Solutions Manual
for Harris'
Quantitative Chemical Analysis

Eighth Edition

Daniel C. Harris
Michelson Laboratory

W. H. Freeman and Company
New York

ISBN: 1-4292-3123-8
EAN: 978-1-4292-3123-7

Sixth printing

W.H. Freeman and Company
41 Madison Avenue
New York, NY 10010
Houndmills, Basingstoke RG21 6XS England
www.whfreeman.com

Contents

Contents

CHAPTER 0
THE ANALYTICAL PROCESS

0-1. Qualitative analysis finds out what is in a sample. Quantitative analysis measures how much is in a sample.

0-2. Steps in a chemical analysis:

(1) Formulate the question: Convert a general question into a specific one that can be answered by a chemical measurement.

(2) Select the appropriate analytical procedure.

(3) Obtain a representative sample.

(4) Sample preparation: Convert the representative sample into a sample suitable for analysis. If necessary, concentrate the analyte and remove or mask interfering species.

(5) Analysis: Measure the unknown concentration in replicate analyses.

(6) Produce a clear report of results, including estimates of uncertainty.

(7) Draw conclusions: Based on the analytical results, decide what actions to take.

0-3. Masking converts an interfering species to a noninterfering species.

0-4. A calibration curve shows the response of an analytical method as a function of the known concentration of analyte in standard solutions. Once the calibration curve is known, then the concentration of an unknown can be deduced from a measured response.

0-5. (a) A homogeneous material has the same composition everywhere. In a heterogeneous material, the composition is not the same everywhere.

(b) In a segregated heterogeneous material, the composition varies on a large scale. There could be large patches with one composition and large patches with another composition. The differences are segregated into different regions. In a random heterogeneous material, the differences occur on a fine scale. If we collect a "reasonable-size" portion, we will capture each of the different compositions that are present.

(c) To sample a *segregated heterogeneous material*, we take representative amounts from each of the obviously different regions. In panel b in Box 0-1, 66% of the area has composition A, 14% is B, and 20% is C. To construct a representative bulk sample, we could take 66 randomly selected samples from region A, 14 from region B, and 20 from region C. To sample a *random heterogeneous material*, we divide the material into imaginary segments and collect random segments with the help of a table of random numbers.

0-6.   We are apparently observing *interference* by Mn^{2+} in the I^- analysis by method A. The result of the I^- analysis is affected by the presence of Mn^{2+}. The greater the concentration of Mn^{2+} in the mineral water, the greater is the apparent concentration of I^- found by method A. Method B is not subject to the same interference, so the concentration of I^- is low and independent of addition of Mn^{2+}. There must be some Mn^{2+} in the original mineral water, which causes method A to give a higher result than method B even when no Mn^{2+} is deliberately added.

CHAPTER 1
MEASUREMENTS

A note from Dan: Don't worry if your numerical answers are slightly different from those in the *Solutions Manual*. You or I may have rounded intermediate results. In general, retain many extra digits for intermediate answers and save your roundoff until the end. We'll study this process in Chapter 3.

1-1. (a) meter (m), kilogram (kg), second (s), ampere (A), kelvin (K), mole (mol)

      (b) hertz (Hz), newton (N), pascal (Pa), joule (J), watt (W)

1-2. Abbreviations above kilo are capitalized: M (mega, 10^6), G (giga, 10^9), T (tera, 10^{12}), P (peta, 10^{15}), E (exa, 10^{18}), Z (zetta, 10^{21}) and Y (yotta, 10^{24}).

1-3.

(a)	mW	=	milliwatt	=	10^{-3} watt
(b)	pm	=	picometer	=	10^{-12} meter
(c)	kΩ	=	kiloohm	=	10^3 ohm
(d)	μF	=	microfarad	=	10^{-6} farad
(e)	TJ	=	terajoule	=	10^{12} joule
(f)	ns	=	nanosecond	=	10^{-9} second
(g)	fg	=	femtogram	=	10^{-15} gram
(h)	dPa	=	decipascal	=	10^{-1} pascal

1-4. (a) 100 fJ or 0.1 pJ   (d) 0.1 nm or 100 pm

      (b) 43.172 8 nF   (e) 21 TW

      (c) 299.79 THz or 0.299 79 PHz   (f) 0.483 amol or 483 zmol

1-5. (a) 5.4 Pg = 5.4×10^{15} g. 5.4×10^{15} g $\times \dfrac{1\,\text{kg}}{1\,000\,\text{g}} = 5.4 \times 10^{12}$ kg of C

      (b) The formula mass of CO_2 is 12.010 7 + 2(15.999 4) = 44.009 5

$$5.4 \times 10^{12}\,\text{kg C} \times \frac{44.009\,5\,\text{kg } CO_2}{12.010\,7\,\text{kg C}} = 2.0 \times 10^{13}\,\text{kg } CO_2$$

      (c) 2.0×10^{13} kg $CO_2 \times \dfrac{1\,\text{ton}}{1\,000\,\text{kg}} = 2.0 \times 10^{10}$ tons of CO_2

$$\frac{2.0 \times 10^{10}\,\text{tons}}{5 \times 10^{9}\,\text{people}} = 4\text{ tons per person}$$

1-6. Table 1-4 tells us that 1 horsepower $= 745.700$ W $= 745.700$ J/s.

$$100.0 \text{ horsepower} = (100.0 \text{ horsepower})\left(\frac{745.700 \text{ J/s}}{\text{horsepower}}\right) = 7.457 \times 10^4 \text{ J/s}.$$

$$\frac{7.457 \times 10^4 \dfrac{\cancel{J}}{\cancel{s}}}{4.184 \dfrac{\cancel{J}}{\text{cal}}} \times 3600 \dfrac{\cancel{s}}{\text{h}} = 6.416 \times 10^7 \dfrac{\text{cal}}{\text{h}}.$$

1-7. (a)

$$\frac{\left(2.2 \times 10^6 \dfrac{\cancel{\text{cal}}}{\cancel{\text{day}}}\right)\left(4.184 \dfrac{\text{J}}{\cancel{\text{cal}}}\right)\left(\dfrac{1 \cancel{\text{day}}}{24 \cancel{\text{h}}}\right)\left(\dfrac{1 \cancel{\text{h}}}{3600 \text{ s}}\right)}{(120 \cancel{\text{pound}})\left(0.453\,6 \dfrac{\text{kg}}{\cancel{\text{pound}}}\right)} = 2.0 \text{ J/(s·kg)}$$

$$= 2.0 \text{ W/kg}$$

Similarly, $3.4 \times 10^3 \dfrac{\text{kcal}}{\text{day}} \Rightarrow 3.0 \text{ J/(s·kg)} = 3.0 \text{ W/kg}.$

(b) The office worker's power output is

$$\left(2.2 \times 10^6 \dfrac{\cancel{\text{cal}}}{\cancel{\text{day}}}\right)\left(4.184 \dfrac{\text{J}}{\cancel{\text{cal}}}\right)\left(\dfrac{1 \cancel{\text{day}}}{24 \cancel{\text{h}}}\right)\left(\dfrac{\cancel{\text{h}}}{3600 \text{ s}}\right) = 1.1 \times 10^2 \dfrac{\text{J}}{\text{s}} = 1.1 \times 10^2 \text{ W}$$

The person's power output is greater than that of the 100 W light bulb.

1-8. $\left(5.00 \times 10^3 \dfrac{\cancel{\text{Btu}}}{\cancel{\text{h}}}\right)\left(1055.06 \dfrac{\text{J}}{\cancel{\text{Btu}}}\right)\left(\dfrac{1 \cancel{\text{h}}}{3600 \text{ s}}\right) = 1.47 \times 10^3 \dfrac{\text{J}}{\text{s}} = 1.47 \times 10^3 \text{ W}$

1-9. (a) $\left(1000 \dfrac{\text{m}}{\text{km}}\right)\left(\dfrac{1 \cancel{\text{inch}}}{0.025\,4 \cancel{\text{m}}}\right)\left(\dfrac{1 \cancel{\text{foot}}}{12 \cancel{\text{inch}}}\right)\left(\dfrac{1 \text{ mile}}{5\,280 \cancel{\text{foot}}}\right) = 0.621\,37 \dfrac{\text{mile}}{\text{km}}$

(b) $\left(\dfrac{100 \cancel{\text{km}}}{4.6 \cancel{\text{L}}}\right)\left(\dfrac{0.621\,37 \text{ miles}}{\cancel{\text{km}}}\right)\left(\dfrac{3.785\,4 \cancel{\text{L}}}{\text{gallon}}\right) = 51 \dfrac{\text{miles}}{\text{gallon}}$

(c) The diesel engine produces 223 g CO_2/km, which we will convert into g/mile:

$$\left(223 \dfrac{\text{g } CO_2}{\cancel{\text{km}}}\right)\left(\dfrac{1 \cancel{\text{km}}}{0.621\,37 \text{ mile}}\right) = 359 \dfrac{\text{g } CO_2}{\text{mile}}$$

In 15 000 miles, $CO_2 = (15\,000 \cancel{\text{miles}})(359 \text{ g}/\cancel{\text{mile}}) = 5.38 \times 10^6$ g or 5.38×10^3 kg $= 5.38$ metric tons. The gasoline engine produces 266 g CO_2/km, which we convert into 428 g/mile or 6.42 metric tons in 15 000 miles.

1-10. Newton = force = mass × acceleration = $kg\left(\dfrac{m}{s^2}\right)$

Joule = energy = force × distance = $kg\left(\dfrac{m}{s^2}\right) \cdot m = kg\left(\dfrac{m^2}{s^2}\right)$

Pascal = pressure = force / area = $kg\left(\dfrac{m}{s^2}\right)/m^2 = \dfrac{kg}{m \cdot s^2}$

1-11. $\left(0.03\dfrac{mg}{m^2 \cdot day}\right)\left(1000\dfrac{m}{km}\right)^2 (535~km^2)\left(\dfrac{1~g}{1000~mg}\right) \times$

$\left(\dfrac{1~kg}{1000~g}\right)\left(\dfrac{1~ton}{1000~kg}\right)\left(365\dfrac{day}{year}\right) = 6\dfrac{ton}{year}$

1-12. (a) molarity = moles of solute / liter of solution

(b) molality = moles of solute / kilogram of solvent

(c) density = grams of substance / milliliter of substance

(d) weight percent = 100 × (mass of substance/mass of solution or mixture)

(e) volume percent = 100 × (volume of substance/volume of solution or mixture)

(f) parts per million = 10^6 × (grams of substance/grams of sample)

(g) parts per billion = 10^9 × (grams of substance/grams of sample)

(h) formal concentration = moles of formula/liter of solution

1-13. Acetic acid (CH_3CO_2H) is a weak electrolyte that is partially dissociated. When we dissolve 0.01 mol in a liter, the concentrations of CH_3CO_2H plus $CH_3CO_2^-$ add to 0.01 M. The concentration of CH_3CO_2H alone is less than 0.01 M.

1-14. 32.0 g / [(22.990 + 35.453) g/mol] = 0.548 mol NaCl
0.548 mol / 0.500 L = 1.10 M

1-15. $\left(1.71\dfrac{mol~CH_3OH}{L~solution}\right)(0.100~L~solution) = 0.171~mol~CH_3OH$

$$(0.171 \text{ mol CH}_3\text{OH})\left(\frac{32.04 \text{ g}}{\text{mol CH}_3\text{OH}}\right) = 5.48 \text{ g}$$

1-16. (a) $19 \text{ mPa} = 19 \times 10^{-3} \text{ Pa}.$ $\quad 19 \times 10^{-3} \text{ Pa} \times \dfrac{1 \text{ bar}}{10^5 \text{ Pa}} = 1.9 \times 10^{-7} \text{ bar}$

(b) $T(\text{K}) = 273.15 + ^\circ\text{C} = 273.15 - 70 = 203 \text{ K}$

$$\frac{n}{V} = \frac{P}{RT} = \frac{1.9 \times 10^{-7} \text{ bar}}{0.08314 \dfrac{\text{L} \cdot \text{bar}}{\text{mol} \cdot \text{K}} \times 203 \text{ K}} = 1.1 \times 10^{-8} \text{ M} = 11 \text{ nM}$$

1-17. $1 \text{ ppm} = \dfrac{1 \text{ g solute}}{10^6 \text{ g solution}}.$ Since 1 L of dilute solution $\approx 10^3$ g,

$1 \text{ ppm} = 10^{-3} \text{ g solute/L}$ ($= 10^{-3}$ g solute / 10^3 g solution).

Since $10^{-3} \text{ g} = 10^3 \mu\text{g}$, $1 \text{ ppm} = 10^3 \mu\text{g/L}$ or $1 \mu\text{g/mL}$.

Since $10^{-3} \text{ g} = 1 \text{ mg}$, $1 \text{ ppm} = 1 \text{ mg/L}$.

1-18. 0.2 ppb means 0.2×10^{-9} g of $C_{20}H_{42}$ per g of rainwater

$$= 0.2 \times 10^{-6} \frac{\text{g } C_{20}H_{42}}{1\,000 \text{ g rainwater}} \approx \frac{0.2 \times 10^{-6} \text{ g } C_{20}H_{42}}{\text{L rainwater}}.$$

$$\frac{0.2 \times 10^{-6} \text{ g / L}}{282.55 \text{ g / mol}} = 7 \times 10^{-10} \text{ M}$$

1-19. $\left(0.705 \dfrac{\text{g HClO}_4}{\text{g solution}}\right)(37.6 \text{ g solution}) = 26.5 \text{ g HClO}_4$

$37.6 \text{ g solution} - 26.5 \text{ g HClO}_4 = 11.1 \text{ g H}_2\text{O}$

1-20. (a) $\left(1.67 \dfrac{\text{g solution}}{\text{mL}}\right)\left(1000 \dfrac{\text{mL}}{\text{L}}\right) = 1.67 \times 10^3 \text{ g solution}$

(b) $\left(0.705 \dfrac{\text{g HClO}_4}{\text{g solution}}\right)(1.67 \times 10^3 \text{ g solution}) = 1.18 \times 10^3 \text{ g HClO}_4$

(c) $(1.18 \times 10^3 \text{ g}) / (100.46 \text{ g /mol}) = 11.7 \text{ mol}$

1-21. molality $= \dfrac{\text{mol KI}}{\text{kg solvent}}$

$20.0 \text{ wt\% KI} = \dfrac{200 \text{ g KI}}{1\,000 \text{ g solution}} = \dfrac{200 \text{ g KI}}{800 \text{ g H}_2\text{O}}$

To find the grams of KI in 1 kg of H$_2$O, we set up a proportion:

$$\frac{200 \text{ g KI}}{800 \text{ g H}_2\text{O}} = \frac{x \text{ g KI}}{1\,000 \text{ g H}_2\text{O}} \Rightarrow x = 250 \text{ g KI}$$

But 250 g KI = 1.51 mol KI, so the molality is 1.51 m.

1-22. (a) $\dfrac{150 \times 10^{-15} \text{ mol/cell}}{2.5 \times 10^4 \text{ vesicles/cell}} = 6.0 \dfrac{\text{amol}}{\text{vessicle}}$

(b) $(6.0 \times 10^{-18} \text{ mol})\left(6.022 \times 10^{23} \dfrac{\text{molecules}}{\text{mol}}\right) = 3.6 \times 10^6$ molecules

(c) Volume $= \dfrac{4}{3}\pi (200 \times 10^{-9} \text{ m})^3 = 3.35 \times 10^{-20} \text{ m}^3$;

$$\frac{3.35 \times 10^{-20} \text{ m}^3}{10^{-3} \text{ m}^3 / \text{L}} = 3.35 \times 10^{-17} \text{ L}$$

(d) $\dfrac{10 \times 10^{-18} \text{ mol}}{3.35 \times 10^{-17} \text{ L}} = 0.30 \text{ M}$

1-23. $\dfrac{80 \times 10^{-3} \text{ g}}{180.2 \text{ g} / \text{mol}} = 4.4 \times 10^{-4} \text{ mol}$; $\quad \dfrac{4.4 \times 10^{-4} \text{ mol}}{100 \times 10^{-3} \text{ L}} = 4.4 \times 10^{-3} \text{ M}$;

Similarly, 120 mg/100 L = 6.7×10^{-3} M.

1-24. (a) Mass of 1.000 L $= 1.046 \dfrac{\text{g}}{\text{mL}} \times 1\,000 \dfrac{\text{mL}}{\text{L}} \times 1.000 \text{ L} = 1\,046 \text{ g}$

Grams of $C_2H_6O_2$ per liter $= 6.067 \dfrac{\text{mol}}{\text{L}} \times 62.07 \dfrac{\text{g}}{\text{mol}} = 376.6 \dfrac{\text{g}}{\text{L}}$

(b) 1.000 L contains 376.6 g of $C_2H_6O_2$ and $1\,046 - 376.6 = 669$ g of H_2O
$= 0.669$ kg

Molality $= \dfrac{6.067 \text{ mol } C_2H_6O_2}{0.669 \text{ kg } H_2O} = 9.07 \dfrac{\text{mol } C_2H_6O_2}{\text{kg } H_2O} = 9.07 \ m$

1-25. Shredded wheat: 1.000 g contains 0.099 g protein + 0.799 g carbohydrate

$0.099 \text{ g} \times 4.0 \dfrac{\text{Cal}}{\text{g}} + 0.799 \text{ g} \times 4.0 \dfrac{\text{Cal}}{\text{g}} = 3.6 \text{ Cal}$

Doughnut: 1.000 g contains 0.046 g protein + 0.514 g carbohydrate + 0.186 g fat

$0.046 \text{ g} \times 4.0 \dfrac{\text{Cal}}{\text{g}} + 0.514 \text{ g} \times 4.0 \dfrac{\text{Cal}}{\text{g}} + 0.186 \text{ g} \times 9.0 \dfrac{\text{Cal}}{\text{g}} = 3.9 \text{ Cal}$

In a similar manner, we find 2.8 $\dfrac{\text{Cal}}{\text{g}}$ for hamburger and 0.48 $\dfrac{\text{Cal}}{\text{g}}$ for apple.

There are 16 ounces in 1 pound, which Table 1-4 says is equal to 453.592 37 g

$$\Rightarrow 28.35 \frac{g}{ounce}.$$

To convert Cal/g to Cal/ounce, multiply by 28.35:

	Shredded Wheat	Doughnut	Hamburger	Apple
Cal/g	3.6	3.9	2.8	0.48
Cal/ounce	102	111	79	14

1-26. Mass of water $= \pi (225 \text{ m})^2 (10.0 \text{ m}) \left(\dfrac{1\,000 \text{ kg}}{\text{m}^3} \right) = 1.59 \times 10^9$ kg

$$1.6 \text{ ppm} = \frac{1.6 \times 10^{-3} \text{ g F}^-}{\text{kg H}_2\text{O}}$$

Mass of F^- required $=$

$$\left(1.6 \times 10^{-3} \frac{\text{g F}^-}{\text{kg H}_2\text{O}} \right) (1.59 \times 10^9 \text{ kg H}_2\text{O}) = 2.5 \times 10^6 \text{ g F}^-.$$

(If we retain three digits for the next calculation, this last number is 2.54×10^6.)

The atomic mass of F is 18.998 and the formula mass of H_2SiF_6 is 144.09. One mole of H_2SiF_6 contains 6 moles of F.

$$\frac{\text{mass of F}^-}{\text{mass of H}_2\text{SiF}_6} = \frac{6 \times 18.998}{144.09} = \frac{2.54 \times 10^6 \text{ g F}}{x \text{ g H}_2\text{SiF}_6} \Rightarrow x = 3.2 \times 10^6 \text{ g H}_2\text{SiF}_6$$

1-27. (a) $PV = nRT$

$$(1.000 \text{ bar})(5.24 \times 10^{-6} \text{ L}) = n \left(0.083\,14 \frac{\text{L} \cdot \text{bar}}{\text{mol} \cdot \text{K}} \right) (298.15 K)$$

$$\Rightarrow n = 2.11 \times 10^{-7} \text{ mol} \Rightarrow 2.11 \times 10^{-7} \text{ M}$$

(b) Ar: 0.934% means 0.009 34 L of Ar per L of air

$$(1.000 \text{ bar})(0.009\,34 \text{ L}) = n \left(0.083\,14 \frac{\text{L} \cdot \text{bar}}{\text{mol} \cdot \text{K}} \right) (298.15 \text{ K})$$

$$\Rightarrow n = 3.77 \times 10^{-4} \text{ mol} \Rightarrow 3.77 \times 10^{-4} \text{ M}$$

Kr: 1.14 ppm $\Rightarrow$ 1.14 μL Kr per L of air $\Rightarrow 4.60 \times 10^{-8}$ M

Xe: 87 ppb $\Rightarrow$ 87 nL Xe per L of air $\Rightarrow 3.5 \times 10^{-9}$ M

1-28. $2.00 \text{ L} \times 0.050\,0 \dfrac{\text{mol}}{\text{L}} \times 61.83 \dfrac{\text{g}}{\text{mol}} = 6.18$ g in a 2 L volumetric flask

1-29. Weigh out $2 \times 0.050\,0$ mol $= 0.100$ mol $= 6.18$ g $B(OH)_3$ and dissolve in 2.00 kg H_2O.

1-30. $M_{con} \cdot V_{con} = M_{dil} \cdot V_{dil}$

$$\left(0.80 \frac{mol}{L}\right)(1.00\ L) = \left(0.25\frac{mol}{L}\right)\ V_{dil} \Rightarrow V_{dil} = 3.2\ L$$

1-31.   We need $1.00\ L \times 0.10\frac{mol}{L} = 0.10$ mol NaOH $= 4.0$ g NaOH

$$\frac{4.0\ \text{g NaOH}}{0.50\dfrac{\text{g NaOH}}{\text{g solution}}} = 8.0\ \text{g solution}$$

1-32.   (a) $V_{con} = V_{dil}\dfrac{M_{dil}}{M_{con}} = 1\,000\ \text{mL}\left(\dfrac{1.00\ M}{18.0\ M}\right) = 55.6\ \text{mL}$

(b) One liter of 98.0% H_2SO_4 contains $(18.0\ mol)(98.079\ g/\ mol) = 1.77 \times 10^3$ g of H_2SO_4. Since the solution contains 98.0 wt% H_2SO_4, and the mass of H_2SO_4 per mL is 1.77 g, the mass of solution per milliliter (the density) is

$$\frac{1.77\ \text{g } H_2SO_4\ /\text{mL}}{0.980\ \text{g } H_2SO_4\ /\text{g solution}} = 1.80\ \text{g solution/mL}$$

1-33.   2.00 L of 0.169 M NaOH $= 0.338$ mol NaOH $= 13.5$ g NaOH

$$\text{density} = \frac{\text{g solution}}{\text{mL solution}}$$

$$= \frac{13.5\ \text{g NaOH}}{(16.7\ \text{mL solution})\left(0.534\dfrac{\text{g NaOH}}{\text{g solution}}\right)} = 1.52\frac{\text{g}}{\text{mL}}$$

1-34.   FM of $Ba(NO_3)_2 = 261.34$ 4.35 g of solid with 23.2 wt% $Ba(NO_3)_2$ contains

$(0.232)(4.35\ g) = 1.01$ g $Ba(NO_3)_2$

$$\text{mol } Ba^{2+} = \frac{(1.01\ \text{g } Ba(NO_3)_2)}{(261.34\ \text{g } Ba(NO_3)_2\ /\text{mol})} = 3.86 \times 10^{-3}\ \text{mol}$$

mol $H_2SO_4 = $ mol $Ba^{2+} = 3.86 \times 10^{-3}$ mol

$$\text{volume of } H_2SO_4 = \frac{(3.86 \times 10^{-3}\ \text{mol})}{(3.00\ \text{mol/L})} = 1.29\ \text{mL}$$

1-35.   25.0 mL of 0.023 6 M Th^{4+} contains

$(0.025\ 0\ L)(0.023\ 6\ M) = 5.90 \times 10^{-4}$ mol Th^{4+}

mol HF required for stoichiometric reaction $= 4 \times$ mol $Th^{4+} = 2.36 \times 10^{-3}$ mol

50% excess $= 1.50(2.36 \times 10^{-3}\ \text{mol}) = 3.54 \times 10^{-3}$ mol HF

Required mass of pure HF $= (3.54 \times 10^{-3}\ \text{mol})(20.01\ \text{g/mol}) = 0.070\,8\ \text{g}$

Mass of 0.491 wt% HF solution $= \dfrac{(0.070\,8\ \text{g HF})}{(0.004\,91\ \text{g HF} / \text{g solution})} = 14.4\ \text{g}$

1-36. Concentrations of reagents used in an analysis are determined either by weighing out supposedly pure primary standards or by reaction with such standards. If the standards are not pure, none of the concentrations will be correct.

1-37. The equivalence point occurs when the exact stoichiometric quantities of reagents have been mixed. The end point, which comes near the equivalence point, is marked by a sudden change in a physical property brought about by the disappearance of a reactant or appearance of a product.

1-38. In a blank titration, the quantity of titrant required to reach the end point in the absence of analyte is measured. By subtracting this quantity from the amount of titrant needed in the presence of analyte, we reduce the systematic error.

1-39. In a direct titration, titrant reacts directly with analyte. In a back titration, a known excess of reagent that reacts with analyte is used. The excess is then measured with a second titrant.

1-40. Primary standards are purer than reagent-grade chemicals. The assay of a primary standard must be very close to the nominal value (such as 99.95–100.05%), whereas the assay on a reagent chemical might be only 99%. Primary standards must have very long shelf lives.

1-41. Since a relatively large amount of acid might be required to dissolve a small amount of sample, we cannot tolerate even modest amounts of impurities in the acid for trace analysis. Otherwise, the quantity of impurity could be greater than quantity of analyte in the sample.

1-42. 40.0 mL of 0.040 0 M $Hg_2(NO_3)_2 = 1.60$ mmol of Hg_2^{2+}, which will require 3.20 mmol of KI. This is contained in volume $= \dfrac{3.20\ \text{mmol}}{0.100\ \text{mmol/mL}} = 32.0\ \text{mL}$.

1-43. 108.0 mL of 0.165 0 M oxalic acid = 17.82 mmol, which requires
$\left(\dfrac{2\ \text{mol MnO}_4^-}{5\ \text{mol H}_2\text{C}_2\text{O}_4} \right) (17.82\ \text{mol H}_2\text{C}_2\text{O}_4) = 7.128\ \text{mmol of MnO}_4^-.$

7.128 mmol / (0.165 0 mmol/mL) = 43.20 mL of $KMnO_4$.

Another way to see this is to note that the reagents are both 0.165 0 M. Therefore, volume of MnO_4^- = $\frac{2}{5}$(volume of oxalic acid).

For second question, volume of oxalic acid = $\frac{5}{2}$(volume of MnO_4^-) = 270.0 mL.

1-44. 1.69 mg of NH_3 = 0.099 2 mmol of NH_3. This will react with $\frac{3}{2}$(0.099 2) =

0.149 mmol of OBr^-. The molarity of OBr^- is 0.149 mmol/1.00 mL = 0.149 M.

1-45. mol sulfamic acid = $\dfrac{0.333\,7\ g}{97.094\ g/mol}$ = 3.436_9 mmol

molarity of NaOH = $\dfrac{3.436_9\ \text{mmol}}{34.26\ \text{mL}}$ = 0.100 3 M

1-46. HCl added to powder = (10.00 mL)(1.396 M) = 13.96 mmol

NaOH required = (39.96 mL)(0.100 4 M) = 4.012 mmol

HCl consumed by carbonate = 13.96 – 4.012 = 9.94_8 mmol

mol $CaCO_3$ = $\frac{1}{2}$ mol HCl consumed = 4.97_4 mmol = 0.497_8 g $CaCO_3$

wt% $CaCO_3$ = $\dfrac{0.497_8\ g\ CaCO_3}{0.541\,3\ g\ limestone} \times 100$ = 92.0 wt%

CHAPTER 2
TOOLS OF THE TRADE

2-1. The primary rule is to familiarize yourself with the hazards of what you are about to do and not to do something you consider to be dangerous.

2-3. Dichromate ($Cr_2O_7^{2-}$) is soluble in water and contains carcinogenic Cr(VI).

Reducing Cr(VI) to Cr(III) decreases the toxicity of the metal. Converting aqueous Cr(III) to solid $Cr(OH)_3$ decreases the solubility of the metal and therefore decreases its ability to be spread by water. Evaporation produces the minimum volume of waste.

2-4. The upper "0" means that the reagent has no fire hazard. The right hand "0" indicates that the reagent is stable. The "3" tells us that the reagent is corrosive or toxic and we should avoid skin contact or inhalation.

2-5. The lab notebook must: (1) state what was done; (2) state what was observed; and (3) be understandable to a stranger.

2-6. See Section 2.3.

2-7. The buoyancy correction is 1 when the substance being weighed has the same density as the weights used to calibrate the balance.

2-8. $$ m = \frac{(14.82 \text{ g}) \left(1 - \dfrac{0.001\,2 \text{ g/mL}}{8.0 \text{ g/mL}}\right)}{\left(1 - \dfrac{0.001\,2 \text{ g/mL}}{0.626 \text{ g/mL}}\right)} = 14.85 \text{ g} $$

2-9. The smallest correction will be for PbO_2, whose density is closest to 8.0 g/mL. The largest correction will be for the least dense substance, lithium.

2-10. $$ m = \frac{4.236\,6 \text{ g} \left(1 - \dfrac{0.001\,2 \text{ g/mL}}{8.0 \text{ g/mL}}\right)}{\left(1 - \dfrac{0.001\,2 \text{ g/mL}}{1.636 \text{ g/mL}}\right)} = 4.239\,1 \text{ g} $$

Without correcting for buoyancy, we would think the mass of primary standard is less than the actual mass and we would think the molarity of base reacting with the standard is also less than the actual molarity. The percentage error would be

$$ \frac{\text{true mass} - \text{measured mass}}{\text{true mass}} \times 100 = \frac{4.239\,1 - 4.236\,6}{4.239\,1} \times 100 = 0.06\%. $$

2-11. (a) One mol of He (= 4.003 g) occupies a volume of

$$V = \frac{nRT}{P} = \frac{(1 \text{ mol})\left(0.083\,14\,\dfrac{L \cdot bar}{mol \cdot K}\right)(293.15\,K)}{1\,bar} = 24.37\,L$$

Density = 4.003 g / 24.37 L = 0.164 g/L = 0.000 164 g/mL

(b) $m = \dfrac{(0.823\,\text{g})\left(1 - \dfrac{0.000\,164\,\text{g/mL}}{8.0\,\text{g/mL}}\right)}{\left(1 - \dfrac{0.000\,164\,\text{g/mL}}{0.97\,\text{g/mL}}\right)} = 0.823\,\text{g}$

2-12. (a) (0.42) (2 330 Pa) = 979 Pa

(b) Air density =

$\dfrac{(0.003\,485)(94\,000) - (0.001\,318)(979)}{293.15} = 1.11\,\text{g/L} = 0.001\,1\,\text{g/mL}$

(c) mass = 1.000 0 g $\left(\dfrac{1 - \dfrac{0.001\,1\,\text{g/mL}}{8.0\,\text{g/mL}}}{1 - \dfrac{0.001\,1\,\text{g/mL}}{1.00\,\text{g/mL}}}\right) = 1.001\,0\,\text{g}$

2-13. $m_b = m_a \dfrac{r_a^2}{r_b^2} = (100.0000\,\text{g})\dfrac{(6\,370\,000\,\text{m})^2}{(6\,370\,030\,\text{m})^2} = 99.999\,1\,\text{g}$

2-14. TD means "to deliver" and TC means "to contain."

2-15. Dissolve (0.250 0 L)(0.150 0 mol/L) = 0.037 50 mol of K_2SO_4 (= 6.535 g, FM 174.26 g/mol) in less than 250 mL of water in a 250-mL volumetric flask. Add more water and mix. Dilute to the 250.0 mL mark and invert the flask many times for complete mixing.

2-16. The plastic flask is needed for trace analysis of analytes at ppb levels that might be lost by adsorption on the glass surface.

2-17. (a) With a suction device, suck liquid up past the 5.00 mL mark. Discard one or two pipet volumes of liquid to rinse the pipet. Take up a third volume past the calibration mark and quickly replace the bulb with your index finger. (Alternatively, use an automatic suction device that remains attached to the pipet.) Wipe excess liquid off the outside of the pipet with a clean tissue. Touch the tip of the pipet to the side of a beaker and drain liquid until the bottom of the meniscus reaches the center of the mark. Transfer the pipet to a receiving vessel and drain it by gravity while holding the tip against the wall. After draining stops, hold the pipet to the wall for a few more seconds to complete draining. Do not blow out the last drop. The pipet should be nearly vertical at the end of delivery.

(b) Transfer pipet.

2-18. (a) Adjust the knob for 50.0 μL. Place a fresh tip tightly on the barrel. Depress the plunger to the first stop, corresponding to 50.0 μL. Hold the pipet vertically, dip it 3–5 mm into reagent solution, and slowly release the plunger to suck up liquid. Leave the tip in the liquid for a few more seconds. Withdraw the pipet vertically. Take up and discard three squirts of reagent to clean and wet the tip and fill it with vapor. To dispense liquid, touch the tip to the wall of the receiver and gently depress the plunger to the first stop. After a few seconds, depress the plunger further to squirt out the last liquid.

(b) The procedure in (a) is called forward mode. For a foaming liquid, use reverse mode. Depress the plunger beyond the 50.0 μL stop and take in more than 50.0 μL. To deliver 50.0 μL, depress the plunger to the first stop and not beyond.

2-19. The trap prevents liquid filtrate from being sucked into the vacuum system. The watchglass keeps dust out of the sample.

2-20. Phosphorus pentoxide

2-21. $20.2144 \text{ g} - 10.2634 \text{ g} = 9.9510 \text{ g}$. Column 3 of Table 2-7 tells us that the true volume is $(9.9510 \text{ g})(1.0029 \text{ mL/g}) = 9.9799 \text{ mL}$.

2-22. Expansion $= \dfrac{0.9991026}{0.9970479} = 1.0020608 \approx 0.2\%$. Densities were taken from Table 2-7. The 0.5000 M solution at 25° would be $(0.5000 \text{ M})/(1.002) = 0.4990$ M.

2-23. Using column 2 of Table 2-7, mass in vacuum $=$
$(50.037 \text{ mL})(0.9982071 \text{ g/mL}) = 49.947 \text{ g}$.

Using column 3, mass in air $= \dfrac{50.037 \text{ mL}}{1.0029 \text{ mL/g}} = 49.892 \text{ g}$.

2-24. When the solution is cooled to 20°C, the concentration will be higher than the concentration at 24°C by a factor of $\frac{\text{density at 20°C}}{\text{density at 24°C}}$. Therefore, the concentration needed at 24° will be lower than the concentration at 20°C.

Desired concentration at 24°C $= (1.000 \text{ M}) \left(\dfrac{0.997\,299\,5 \text{ g/mL}}{0.998\,207\,1 \text{ g/mL}} \right) = 0.999\,1$ M

(using the quotient of densities from Table 2-7). The true mass of KNO_3 needed is $\left(0.5000 \text{ L}\right) \left(0.9991 \dfrac{\text{mol}}{\text{L}}\right) \left(101.103 \dfrac{\text{g}}{\text{mol}}\right) = 50.506$ g.

$$m' = \frac{(50.506 \text{ g}) \left(1 - \dfrac{0.001\,2 \text{ g/mL}}{2.109 \text{ g/mL}}\right)}{\left(1 - \dfrac{0.001\,2 \text{ g/mL}}{8.0 \text{ g/mL}}\right)} = 50.484 \text{ g}$$

2-25. (a) Fraction within specifications $= e^{-t(\ln 2)/t_m}$. If $t_m = 2$ yr and $t = 2$ yr, then fraction within specifications $= e^{-2(\ln 2)/2} = e^{-\ln 2} = \frac{1}{2}$.

(b) Fraction within specifications $= 0.95 = e^{-t(\ln 2)/2 \text{ yr}}$

$\ln(0.95) = -t(\ln 2)/2 \Rightarrow t = -2\ln(0.95)/\ln 2 = 0.148$ yr $= 54$ days ≈ 8 weeks

To solve for t, take the natural logarithm of both sides:

2-26. Al extracted from glass $= (0.200 \text{ L})(5.2 \times 10^{-6} \text{ M}) = 1.04 \times 10^{-6}$ mol

mass of Al $= (1.04 \times 10^{-6} \text{ mol})(26.98 \text{ g/mol}) = 28.1$ μg

This much Al was extracted from 0.50 g of glass, so

wt% Al extracted $= 100 \times \dfrac{28.1 \times 10^{-6} \text{ g}}{0.50 \text{ g}} = 0.005\,6_2$ wt%

Fraction of Al extracted $= \dfrac{0.005\,6_2 \text{ wt\%}}{0.80 \text{ wt\%}} = 0.007\,0$ (or 0.70% of the Al)

2-27.

Graph of van Deemter Equation		
	Flow rate	Plate height
Constants	(mL/min)	(mm)
A =	4	8.194
1.65	6	6.092
B =	8	5.064
25.8	10	4.466
C =	20	3.412
0.0236	30	3.218
	40	3.239
	50	3.346
	60	3.496
	70	3.671
	80	3.861
	90	4.061
	100	4.268
Formula:		
C5 = $A$6+$A$8/B5+$A$10*B5		

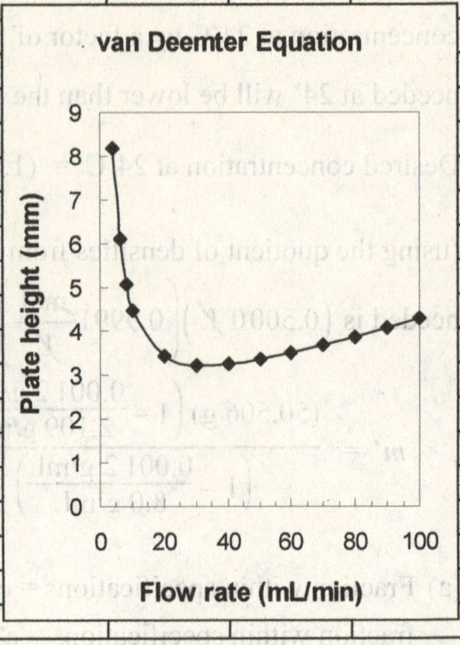

van Deemter Equation

Plate height (mm) vs Flow rate (mL/min)

CHAPTER 3
EXPERIMENTAL ERROR

3-1.   (a) 5   (b) 4   (c) 3

3-2.   (a) 1.237   (b) 1.238   (c) 0.135   (d) 2.1   (e) 2.00

3-3.   (a) 0.217   (b) 0.216   (c) 0.217

3-4.   (b) 1.18 (3 significant figures)   (c) 0.71 (2 significant figures)

3-5.   (a) 3.71   (b) 10.7   (c) 4.0×10^1 (d) 2.85×10^{-6}

       (e) 12.625 1 (f) 6.0×10^{-4} (g) 242

3-6.   (a) $BaF_2 = 137.327 + 2(18.998\ 403\ 2) = 175.324$ because the atomic mass of Ba

        has only 3 decimal places.

      (b) $C_6H_4O_4 = 6(12.010\ 7) + 4(1.007\ 94) + 4(15.999\ 4) = 140.093\ 6$

        (The fourth-decimal place in the atomic mass of C has an uncertainty of ± 8

        and the fourth-decimal place of O has an uncertainty of ± 3. These uncer-

        tainties are large enough to make the fourth-decimal place in molecular mass

        of $C_6H_4O_4$ insignificant. Therefore, another good answer is 140.094.)

3-7.   (a) 12.3   (b) 75.5   (c) 5.520×10^3 (d) 3.04

       (e) 3.04×10^{-10} (f) 11.9   (g) 4.600   (h) 4.9×10^{-7}

3-9.   All measurements have some uncertainty, so there is no way to know the true
value.

3-10. Systematic error is always above or always below the "true value" if you make
replicate measurements. In principle, you can find the source of this error and
eliminate it in a better experiment so the measured mean equals the true mean.
Random error is equally likely to be positive or negative and cannot be
eliminated. Random error can be reduced in a better experiment.

3-11. The apparent mass of product is systematically low because the initial mass of
the (crucible plus moisture) is higher than the true mass of the crucible. The error
is systematic. There is also always some random error superimposed on the
systematic error.

3-12. (a) 25.031 mL is a systematic error. The pipet always delivers more than it is rated for. The number ± 0.009 is the random error in the volume delivered. The volume fluctuates around 25.031 by ± 0.009 mL.

(b) The numbers 1.98 and 2.03 mL are systematic errors. The buret delivers too little between 0 and 2 mL and too much between 2 and 4 mL. The observed variations ± 0.01 and ± 0.02 are random errors.

(c) The difference between 1.9839 and 1.9900 g is random error. The mass will probably be different the next time I try the same procedure.

(d) Differences in peak area are random error based on inconsistent injection volume, inconsistent detector response, and probably other small variations in the condition of the instrument from run to run.

3-13.   (a) Carmen   (b) Cynthia   (c) Chastity   (d) Cheryl

3-14.   3.124 (± 0.005), 3.124 (± 0.2%). It would also be reasonable to keep an additional digit: 3.123_6 ($\pm 0.005_2$), 3.123_6 ($\pm 0.1_7$%)

3-15.   (a)   6.2 (± 0.2)

$\underline{-\ 4.1\ (\pm 0.1)}$

$2.1 \pm e$ $e^2 = 0.2^2 + 0.1^2 \Rightarrow e = 0.2_{24}$ Answer: 2.1 ± 0.2 (or 2.1 ± 11%)

(b)   9.43 (± 0.05)   9.43 (± 0.53%)

$\underline{\times\ 0.016\ (\pm 0.001)}$ $\Rightarrow$ $\underline{\times\ 0.016\ (\pm 6.25\%)}$ $\%e^2 = 0.53^2 + 6.25^2$

0.150 88 ($\pm\ \%e$) $\Rightarrow$ $\%e = 6.272$

Relative uncertainty = 6.27%; Absolute uncertainty = $0.150\ 88 \times 0.062\ 7$

= 0.009 46;   Answer: 0.151 ± 0.009 (or 0.151 ± 6%)

(c) The first term in brackets is the same as part (a), so we can rewrite the problem as $2.1\ (\pm 0.2_{24}) \div 9.43\ (\pm 0.05) = 2.1\ (\pm 10.7\%) \div 9.43\ (\pm 0.53\%)$

$\%e = \sqrt{10.7^2 + 0.53^2} = 10.7\%$

Absolute uncertainty = $0.107 \times 0.223 = 0.023\ 9$

Answer: $0.22_3 \pm 0.02_4$ (± 11%)

(d) The term in brackets is

$6.2\ (\pm 0.2) \times 10^{-3}$ $e = \sqrt{0.2^2 + 0.1^2} \Rightarrow e = 0.2_{24}$

$\underline{+\ 4.1\ (\pm 0.1) \times 10^{-3}}$

$10.3\ (\pm 0.2_{24}) \times 10^{-3} = 10.3 \times 10^{-3}\ (\pm 2.17\%)$

$9.43\ (\pm 0.53\%) \times 0.010\ 3\ (\pm 2.17\%) = 0.097\ 13 \pm 2.23\% = 0.097\ 13 \pm 0.002\ 17$

Answer: $0.097_1 \pm 0.002_2$ ($\pm 2._2$%)

3-16.   (a)   Uncertainty $= \sqrt{0.03^2 + 0.02^2 + 0.06^2} = 0.07$

Answer: 10.18 (±0.07) (±0.7%)

(b)   91.3 (±1.0) × 40.3 (±0.2)/21.1 (±0.2)

$= 91.3\ (\pm 1.10\%) \times 40.3\ (\pm 0.50\%)/21.1\ (\pm 0.95\%)$

% uncertainty $= \sqrt{1.10^2 + 0.50^2 + 0.95^2} = 1.54\%$

Answer: 174 (±3) (±2%)

(c)   [4.97 (±0.05) − 1.86 (±0.01)]/21.1 (±0.2)

$= [3.11\ (\pm 0.0510)]/21.1\ (\pm 0.2) = [3.11\ (\pm 1.64\%)]/21.1\ (\pm 0.95\%)$

$= 0.147\ (\pm 1.90\%) = 0.147\ (\pm 0.003)\ (\pm 2\%)$

(d) $\begin{array}{r} 2.016\,4 \quad (\pm 0.000\,8) \\ 1.233 \quad (\pm 0.002) \\ + \; 4.61 \quad (\pm 0.01) \\ \hline 7.85_{94} \; \sqrt{(0.000\,8)^2 + (0.002)^2 + (0.01)^2} = 0.01_{02} \end{array}$

Answer: 7.86 (±0.01)(±0.1%)

(e) $\begin{array}{r} 2\,016.4 \quad (\pm 0.8) \\ + \; 123.3 \quad (\pm 0.2) \\ + \quad 46.1 \quad (\pm 0.1) \\ \hline 2\,185.8 \; \sqrt{(0.8)^2 + (0.2)^2 + (0.1)^2} = 0.8 \end{array}$

Answer: 2 185.8 (±0.8) (±0.04%)

(f)   For $y = x^a$, $\%e_y = a\%e_x$

$x = 3.14 \pm 0.05 \Rightarrow \%e_x = (0.05 / 3.14) \times 100 = 1.592\%$

$\%e_y = \frac{1}{3}(1.592\%) = 0.531\%$

Answer: $1.464_3 \pm 0.007_8\ (\pm 0.5_3\%)$

(g)   For $y = \log x$, $e_y = 0.434\,29\,\dfrac{e_x}{x}$

$x = 3.14 \pm 0.05 \Rightarrow e_y = 0.434\,29\left(\dfrac{0.05}{3.14}\right) = 0.006\,915$

Answer: $0.496_9 \pm 0.006_9\ (\pm 1.3_9\%)$

3-17.   (a) $y = x^{1/2} \Rightarrow \%e_y = \frac{1}{2}\left(100 \times \dfrac{0.001\,1}{3.141\,5}\right) = 0.017\,5\%$

$(1.75 \times 10^{-4})\sqrt{3.141\,5} = 3.1 \times 10^{-4}$ Answer: $1.772\,4_3 \pm 0.000\,3_1$

(b) $y = \log x \Rightarrow e_y = 0.434\,29\left(\dfrac{0.001\,1}{3.141\,5}\right) = 1.52 \times 10^{-4}$

Answer: $0.497\,1_4 \pm 0.000\,1_5$

(c) $y = $ antilog $x = 10^x \Rightarrow e_y = y \times 2.302\,6\,e_x$

$= (10^{3.141\,5})(2.302\,6)(0.001\,1) = 3.51$ Answer: $1.385_2 \pm 0.003_5 \times 10^3$

(d) $y = \ln x \Rightarrow e_y = \dfrac{0.001\,1}{3.141\,5} = 3.5 \times 10^{-4}$ Answer: $1.144\,7_0 \pm 0.000\,3_5$

(e) Numerator of log term: $y = x^{1/2} \Rightarrow e_y = \dfrac{1}{2}\left(\dfrac{0.006}{0.104} \times 100\right) = 2.88\%$

$\dfrac{0.3225 \pm 2.88\%}{0.0511 \pm 0.0009} = \dfrac{0.3225 \pm 2.88\%}{0.0511 \pm 1.76\%}$

$= 6.311 \pm 3.375\% = 6.311 \pm 0.213$

For $y = \log x$, $e_y = 0.434\,29\,\dfrac{e_x}{x} = 0.434\,29\left(\dfrac{0.213}{6.311}\right) = 0.015$

Answer: $0.80_0 \pm 0.01_5$

3-18. (a) Standard uncertainty in atomic mass is equal to the uncertainty listed in the periodic table divided by $\sqrt{3}$ because atomic mass has a rectangular distribution of values.

Na = 22.989 769 28 $\pm 0.000\,000\,02/\sqrt{3}$ g/mol

Cl = 35.453 $\pm 0.002/\sqrt{3}$ g/mol

_____

58.442 770 $\sqrt{[(2 \times 10^{-8})^2]/3 + [(2 \times 10^{-3})^2]/3} = 1.2 \times 10^{-3}$

58.443 $\pm 0.001_2$ g/mol

(b) molarity $= \dfrac{\text{mol}}{\text{L}} = \dfrac{[2.634\,(\pm 0.002)\text{g}] / [58.443\,(\pm 0.001_2)\text{g/mol}]}{0.100\,00\,(\pm 0.000\,08)\,\text{L}}$

$= \dfrac{2.634\,(\pm 0.076\%) / [58.443\,(\pm 0.002\,1\%)]}{0.100\,00\,(\pm 0.08\%)}$

relative error $= \sqrt{(0.076\%)^2 + (0.002\,1\%)^2 + (0.08\%)^2} = 0.11\%$

molarity $= 0.450\,7\,(\pm 0.000\,5)$ M

3-19. $m = \dfrac{m'\left(1 - \dfrac{d_a}{d_w}\right)}{1 - \dfrac{d_a}{d}}$

$m = \dfrac{[1.034\,6\,(\pm 0.000\,2)\,\text{g}]\left(1 - \dfrac{0.001\,2(\pm 0.000\,1)\,\text{g/mL}}{8.0\,(\pm 0.5)\,\text{g/mL}}\right)}{1 - \dfrac{0.001\,2(\pm 0.000\,1)\,\text{g/mL}}{0.997\,299\,5\,\text{g/mL}}}$

$$m = \frac{[1.034\,6\,(\pm0.019\,3\%)]\left(1 - \dfrac{0.001\,2\,(\pm8.33\%)}{8.0\,(\pm6.25\%)}\right)}{1 - \dfrac{0.001\,2\,(\pm8.33\%)}{0.997\,299\,5\,(\pm0\%)}}$$

$$m = \frac{[1.034\,6\,(\pm0.019\,3\%)][1 - 0.000\,150\,(\pm10.4\%)]}{[1 - 0.001\,203\,(\pm8.33\%)]}$$

$$m = \frac{[1.034\,6\,(\pm0.019\,3\%)]\,[1 - 0.000\,150\,(\pm0.000\,015\,6)]}{[1 - 0.001\,203\,(\pm0.000\,100)]}$$

$$m = \frac{[1.034\,6\,(\pm0.019\,3\%)]\,[0.999\,850\,0\,(\pm0.000\,015\,6)]}{[0.998\,797\,(\pm0.000\,100)]}$$

$$m = \frac{[1.034\,6\,(\pm0.019\,3\%)]\,[0.999\,850\,0\,(\pm0.001\,56\%)]}{[0.998\,797\,(\pm0.010\,0\%)]}$$

$$m = 1.035\,7\,(\pm0.021\,8\%) = 1.035\,7\,(\pm0.000\,2)\ g$$

3-20. $\text{mol Fe}_2\text{O}_3 = \dfrac{0.277_4 \pm 0.001_8\ g}{159.688\ g/mol} = \dfrac{0.277_4}{159.688} \pm \dfrac{0.001_8}{159.688}$

$\qquad\qquad = 1.73_{71} \pm 0.01_{13}\ \text{mmol Fe}_2\text{O}_3;$

$\text{mass of Fe} = 2[1.73_{71}\,(\pm0.01_{13}) \times 10^{-3}\ mol][55.845\ g/mol] = 0.194_{02} \pm 0.001_{26}\ g$

$\text{mass of Fe per tablet} = (0.194_{02} \pm 0.001_{26}\ g)/12 = 16.1_{68} \pm 0.1_{05}\ mg$

$\qquad\qquad = 16.2 \pm 0.1\ mg$

3-21. $\text{mol H}^+ = 2 \times \text{mol Na}_2\text{CO}_3$

$\text{mol Na}_2\text{CO}_3 = \dfrac{0.967\,4\,(\pm0.000\,9)\ g}{105.988\,4\,(\pm0.000\,7)\ g/mol} = \dfrac{0.967\,4\,(\pm0.093\%)\ g}{105.988\,(\pm0.000\,66\%)\ g/mol}$

$\qquad\qquad = 0.009\,127\,4\,(\pm0.093\%)\ mol$

$\text{mol H}^+ = 2(0.009\,127\,4\,(\pm0.093\%)) = 0.018\,255\,(\pm0.093\%)\ mol$

(Relative error is not affected by the multiplication by 2 because mol H^+

and uncertainty in mol H^+ are both multiplied by 2.)

$\text{molarity of HCl} = \dfrac{0.018\,255\,(\pm0.093\%)\ mol}{0.027\,35\,(\pm0.000\,04)\ L} = \dfrac{0.018\,255\,(\pm0.093\%)\ mol}{0.027\,35\,(\pm0.146\%)\ L}$

$\qquad\qquad = 0.66746\,(\pm0.173\%) = 0.667\,46\,(\pm0.001\,155)$

$\qquad\qquad = 0.667 \pm 0.001\ M$

3-22. To find the uncertainty in $c_0{}^3$, we use the function $y = x^a$ in Table 3-1,

where $x = c_0$ and $a = 3$. The uncertainty in $c_0{}^3$ is

$$\%e_y = a\,\%e_x = 3 \times \frac{0.000\,000\,33}{5.431\,020\,36} \times 100 = 1.823 \times 10^{-5}\%$$

So $c_0{}^3 = (5.431\,020\,36 \times 10^{-8}\ cm)^3 = 1.601\,932\,796\,0 \times 10^{-22}\ cm^3$ with a relative uncertainty of $1.823 \times 10^{-5}\%$. We retain extra digits for now and round off at the end of the calculations. (If your calculator cannot hold as many digits as we need for this arithmetic, you can do the math with a spreadsheet set to display 10 decimal places.)

The value of Avogadro's number is computed as follows:

$$N_A = \frac{m_{Si}}{(\rho c_o^3)/8} = \frac{28.085\ 384\ 2\ \text{g/mol}}{(2.329\ 031\ 9\ \text{g/cm}^3 \times 1.601\ 932\ 79_{60} \times 10^{-22}\ \text{cm}^3)/8}$$

$$= 6.022\ 136\ 936\ 1 \times 10^{23}\ \text{mol}^{-1}$$

The relative uncertainty in Avogadro's number is found from the relative uncertainties in m_{Si}, ρ, and c_o^3. (There is no uncertainty in the number 8 atoms/unit cell.)

percent uncertainty in m_{Si} = 100 (0.000 003 5/28.085 384 2) = $1.246 \times 10^{-5}\%$

percent uncertainty in ρ = 100 (0.000 001 8/2.329 031 9) = $7.729 \times 10^{-5}\%$

percent uncertainty in c_o^3 = $1.823 \times 10^{-5}\%$ (calculated before)

percent uncertainty in N_A = $\sqrt{\%e_{m_{Si}}^2 + \%e_r^2 + (\%e_{c_o^3})^2}$ =

$$= \sqrt{(1.246 \times 10^{-5})^2 + (7.729 \times 10^{-5})^2 + (1.823 \times 10^{-5})^2} = 8.038 \times 10^{-5}\%$$

The absolute uncertainty in N_A is $(8.038 \times 10^{-5}\%)(6.022\ 136\ 936\ 1 \times 10^{23})/100$ = $0.000\ 004\ 841 \times 10^{23}$. Now we will round off N_A to the second digit of its uncertainty to express it in a manner consistent with the other data in this problem:

$$N_A = 6.022\ 136\ 9\ (\pm 0.000\ 004\ 8) \times 10^{23} \text{ or } 6.022\ 136\ 9\ (48) \times 10^{23}$$

3-23. C: $12.010\ 7 \pm 0.000\ 8/\sqrt{3}$; H: $1.007\ 94 \pm 0.000\ 07/\sqrt{3}$
O: $15.999\ 4 \pm 0.000\ 3/\sqrt{3}$; N: $14.006\ 7 \pm 0.000\ 2/\sqrt{3}$

+9C:	$9(12.010\ 7 \pm 0.000\ 4_6)$	=	$108.096\ 3 \pm 0.004\ 2$
+9H:	$9(1.007\ 94 \pm 0.000\ 04_0)$	=	$9.071\ 46 \pm 0.000\ 36$
+6O:	$6(15.999\ 4 \pm 0.000\ 1_7)$	=	$95.996\ 4 \pm 0.001\ 0$
+3N:	$3(14.006\ 7 \pm 0.000\ 1_2)$	=	$42.020\ 1 \pm 0.000\ 35$

$C_9H_9O_6N_3$: $255.184\ 26 \pm ?$

Uncertainty = $\sqrt{0.004\ 2^2 + 0.000\ 36^2 + 0.001\ 0^2 + 0.000\ 35^2}$ = 0.004

Answer: 255.184 ± 0.004

3-24. Relative uncertainties:

Large volume: $0.000\ 5$ L$/5.013\ 8_2$ L $= 0.010\%$

Small volume: $0.000\ 9$ mL$/3.793\ 0$ mL $= 0.024\%$

Pressure: 0.03 mm$/400$ mm $= 0.008\%$

Temperature: 0.03 K$/300$ K $= 0.01\%$

The largest uncertainty is in the volume of the small vessel $= 0.024\%$.

Uncertainty in $CO_2 \approx 0.024\%$ of 400 ppm $= 0.000\ 24 \times 400$ ppm $= 0.1$ ppm.

CHAPTER 4
STATISTICS

4-1. The smaller the standard deviation, the greater the precision. There is no necessary relationship between standard deviation and accuracy. The statistics that we do in this chapter pertains to precision, not accuracy.

4-2. (a) $\mu \pm \sigma$ corresponds to $z = -1$ to $z = +1$. The area from $z = 0$ to $z = +1$ is 0.341 3. The area from $z = 0$ to $z = -1$ is also 0.341 3.

Total area (= fraction of population) from $z = -1$ to $z = +1 = 0.682\,6$.

(b) $z = -2$ to $z = +2 \Rightarrow$ area $= 2 \times 0.477\,3 = 0.954\,6$

(c) $z = 0$ to $z = +1 \Rightarrow$ area $= 0.341\,3$

(d) $z = 0$ to $z = 0.5 \Rightarrow$ area $= 0.191\,5$

(e) Area from $z = -1$ to $z = 0$ is 0.341 3. Area from $z = -0.5$ to $z = 0$ is 0.191 5. Area from $z = -1$ to $z = -0.5$ is $0.341\,3 - 0.191\,5 = 0.149\,8$.

4-3. (a) Mean $= \frac{1}{8}(1.526\,60 + 1.529\,74 + 1.525\,92 + 1.527\,31 + 1.528\,94 +$

$1.528\,04 + 1.526\,85 + 1.527\,93) = 1.527\,67$

(b) Standard deviation $=$

$$\sqrt{\frac{(1.526\,60 - 1.527\,67)^2 + \cdots + (1.527\,93 - 1.527\,67)^2}{8 - 1}} = 0.001\,26$$

(c) Variance $= (0.001\,26)^2 = 1.59 \times 10^{-6}$

(d) Significant figures: $\bar{x} \pm s = 1.527_7 \pm 0.001_3$ or 1.528 ± 0.001.

4-4. (a) 1005.3 hours corresponds to $z = (1005.3 - 845.2)/94.2 = 1.700$.

In Table 4-1, the area from the mean to $z = 1.700$ is 0.455 4. The area above $z = 1.700$ is therefore $0.5 - 0.455\,4 = 0.044\,6$.

(b) 798.1 corresponds to $z = (798.1 - 845.2)/94.2 = -0.500$.

The area from the mean to $z = -0.500$ is the same as the area from the mean to $z = +0.500$, which is 0.191 5 in Table 4-1.

901.7 corresponds to $z = (901.7 - 845.2)/94.2 = 0.600$.

The area from the mean to $z = 0.600$ is 0.225 8 in Table 4-1.

The area between 798.1 and 901.7 is the sum of the two areas:

$0.191\,5 + 0.225\,8 = 0.417\,3$

(c) The following spreadsheet shows that the area from $-\infty$ to 800 h is 0.315 7 and the area from $-\infty$ to 900 h is 0.719 6. Therefore, the area from 800 to 900 h is 0.719 6 – 0.315 7 = 0.404 0.

	A	B	C
1	Mean =	Std dev =	
2	845.2	94.2	
3			
4	Area from -∞ to 800 =		0.3157
5	Area from -∞ to 900 =		0.7196
6	Area from 800 to 900 =		0.4040
7			
8	C4 = NORMDIST(800,$A$2,$B$2,TRUE)		
9	C5 = NORMDIST(900,$A$2,$B$2,TRUE)		
10	C6 = C5-C4		

4-5.   (a)   Half the people with tumors have $K < 0.92$ and would not be identified by the test. The false negative rate is 50%.

(b)   The false positive rate is the fraction of healthy people with $K \geq 0.92$. To use Table 4-1, we need to convert $x = 0.92$ to a z value defined as

$$z = \frac{x - \mu}{s} = \frac{0.92 - 0.75}{0.07} = 2.43$$

In Table 4-1, area from mean ($z = 0$) to $z = 2.4$ is 0.491 8. Area from mean to $z = 2.5$ is 0.493 8. We estimate that area from mean to $z = 2.43$ is a little greater than 0.492. Area above $z = 2.43$ is therefore $0.5 - 0.492 = 0.008$. That is, 0.8% of healthy people will have a false positive indication of cancer.

In the following spreadsheet, cell E5 computes the area below $K = 0.92$ with the formula NORMDIST(0.92, $B$4,$B$5,True), where B4 contains K and B5 contains the standard deviation. The area below 0.92 is found in cell E5 to be 0.992 4. The area above $K = 0.92$ is therefore $1 - 0.002 4 = 0.007 6$.

	A	B	C	D	E	F	G	H
1	Gaussian distribution for phase partitioning of plasma proteins							
2								
3	Healthy patients			For healthy people,			Area below cutoff	
4	Mean K =	0.75		area below 0.92 =			for people with tumors	
5	s =	0.07			0.992421		Cutoff (K)	Area
6	Malignant tumor			area above 0.92 =			0.8	0.137656
7	Mean K =	0.92			0.007579		0.81	0.158655
8	s =	0.11					0.82	0.181651
9				area below 0.845 =			0.83	0.206627
10					0.912632		0.84	0.233529
11				area above 0.845 =			0.85	0.26227
12					0.087368		0.845	0.247677
13								
14	E5 = NORMDIST(0.92,$B$4,$B$5,TRUE)				H6 = NORMDIST(G6,$B$7,$B$8,TRUE)			
15	E7 = 1 - E5							

(c) In column G, we vary the value of K and compute the area above K under the curve for people with malignant tumors in column H. We search for the value of K that gives an area of 0.25, which means that 25% of people with tumors will not be identified. The value 0.84 gives an area of 0.233 5 and the value 0.85 gives an area of 0.262 3. By trial and error, we find that $K = 0.845$ gives an area near 0.25.

In cell E10, we insert $K = 0.845$ into the NORMDIST function for healthy people and find that the area below $K = 0.845$ is 0.912 6. The area above $K = 0.845$ is $1 - 0.912\ 6 = 0.087\ 4$. That is, 8.7% of healthy people will produce a false positive result, indicating the presence of a tumor.

4-6.

	A	B	C	D	E	F	G	H
1	Gaussian curve for light bulbs							
2								
3	mean =	x (hours)	y (bulbs)	Formula for cell C2 =				
4	845.2	500	0.49	($A$8*$A$10/($A$6*$A$12))*				
5	std dev =	525	1.25	EXP(-((B4-$A$4)^2)/(2*$A$6^2))				
6	94.2	550	2.98					
7	total bulbs =	600	13.64					
8	4768	625	26.28					
9	bulbs per bar	650	47.19					
10	20	675	78.95					
11	sqrt(2 pi) =	700	123.11					
12	2.506628	725	178.92					
13		750	242.35					
14		775	305.94					
15		800	359.94					
16		825	394.68					
17		845.2	403.85					
18		850	403.33					
19		875	384.14					
20		900	340.99					
21		925	282.09					
22		950	217.50					
23		975	156.29					
24		1000	104.67					
25		1025	65.33					
26		1050	38.00					
27		1075	20.60					
28		1100	10.41					
29		1125	4.90					
30		1150	2.15					
31		1175	0.88					
32		1200	0.34					

4-7. Use the same spreadsheet as in the previous problem, but vary the standard deviation. Here are the results:

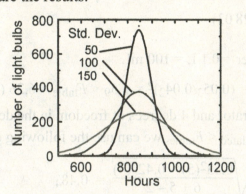

4-8. A confidence interval is a region around the measured mean in which the true mean is likely to lie: If we were to repeat a set of n measurements many times and compute the mean and standard deviation for each set, the 95% confidence interval would include the true population mean (whose value we do not know) in 95% of the sets of n measurements.

4-9. Since the bars are drawn at a 50% confidence level, 50% of them ought to include the mean value if many experiments are performed. 90% of the 90% confidence bars must reach the mean value if we do enough experiments. The 90% bars must be longer than the 50% bars because more of the 90% bars must reach the mean.

4-10. Case 1: Comparing a measured result to a "known" value. See if the known value is included within the 95% confidence interval computed as in Equation 4-7.

Case 2: Comparing replicate measurements. Use Equations 4-8 and 4-9 if the two standard deviations are not significantly different from each other. Use Equations 4-8a and 4-9a if the standard deviations are significantly different. Use the F test to decide if the two standard deviations are significantly different.

Case 3: Comparing individual differences. (Use Equations 4-10 and 4-11.)

4-11. $\bar{x} = 0.14_8$, $s = 0.03_4$

$$90\% \text{ confidence interval} = 0.14_8 \pm \frac{(2.015)(0.03_4)}{\sqrt{6}} = 0.14_8 \pm 0.02_8$$

$$99\% \text{ confidence interval} = 0.14_8 \pm \frac{(4.032)(0.03_4)}{\sqrt{6}} = 0.14_8 \pm 0.05_6$$

4-12. 99% confidence interval $= \bar{x} \pm \dfrac{(3.707)(0.000\,07)}{\sqrt{7}} = \bar{x} \pm 0.000\,10$

(1.527 83 to 1.528 03)

4-13. (a) dL = deciliter = 0.1 L = 100 mL

(b) $F_{calculated} = (0.05_3/0.04_2)^2 = 1.5_9 < F_{table} = 6.26$ (for 5 degrees of freedom in the numerator and 4 degrees of freedom in the denominator).

Since $F_{calculated} < F_{table}$, we can use the following equations:

$$s_{pooled} = \sqrt{\frac{0.53^2(5) + 0.42^2(4)}{6 + 5 - 2}} = 0.48_4$$

$$t = \frac{|14.5_7 - 13.9_5|}{0.48_4}\sqrt{\frac{6 \cdot 5}{6 + 5}} = 2.12 < 2.262 \text{ (listed for 95\% confidence and}$$

9 degrees of freedom). The results agree and the trainee should be released.

4-14.

	A	B	C	D	E	F
1	Comparison of two methods					
2						
3	Sample	Method 1	Method 2	d_i		
4	A	0.88	0.83	0.05	= B4-C4	
5	B	1.15	1.04	0.11		
6	C	1.22	1.39	-0.17		
7	D	0.93	0.91	0.02		
8	E	1.17	1.08	0.09		
9	F	1.51	1.31	0.20		
10			mean =	0.050	= AVERAGE(D4:D9)	
11			stdev =	0.124	= STDEV(D4:D9)	
12			$t_{calculated}$ =	0.987	= D10/D11*SQRT(6)	
13			t_{table} =	2.571	= TINV(0.05,5)	

$t_{calculated} = 0.987 < 2.571$ (Student's t for 95% confidence and 5 deg of freedom)
The difference is <u>not</u> significant.

4-15. In the following spreadsheet, we find $t_{calculated}$ (which is labeled t Stat in cell F10) is less than t_{table} (t Critical two-tail in cell F14). Therefore, the difference between the methods is *not* significant.

The probability P(T<=t) two-tail in cell F13 is 0.37. There is a 37% chance of finding the observed difference between equivalent methods by random variations in results. The probability would have to be ≤0.05 for us to conclude that the methods differ.

	A	B	C	D	E	F	G
1	Paired t test				t-Test: Paired Two Sample for Means		
2							
3	Sample	Method 1	Method 2			Variable 1	Variable 2
4	A	0.88	0.83		Mean	1.14333333	1.09333333
5	B	1.15	1.04		Variance	0.05118667	0.04818667
6	C	1.22	1.39		Observations	6	6
7	D	0.93	0.91		Pearson Correlation	0.84541418	
8	E	1.17	1.08		Hypothesized Mean Difference	0	
9	F	1.51	1.31		df	5	
10					t Stat	0.98692754	
11	Calculated t Statistic in cell F10 is				P(T<=t) one-tail	0.18449929	
12	less than critical t in cell F14.				t Critical one-tail	2.01504918	
13	Therfore, the difference between the				P(T<=t) two-tail	0.36899857	
14	methods is *not significant*.				t Critical two-tail	2.57057764	

4-16. $F_{calculated} = s_2^2/s_1^2 = (0.039)^2/(0.025)^2 = 2.43$

$F_{table} = 9.28$ for 3 degrees of freedom in the numerator and denominator

Since $F_{calculated} < F_{table}$, the difference in standard deviation is not significant and we use Equations 4-8 and 4-9.

$$s_{pooled} = \sqrt{\frac{s_1^2(n_1-1) + s_2^2(n_2-1)}{n_1+n_2-2}} = \sqrt{\frac{0.025^2(4-1) + 0.039^2(4-1)}{4+4-2}} = 0.032\,8$$

$$t_{calculated} = \frac{|\bar{x}_1 - \bar{x}_2|}{s_{pooled}}\sqrt{\frac{n_1 n_2}{n_1+n_2}} = \frac{|1.382 - 1.346|}{0.032\,8}\sqrt{\frac{4\cdot 4}{4+4}} = 1.55$$

t_{table} (4 + 4 − 2 = 6 degrees of freedom) = 2.447

Since $t_{calculated} < t_{table}$, the difference is <u>not</u> significant.

4-17. For Method 1, we compute $\bar{x}_1 = 0.082\,605_2$, $s_1 = 0.000\,013_4$.

For Method 2, $\bar{x}_2 = 0.082\,00_5$, $s_2 = 0.000\,12_9$.

The two standard deviations differ by approximately a factor of 10. We should use the F test to compare the two standard deviations:

$F_{calculated} = s_2^2/s_1^2 = (0.000\,12_9)^2/(0.000\,013_4)^2 = 92.7$

$F_{table} = 6.26$. Since $F_{calculated} > F_{table}$, we use Equations 4-8a and 4-9a.

The following spreadsheet shows that $t_{calculated} = 11.3$ and $t_{table} = 2.57$. $t_{calculated} > t_{table}$, so the difference <u>is</u> significant at the 95% confidence level.

Paired t test		t-Test: Two-Sample Assuming Unequal Variances		
Method 1	Method 2		Variable 1	Variable 2
0.082601	0.08183	Mean	0.082605	0.082005
0.082621	0.08186	Variance	1.8E-10	1.67E-08
0.082589	0.08205	Observations	5	6
0.082617	0.08206	Hypothesized Mean Difference	0	
0.082598	0.08215	df	5	
	0.08208	t Stat	11.31371	$\leftarrow t_{calculated} = 11.3$
		P(T<=t) one-tail	4.72E-05	
		t Critical one-tail	2.015049	
		P(T<=t) two-tail	9.43E-05	
		t Critical two-tail	2.570578	$\leftarrow t_{table} = 2.57$

4-18. 90% confidence interval $= \bar{x} \pm \dfrac{(2.353)(1\%)}{\sqrt{4}} = \bar{x} \pm 1.1_8\% < 1.2\%$.

The answer is yes.

4-19. For indicators 1 and 2: $F_{calculated} = (0.002\,25/0.000\,98)^2 = 5.2_7 > F_{table} \approx 2.2$
(for 27 degrees of freedom in the numerator and 17 degrees of freedom in the
denominator). Since $F_{calculated} > F_{table}$, we use the following equations:

$$\text{Degrees of freedom} = \frac{(s_1^2/n_1 + s_2^2/n_2)^2}{\dfrac{(s_1^2/n_1)^2}{n_1 - 1} + \dfrac{(s_2^2/n_2)^2}{n_2 - 1}}$$

$$= \frac{(0.002\,25^2/28 + 0.000\,98^2/18)^2}{\dfrac{(0.002\,25^2/28)^2}{28 - 1} + \dfrac{(0.000\,98^2/18)^2}{18 - 1}} = 39.8 \approx 40$$

$$t_{calculated} = \frac{|\bar{x}_1 - \bar{x}_2|}{\sqrt{s_1^2/n_1 + s_2^2/n_2}} = \frac{|0.095\,65 - 0.086\,86|}{\sqrt{0.002\,25^2/28 + 0.000\,98^2/18}} = 18.2$$

This is much greater than t for 40 degrees of freedom, which is 2.02.
The difference <u>is</u> significant.

For indicators 2 and 3: $F_{calculated} = (0.001\,13/0.000\,98)^2 = 1.3_3 < F_{table} \approx 2.2$ (for
28 degrees of freedom in the numerator and 17 degrees of freedom in the
denominator). Since $F_{calculated} < F_{table}$, we use the following equations:

$$s_{pooled} = \sqrt{\frac{s_1^2\,(n_1 - 1) + s_2^2\,(n_2 - 1)}{n_1 + n_2 - 2}} = 0.001\,075\,8$$

$$t_{calculated} = \frac{|\bar{x}_1 - \bar{x}_2|}{s_{pooled}} \sqrt{\frac{n_1 n_2}{n_1 + n_2}} = 1.39 < 2.02 \Rightarrow \text{difference is } \underline{not} \text{ significant.}$$

4-20. $\quad s_{pooled} = \sqrt{\dfrac{30.0^2(31) + 29.8^2(31)}{32 + 32 - 2}} = 29.9$

$t = \dfrac{52.9 - 31.4}{29.9}\sqrt{\dfrac{32 \cdot 32}{32 + 32}} = 2.88.$ The table gives t for 60 degrees of freedom,

which is close to 62. The difference <u>is</u> significant at the 95 and 99% levels.

4-21. $\quad \bar{x} = 97.0_0, \quad s = 1.6_6$

95% confidence interval $= \bar{x} \pm \dfrac{ts}{\sqrt{n}} = 97.0_0 \pm \dfrac{(2.776)(1.6_6)}{\sqrt{5}} = 97.0_0 \pm 2.0_6$

Range $= 94.9_4$ to 99.0_6

The 95% confidence interval does not include the certified value of 94.6 ppm, so
the difference <u>is</u> significant at the 95% confidence level.

If we make one more measurement, the results are $\bar{x} = 96.5_8$, $s = 1.8_0$

95% confidence interval $= 96.5_8 \pm \dfrac{(2.571)(1.8_0)}{\sqrt{6}} = 96.5_8 \pm 1.8_9$

Range $= 94.6_9$ to 98.4_7

The 95% confidence interval still does not include the certified value of 94.6 ppm,
so the difference <u>is</u> still significant at the 95% confidence level.

4-22.   (a)   <u>Rainwater</u>:

$F_{calculated} = (0.008/0.005)^2 = 2._{56} < F_{table} = 4.53$ (for 4 degrees of freedom in
the numerator and 6 degrees of freedom in the denominator). Since
$F_{calculated} < F_{table}$, we use the following equations:

$s_{pooled} = \sqrt{\dfrac{0.005^2(6) + 0.008^2(4)}{7 + 5 - 2}} = 0.006_{37}$

$t_{calculated} = \dfrac{0.069 - 0.063}{0.006_{37}}\sqrt{\dfrac{7 \cdot 5}{7 + 5}} = 1._{61} < t_{table} = 2.228$

The difference is <u>not</u> significant.

<u>Drinking water</u>:

$F_{calculated} = (0.008/0.007)^2 = 1._{31} < F_{table} = 6.39$ (for 4 degrees of

freedom in the numerator and 4 degrees of freedom in the denominator).
Since $F_{calculated} < F_{table}$, we use the following equations:

$s_{pooled} = \sqrt{\dfrac{0.007^2(4) + 0.008^2(4)}{5 + 5 - 2}} = 0.007_{52}$

$t = \dfrac{0.087 - 0.078}{0.007_{52}}\sqrt{\dfrac{5 \cdot 5}{5 + 5}} = 1._{89} < 2.306.$ The difference is <u>not</u> significant.

(b) Gas chromatography:

$$s_{pooled} = \sqrt{\frac{0.005^2(6) + 0.007^2(4)}{7 + 5 - 2}} = 0.005_{88}$$

$$t = \frac{0.078 - 0.069}{0.005_{88}}\sqrt{\frac{7 \cdot 5}{7 + 5}} = 2._{61} > 2.228. \text{ The difference } \underline{is} \text{ significant.}$$

Spectrophotometry:

$$s_{pooled} = \sqrt{\frac{0.008^2(4) + 0.008^2(4)}{5 + 5 - 2}} = 0.008_{00}$$

$$t = \frac{0.087 - 0.063}{0.008_{00}}\sqrt{\frac{5 \cdot 5}{5 + 5}} = 4._{74} > 2.306. \text{ The difference } \underline{is} \text{ significant.}$$

4-23. $\bar{x} = 201.8$; $s = 9.34$

$G_{calculated} = |216 - 201.8| / 9.34 = 1.52$

$G_{table} = 1.672$ for five measurements

Because $G_{calculated} < G_{table}$, we should retain 216.

4-24. Slope $= -1.298\,72 \times 10^4\ (\pm 0.001\,319\,0 \times 10^4)$

$\quad\quad\quad\quad = -1.299\ (\pm 0.001) \times 10^4$ or $-1.298_7\ (\pm 0.001_3) \times 10^4$

Intercept $= 256.695\ (\pm 323.57) = 3\ (\pm 3) \times 10^2$

4-25.

x_i	y_i	x_iy_i	x_i^2	d_i	d_i^2
0	1	0	0	0.071 43	0.005 10
2	2	4	4	−0.214 29	0.045 92
3	3	9	9	0.142 86	0.020 41
sums: 5	6	13	13	0	0.071 43

$$m = \frac{n\sum(x_iy_i) - \sum x_i \sum y_i}{n\sum(x_i^2) - (\sum x_i)^2} = \frac{3 \times 13 - 5 \times 6}{3 \times 13 - 5^2} = \frac{9}{14} = 0.642\,86$$

$$b = \frac{\sum(x_i^2)\sum y_i - \sum(x_iy_i)\sum x_i}{n\sum(x_i^2) - (\sum x_i)^2} = \frac{13 \times 6 - 13 \times 5}{3 \times 13 - 5^2} = \frac{13}{14} = 0.928\,57$$

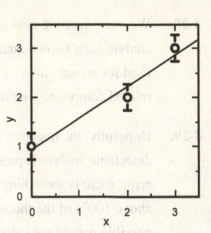

$$s_y = \sqrt{\frac{\sum(d_i^2)}{n-2}} = \sqrt{\frac{0.071\,43}{3-2}} = 0.267\,26$$

$$s_m = s_y\sqrt{\frac{n}{D}} = (0.267\,26)\sqrt{\frac{3}{14}} = 0.123\,71$$

$$s_b = s_y\sqrt{\frac{\sum(x_i^2)}{D}} = (0.267\,26)\sqrt{\frac{13}{14}}$$
$$= 0.257\,54$$

$$\text{slope} = 0.6_4 \pm 0.1_2 \quad \text{intercept} = 0.9_3 \pm 0.2_6$$

4-26.

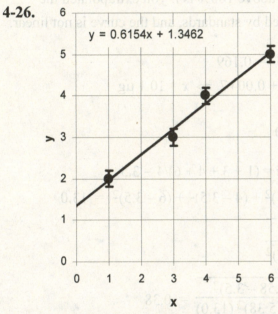

$y = 0.6154x + 1.3462$

4-27.

	A	B	C	D	E	F	G	H	I
1	x	y							
2	3.0	-0.074							
3	10.0	-1.411							
4	20.0	-2.584							
5	30.0	-3.750							
6	40.0	-5.407							
7									
8		LINEST output							
9	m	-0.13789	0.195343	b					
10	s_m	0.006635	0.162763	s_b					
11	R^2	0.993102	0.197625	s_y					
12									
13	Highlight cells B9:C11								
14	Type								
15	"=LINEST(B2:B6,A2:A6,TRUE,TRUE)"								
16	Press CTRL+SHIFT+ENTER (on PC)								
17	Press COMMAND+RETURN (on Mac)								

4-28. We must measure how an analytical procedure responds to a known quantity of analyte (or a known quantity of a related compound) before the procedure can be used for an unknown. Therefore, we must be able to measure out the analyte (or a related compound) in pure form to use as a calibration standard.

4-29. Hopefully, the negative value is within experimental error of 0. If so, no detectable analyte is present. If the negative concentration is beyond experimental error, there is something wrong with your analysis. The same is true for a value above 100% of the theoretical maximum concentration of an analyte. Another possible way to get values below 0 or above 100% is if you extrapolated the calibration curve past the range covered by standards, and the curve is not linear.

4-30. Corrected absorbance $= 0.264 - 0.095 = 0.169$

Equation of line: $0.169 = 0.016\,30\,x + 0.004\,7 \Rightarrow x = 10.1\ \mu g$

4-31. (a) $x = \dfrac{y - b}{m} = \dfrac{2.58 - 1.3_5}{0.61_5} = 2.00$

$\bar{y} = (2 + 3 + 4 + 5)/4 = 3.5$ $\qquad \bar{x} = (1 + 3 + 4 + 6)/4 = 3.5$

$\sum (x_i - \bar{x})^2 = (1 - 3.5)^2 + (3 - 3.5)^2 + (4 - 3.5)^2 + (6 - 3.5)^2 = 13.0$

$$s_x = \frac{s_y}{|m|} \sqrt{\frac{1}{k} + \frac{1}{n} + \frac{(y - \bar{y})^2}{m^2\, \Sigma (x_i - \bar{x})^2}}$$

$$= \frac{0.196\,12}{|0.615\,38|} \sqrt{\frac{1}{1} + \frac{1}{4} + \frac{(2.58 - 3.5)^2}{(0.615\,38)^2\,(13.0)}} = 0.38$$

Answer: $2.0_0 \pm 0.3_8$

(b) For $k = 4$ replicate measurements,

$$s_x = \frac{s_y}{|m|} \sqrt{\frac{1}{k} + \frac{1}{n} + \frac{(y - \bar{y})^2}{m^2\, \Sigma (x_i - \bar{x})^2}}$$

$$= \frac{0.196\,12}{|0.615\,38|} \sqrt{\frac{1}{4} + \frac{1}{4} + \frac{(2.58 - 3.5)^2}{(0.615\,38)^2\,(13.0)}} = 0.26$$

Answer: $2.0_0 \pm 0.2_6$

4-32.   Mean absorbance = $(0.265 + 0.269 + 0.272 + 0.258)/4 = 0.266_0$

Mean blank: $(0.099 + 0.091 + 0.101 + 0.097)/4 = 0.097_0$

$\Rightarrow$ Corrected absorbance $= 0.266_0 - 0.097_0 = 0.169_0$

Cells B30 and B31 of the spreadsheet show that there are 10.1 ± 0.2 µg protein.

	A	B	C	D	E	F	G	H	I
1	Least-Squares Spreadsheet								
2									
3		x	y						
4		0	-0.0003						
5		0	-0.0003						
6		0	0.0007						
7		5	0.0857						
8		5	0.0877						
9		5	0.0887						
10		10	0.1827						
11		10	0.1727						
12		10	0.1727						
13		15	0.2457						
14		15	0.2477						
15		20	0.3257						
16		20	0.3257						
17		20	0.3307						
18									
19		LINEST output:							
20	m	0.01630	0.00470	b					
21	s_m	0.00022	0.00263	s_b					
22	R^2	0.99785	0.00588	s_y					
23									
24	n =		14	B24 = COUNT(B4:B17)					
25	Mean y =		0.16184	B25 = AVERAGE(C4:C17)					
26	$\Sigma(x_i - \text{mean } x)^2$ =		723.214	B26 = DEVSQ(B4:B17)					
27									
28	Measured y =	0.169	Input						
29	Number of replicate measurements of y (k) =	4	Input						
30	Derived x	10.082	B21 = (B28-C20)/B20						
31	s_x =	0.2045	B31 = ((C22/B20)*SQRT((1/B29)+(1/B24)+((B28-B25)^2)/(B20^2*B26))						

Chart: y = 0.0163x + 0.0047; Absorbance vs. Protein (µg)

4-33. (a)

	A	B	C	D	E	F	G	H	I
1	Least-Squares Spreadsheet								
2									
3		x	$y_{corrected}$	y					
4		0	0	9.1					
5		0.062	38.4	47.5					
6		0.122	86.5	95.6					
7		0.245	184.7	193.8					
8		0.486	378.4	387.5					
9		0.971	803.4	812.5					
10		1.921	1662.8	1671.9					
11									
12		LINEST output:							
13	m	869.13	-22.0852	b					
14	s_m	10.6422	8.9474	s_b					
15	R^2	0.9993	18.0527	s_y					
16									
17	n =		7	B17 = COUNT(B4:B10)					
18	Mean y =		450.6	B18 = AVERAGE(C4:C10)					
19	$\Sigma(x_i - \text{mean } x)^2 =$		2.87757	B19 = DEVSQ(B4:B10)					
20									
21	Measured y =		145.0	Input					
22	Number of replicate measurements of y (k) =		4	Input					
23	Derived x		0.19224	B23 = (B21-C13)/B13					
24	$s_x =$		0.0137	B24 = (C15/B13)*SQRT((1/B22)+(1/B17)+((B21-B18)^2)/(B13^2*B19))					

Chart (columns F–I): plot of Signal (mV) vs Methane (vol%), with fit line y = 869.13x − 22.085

(b) Corrected signal = 154.0 − 9.0 = 145.0 mV

(c) Cells B23 and B24 give $[CH_4] = 0.19_2 \ (\pm 0.01_4)$ vol%

4-34. $0.350 = -1.17 \times 10^{-4} x^2 + 0.018\,58\,x - 0.000\,7$

$1.17 \times 10^{-4} x^2 - 0.018\,58\,x + 0.350\,7 = 0$

$$x = \frac{+0.018\,58 \pm \sqrt{0.018\,58^2 - 4\,(1.17 \times 10^{-4})\,(0.350\,7)}}{2\,(1.17 \times 10^{-4})} = 137 \text{ or } 21.9 \ \mu g$$

Correct answer is 21.9 μg.

4-35. (a) The logarithmic graph spreads out the data and is linear over the entire range.

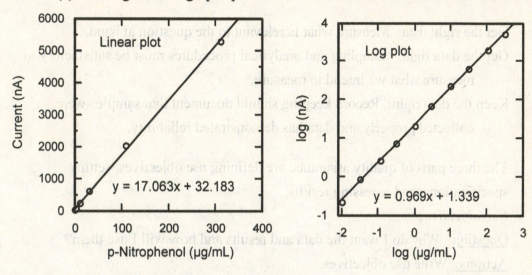

(b) log (current, nA) = 0.969 2 log (concentration, µg/mL) + 1.339

(c) log (99.9) = 0.969 2 log [X] + 1.339

$\Rightarrow$ log [X] = 0.681 6 $\Rightarrow$ [X] = 4.80 µg/mL

4-36. For 8 degrees of freedom, $t_{90\%}$ = 1.860 and $t_{99\%}$ = 3.355.

90% confidence interval: 15.2_2 ($\pm 1.860 \times 0.4_6$) = $15.2_2 \pm 0.8_6$ µg

99% confidence interval: 15.2_2 ($\pm 3.355 \times 0.4_6$) = $15._2 \pm 1._5$ µg

5-1. Get the right data: Measure what is relevant to the question at hand.

Get the data right: Sampling and analytical procedures must be satisfactory to measure what we intend to measure.

Keep the data right: Record keeping should document that samples were collected properly and data has demonstrated reliability.

5-2. The three parts of quality assurance are defining use objectives, setting specifications, and assessing results.

Use objectives:

<u>Question:</u> Why do I want the data and results and how will I use them?

<u>Actions:</u> Write use objectives.

Specifications:

<u>Question:</u> How good do the numbers have to be?

<u>Actions:</u> Write specifications and pick an analytical method to meet the specifications. Consider requirements for sampling, precision, accuracy, selectivity, sensitivity, detection limit, robustness, and allowed rate of false results. Plan to employ blanks, fortification, calibration checks, quality control samples, and control charts. Write and follow standard operating procedures.

Assessment:

<u>Question:</u> Did I meet the specifications?

<u>Actions:</u> Compare data and results with specifications. Document procedures and keep records suitable for meeting use objectives. Verify that the use objectives were met.

5-3. *Precision* is demonstrated by the repeatability of analyses of replicate samples and replicate portions of the same sample. Accuracy is demonstrated by analyzing certified reference materials, by comparing results from different analytical methods, by fortification (spike) recovery, by standard additions, by calibration checks, blanks, and quality control samples (blind samples).

5-4. *Raw data* are directly measured quantities, such as peak area in a chromatogram or volume from a buret. *Treated data* are concentrations or amounts found by applying a calibration method to the raw data. *Results*, such as the mean and standard deviation, are what we ultimately report after applying statistics to treated data.

5-5. A calibration check is an analysis of a solution *formulated by the analyst* to contain a known concentration of analyte. It is the analyst's own check that procedures and instruments are functioning correctly. A performance test sample is an analysis of a solution *formulated by someone other than the analyst* to contain a known concentration of analyte. It is a test to see if the analyst gets the right answer when he or she does not know the right answer.

5-6. A blank is a sample intended to contain no analyte. Positive response to the blank arises from analyte impurities in reagents and equipment and from interference by other species. A method blank is taken through all steps in a chemical analysis. A reagent blank is the same as a method blank, but it has not been subjected to all sample preparation procedures. A field blank is similar to a method blank, but it has been taken into the field and exposed to the same environment as samples collected in the field and transported to the lab.

5-7. Linear range is the analyte concentration interval over which the analytical signal is proportional to analyte concentration. Dynamic range is the concentration range over which there is a useable response to analyte, even if it is not linear. Range is the analyte concentration interval over which an analytical method has specified linearity, accuracy, and precision.

5-8. A false positive is a conclusion that analyte exceeds a certain limit when, in fact, it is below the limit. A false negative is a conclusion that analyte is below a certain limit when, in fact, it is above the limit.

5-9. ~1% of the area under the curve for blanks lies to the right of the detection limit. Therefore, ~1% of samples containing no analyte will give a signal above the detection limit. 50% of the area under the curve for samples containing analyte at the detection limit lies below the detection limit. Therefore, 50% of samples containing analyte at the detection limit will be reported as not containing analyte at a level above the detection limit.

5-10. A control chart tracks the performance of a process to see if it remains within expected bounds. Six indications that a process might be out of control are (1) a reading outside the action lines, (2) 2 out of 3 consecutive readings between the warning and action lines, (3) 7 consecutive measurements all above or all below the center line, (4) six consecutive measurements, all steadily increasing or all steadily decreasing, wherever they are located, (5) 14 consecutive points alternating up and down, regardless of where they are located, and (6) an obvious nonrandom pattern.

5-11. Statement (c) is correct. The purpose of the analysis is to see if concentrations are in compliance with (in other words, do not exceed) levels set by a certain rule.

5-12. The *instrument detection limit* is obtained by replicate measurements of aliquots from one sample. The *method detection limit* is obtained by preparing and analyzing many independent samples. There is more variability in the latter procedure, so the method detection limit should be higher than the instrument detection limit.

Robustness is the ability of an analytical method to be unaffected by small, deliberate changes in operating parameters. *Intermediate precision* is the variation observed when an assay is performed by different people on different instruments on different days in the same lab. Each analysis might incorporate independently prepared reagents and different lots of the same chromatography column from one manufacturer. When demonstrating intermediate precision, experimental conditions are intended to be the same in each analysis. When measuring robustness, conditions are intentionally varied by small amounts.

5-13. *Instrument precision*, also called *injection precision*, is the reproducibility observed when the same quantity of one sample is repeatedly introduced into an instrument.

Intra-assay precision is evaluated by analyzing aliquots of a homogeneous material several times by one person on one day with the same equipment.

Intermediate precision is the variation observed when an assay is performed by different people on different instruments on different days in the same lab.

Interlaboratory precision is the reproducibility observed when aliquots of the same sample are analyzed by different people in different laboratories at different times using equipment and reagents belonging to each lab.

5-14. Criteria:
- Observations outside action lines — no
- 2 out of 3 consecutive measurements between warning and action lines — no

- 7 consecutive measurements all above or all below the center line — YES: observations 2-10 (starting from the left side) are all below the center line
- 6 consecutive measurements steadily increasing or steadily decreasing — no
- 14 consecutive points alternating up and down — no
- Obvious nonrandom pattern — no

5-15. LINEST gives m, b, s_m, s_b, R^2, and s_y in cells B19:C21. TRENDLINE produces the same value of m, b, and R^2, which are printed inside the graph. The 95% confidence interval for y is computed in cell C24.

	A	B	C	D	E	F	G	H	I
1	Graphing data with random Gaussian noise								
2	Generating equation: y = 26.4x+1.37								
3	Gaussian noise =		80						
4									
5	x	y							
6	0	14							
7	10	350							
8	20	566							
9	30	957							
10	40	1067							
11	50	1354							
12	60	1573							
13	70	1732							
14	80	2180							
15	90	2330							
16	100	2508							
17									
18		LINEST output							
19	m	24.80727	89.72727	b					
20	s_m	0.683532	40.43827	s_b					
21	R^2	0.993214	71.6894	s_y					
22									
23	Student's t =		2.262159	=TINV(0.05,9)					
24		t^*s_y	162.1728	=C23*C21					

Inside the graph:
$$y = 24.807x + 89.727$$
$$R^2 = 0.9932$$
Error bars are $\pm t^*s_y$ in cell C24

5-16. (a) For the fortification level of 22.2 ng/mL, the mean of the 5 values is 23.6_6 ng/mL and the standard deviation is 5.6_3 ng/mL.

$$\text{Precision} = 100 \times \frac{5.63}{23.66} = 23.8\%.$$

$$\text{Accuracy} = 100 \times \frac{23.66 - 22.2}{22.2} = 6.6\%$$

For the fortification level of 88.2 ng/mL, the mean of the 5 values is 82.4_8 ng/mL and the standard deviation is 11.4_9 ng/mL.

$$\text{Precision} = 100 \times \frac{11.49}{82.48} = 13.9\%.$$

$$\text{Accuracy} = 100 \times \frac{82.48 - 88.2}{88.2} = -6.5\%$$

For the fortification level of 314 ng/mL, the mean of the 5 values is $302._8$ ng/mL and the standard deviation is 23.5_1 ng/mL.

$$\text{Precision} = 100 \times \frac{23.51}{302.8} = 7.8\%.$$

$$\text{Accuracy} = 100 \times \frac{302.8 - 314}{314} = -3.6\%$$

(b) Standard deviation of 10 samples: $s = 28._2$; mean blank: $y_{blank} = 45._0$

Signal detection limit $= y_{blank} + 3s = 45._0 + (3)(28._2) = 129.6$

$$\text{Concentration detection limit} = \frac{3s}{m} = \frac{(3)(28._2)}{1.75 \times 10^9 \text{ M}^{-1}} = 4.8 \times 10^{-8} \text{ M}$$

$$\text{Lower limit of quantitation} = \frac{10s}{m} = \frac{(10)(28._2)}{1.75 \times 10^9 \text{ M}^{-1}} = 1.6 \times 10^{-7} \text{ M}$$

5-17. (a) 1 wt% $\Rightarrow C = 0.01$: $CV(\%) \approx 2^{(1-0.5\log 0.01)} = 2^2 = 4\%$

If $C = 10^{-12}$, $CV(\%) \approx 2^7 = 128\%$.

(b) If class CV is 50% of the value given by the Horwitz curve, it would be $0.5 \times 2^{(1-0.5\log 0.1)} = 1.4\%$.

5-18. Mean $= 0.38_3$ µg/L and standard deviation $= 0.021_4$ µg/L

$$\% \text{ recovery} = \frac{0.38_3 \text{ µg/L}}{0.40 \text{ µg/L}} \times 100 = 96\%$$

The measurements are already expressed in concentration units. The concentration detection limit is 3 times the standard deviation $= 3(0.021_4 \text{ µg/L}) = 0.064$ µg/L.

5-19. The low concentration of Ni-EDTA has a standard deviation of 28.2 counts for 10 measurements. The detection limit is estimated to be

$$y_{dl} = y_{blank} + 3s = 45 + 3(28.2) = 129.6 \text{ counts}$$

To convert counts to molarity, we note that a 1.00 µM solution gave a net signal of $1797 - 45 = 1752$ counts. The slope of the calibration curve is therefore estimated to be

$$m = \frac{y_{sample} - y_{blank}}{\text{sample concentration}} = \frac{1797 - 45}{1.00 \text{ µM}} = 1.75_2 \times 10^9 \frac{\text{counts}}{\text{M}}$$

The minimum detectable concentration is

$$\frac{3\,s}{m} = \frac{(3)(28.2)\ \text{counts}}{1.75_2 \times 10^9\ \text{counts/M}} = 4.8 \times 10^{-8}\ \text{M}$$

5-20. For a concentration of 0.2 μg/L, the relative standard deviation of 14.4% corresponds to $(0.144)(0.2\ \mu\text{g/L}) = 0.028\ 8\ \mu\text{g/L}$. The detection limit is $3(0.028\ 8\ \mu\text{g/L}) = 0.086\ \mu\text{g/L}$. Here are the results for the other concentrations:

Concentration (μg/L)	Relative standard deviation (%)	Concentration standard deviation (μg/L)	Detection limit (μg/L)
0.2	14.4	0.028 8	0.086
0.5	6.8	0.034 0	0.102
1.0	3.2	0.032 0	0.096
2.0	1.9	0.038 0	0.114
		mean detection limit:	0.10

5-21. If an athlete tests positive for drugs, the test should be repeated with a second sample that was drawn at the same time as the first sample and preserved in an appropriate manner. If there is a 1% chance of a false positive in each test, the chances of observing a false positive twice in a row are 1% of 1% or 0.01%. Instead of falsely accusing 1% of innocent athletes, we would be falsely accusing 0.01% of innocent athletes.

5-22. *Comparison of Lab C with Lab A:*

First, use the F test to see if the standard deviations are significantly different:
$F_{\text{calculated}} = s_C^2/s_A^2 = 0.78^2/0.14^2 = 31._0 > F_{\text{table}} = 3.88$ (with 2 degrees of freedom for s_C and 12 degrees of freedom for s_A)

Standard deviations are not equivalent, so use the following t test:

$$\text{Degrees of freedom} = \frac{(s_1^2/n_1 + s_2^2/n_2)^2}{\dfrac{(s_1^2/n_1)^2}{n_1 - 1} + \dfrac{(s_2^2/n_2)^2}{n_2 - 1}} = \frac{(0.14^2/13 + 0.78^2/3)^2}{\dfrac{(0.14^2/13)^2}{13 - 1} + \dfrac{(0.78^2/3)^2}{3 - 1}} = 2.03 \approx 2$$

$$t_{\text{calculated}} = \frac{|\bar{x}_1 - \bar{x}_2|}{\sqrt{s_1^2/n_1 + s_2^2/n_2}} = \frac{|1.59 - 2.68|}{\sqrt{0.14^2/13 + 0.78^2/3}} = 2.4_1$$

For 2 degrees of freedom, $t_{\text{table}} = 4.303$ for 95% confidence. Since $t_{\text{calculated}} < t_{\text{table}}$, we conclude that the difference between Lab C and Lab A <u>is not</u> significant.

Comparison of Lab C with Lab B:

$F_{calculated} = s_C^2/s_B^2 = 0.78^2/0.56^2 = 1.9_4 < F_{table} = 4.74$ (with 2 degrees of

freedom for s_C and 7 degrees of freedom for s_A). The standard deviations are not

significantly different, so we use the following t test:

$$s_{pooled} = \sqrt{\frac{0.56^2 (8 - 1) + 0.78^2 (3 - 1)}{8 + 3 - 2}} = 0.61_6$$

$$t_{calculated} = \frac{|1.65 - 2.68|}{0.61_6} \sqrt{\frac{8 \cdot 3}{8 + 3}} = 2.4_7$$

$t_{table} = 2.262$ for 95% confidence and $8 + 3 - 2 = 9$ degrees of freedom.

$t_{calculated} > t_{table}$, so the difference is significant at the 95% confidence level.

It makes no sense to conclude that Lab C [2.68 ± 0.78 (3)] > Lab B [1.65 ± 0.56
(8)], but Lab C = Lab A [1.59 ± 0.14 (13)]. The problem with the comparison of
Labs C and A is that the standard deviation of C is much greater than the standard
deviation of A and the number of replicates for C is much less than the number of
replicates for A. The result is that we used a large composite standard deviation
and a small composite number of degrees of freedom. The conclusion is biased
by a large standard deviation and a small number of degrees of freedom. I would
tentatively conclude that results from Lab C are greater than results from Labs B
and A. I would also ask for more replicate results from Lab C. With just 3
replications, it is hard to reach any statistically significant conclusions.

5-23.   A small volume of standard will not change the sample matrix very much, so
matrix effects remain nearly constant. If large, variable volumes of standard are
used, the matrix is different in every mixture and the matrix effects will be
different in every sample.

5-24.   (a) $[Cu^{2+}]_f = [Cu^{2+}]_i \frac{V_i}{V_f} = 0.950 [Cu^{2+}]_i$

(b) $[S]_f = [S]_i \frac{V_i}{V_f} = (100.0 \text{ ppm}) \left(\frac{1.00 \text{ mL}}{100.0 \text{ mL}}\right) = 1.00 \text{ ppm}$

(c) $\frac{[Cu^{2+}]_i}{1.00 \text{ ppm} + 0.950[Cu^{2+}]_i} = \frac{0.262}{0.500} \Rightarrow [Cu^{2+}]_i = 1.04 \text{ ppm}$

5-25. (a) All solutions were made up to the same final volume. Therefore, we prepare a graph of signal versus concentration of added standard. The line in the graph was drawn by the method of least squares with the following spreadsheet. The x-intercept, 8.72 ppb, is the concentration of unknown in the 10-mL solution. In cell B27 of the spreadsheet (on the next page), we find the standard deviation of the x-intercept to be 0.427 ppm. A reasonable answer is $8.7_2 \pm 0.4_3$ ppb.

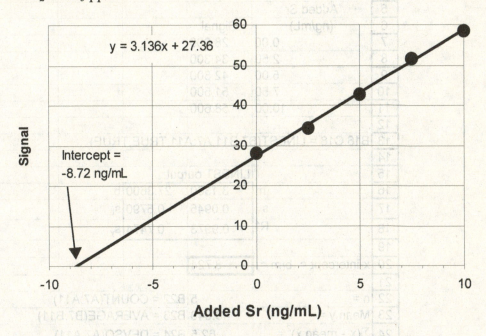

(b) Unknown solution volume = 10.0 mL with Sr = 8.72 ppb = 8.72 ng/mL. In 10.0 mL, there are (10 mL)(8.72 ng/mL) = 87.2 ng. Solution was made from 0.750 mg of tooth enamel. Sr (ppm) in tooth enamel is

$$\text{Concentration (ppm)} = \frac{\text{mass of Sr}}{\text{mass of enamel}} \times 10^6$$

$$= \frac{87.2 \times 10^{-9} \text{ g}}{0.750 \times 10^{-3} \text{ g}} \times 10^6 = 116 \text{ ppm}$$

(c) Relative uncertainty of intercept is $100 \times 0.43/8.72 = 4.9\%$, which leads to a 4.9% uncertainty in the concentration of Sr in the tooth enamel. 0.049×116 ppm = 5.7 ppm. Final answer: 116 ± 6 ppm.

(d) Student's t for $n - 2 = 5 - 2 = 3$ degrees of freedom and 95% confidence is 3.182. We found standard deviation = 5.7 ppm. 95% confidence interval is $\pm ts = (3.182)(5.7 \text{ ppm}) = 18.1$ ppm. Answer: 116 ± 18 ppm.

Spreadsheet for 5-25 (a). To execute LINEST, highlight cells B16-C18, enter "=LINEST(B7:B11,A7:A11,TRUE,TRUE", and press CONTROL + SHIFT + ENTER on a PC or COMMAND(⌘) + RETURN on a Mac.

	A	B	C	D	E
1	Standard Addition Constant Volume Least-Squares Spreadsheet				
2					
3					
4	x	y			
5	Added Sr				
6	(ng/mL)	Signal			
7	0.00	28.000			
8	2.50	34.300			
9	5.00	42.800			
10	7.50	51.500			
11	10.00	58.600			
12					
13	B16:C18 = LINEST(B7:B11,A7:A11,TRUE,TRUE)				
14					
15		LINEST output:			
16	m	3.1360	27.3600	b	
17	s_m	0.0945	0.5790	s_b	
18	R^2	0.9973	0.7474	s_y	
19					
20	x-intercept = -b/m =	-8.724			
21					
22	n =	5	B22 = COUNT(A7:A11)		
23	Mean y =	43.040	B23 = AVERAGE(B7:B11)		
24	$\Sigma(x_i - \text{mean } x)^2 =$	62.5	B24 = DEVSQ(A7:A11)		
25					
26	Std deviation of				
27	x-intercept =	0.427			
28	B27 =(C18/ABS(B16))*SQRT((1/B22) + B23^2/(B16^2*B24))				

5-26. (a) The intercept for tap water is –6.0 mL, corresponding to an addition of (6.0 mL)(0.152 ng/mL) = 0.91_2 ng Eu(III). This much Eu(III) is in 10.00 mL of tap water, so the concentration is 0.91_2 ng/10.00 mL = 0.091 ng/mL. For pond water, the intercept of –14.6 mL corresponds to an addition of (14.6 mL)(15.2 ng/mL) = 2.22×10^2 ng/10.00 mL pond water = 22.2 ng/mL.

(b) Added standard Eu(III) gives a response of 3.03 units/ng for tap water and 0.0822 units/ng for pond water. The relative response is 3.03/0.0822 = 36.9 times greater in tap water than in pond water. There is probably a *matrix effect* in which something in pond water decreases the Eu(III) emission. By using standard addition, we measure the response in the actual sample matrix. Even though Eu(III) in pond water and tap water do not give equal signals, we measure the actual signal in each matrix and can therefore carry out accurate analyses.

5-27.

	A	B	C	D
1	Standard Addition Constant Volume Least-Squares Spreadsheet			
2				
3	V_{total} =	V_s (mL) =	x	y
4	50	NaCl	Concentration of	
5	$[S]_i$ (M) =	standard	added NaCl	I(s+x) =
6	2.64	added	$[S]_f$	signal
7	Vo =	0.000	0	3.13
8	25.00	1.000	0.0528	5.40
9		2.000	0.1056	7.89
10		3.000	0.1584	10.30
11		4.000	0.2112	12.48
12	C7 = $A$6*B7/$A$4			
13	B16:D18 = LINEST(E7:E11,C7:C11,TRUE,TRUE)			
14				
15		LINEST output:		
16	m	44.6970	3.1200	b
17	s_m	0.5511	0.0713	s_b
18	R^2	0.9995	0.0920	s_y
19				
20	x-intercept = -b/m =	-0.06980		
21				
22	n =	5	B22 = COUNT(B7:B11)	
23	Mean y =	7.84	B23 = AVERAGE(E7:E11)	
24	$\Sigma(x_i$ - mean x$)^2$ =	0.0278784	B24 = DEVSQ(C7:C11)	
25				
26	Std deviation of			
27	x-intercept =	0.00235		
28	B27 =(C18/ABS(B16))*SQRT((1/B22) + B23^2/(B16^2*B24))			

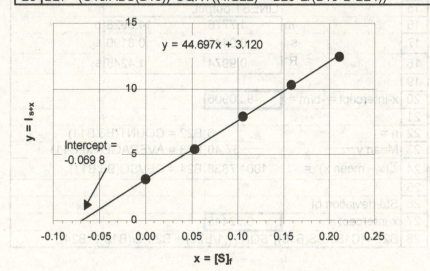

$y = 44.697x + 3.120$

Intercept = -0.069 8

(a) All solutions are made to constant volume, so we plot I_{s+x} vs. $[S]_f$. The negative intercept is $[X]_f = 0.069\ 8$ M. The initial concentration of NaCl is larger by the dilution factor of 2 (50.00 mL/25.00 mL). The initial concentration of NaCl in serum was $2.000 \times 0.069\ 8$ M = 0.013 96 M.

(b) The x-intercept is computed in cell B20 and its uncertainty in cell B27. The relative uncertainty is $(0.002\ 35)/(0.069\ 8) = 3.37\%$. This uncertainty is much larger than the relative uncertainties in volume measurement, so the uncertainty in the original concentration of Na^+ should be 3.37%. A reasonable expression of $[Na^+]$ in the original serum is 0.140 (±3.37%) M = 0.140 (±0.004$_7$) M.

95% confidence interval = ± ts = ± (3.182)(0.004$_7$ M) = ± 0.015 M, where t is taken for 5 − 2 = 3 degrees of freedom.

5-28.

	A	B	C	D
1	Standard Addition Constant Volume Least-Squares Spreadsheet			
2				
3		x	y	
4		Spike (mg/g)	I(s+x) =	
5		$[S]_f$	signal	
6		0.00	15.6	
7		3.12	21.1	
8		7.18	25.5	
9		8.48	30.0	
10		20.0	48.8	
11		38.2	83.4	
12				
13	B16:D18 = LINEST(C6:C11,B6:B11,TRUE,TRUE)			
14				
15		LINEST output:		
16	m	1.7776	14.5928	b
17	s_m	0.0449	0.8190	s_b
18	R^2	0.9974	1.4246	s_y
19				
20	x-intercept = -b/m =	-8.20906		
21				
22	n =	6	B22 = COUNT(B6:B11)	
23	Mean y =	37.40	B23 = AVERAGE(C6:C11)	
24	$\Sigma(x_i$ - mean x$)^2$ =	1004.7838	B24 = DEVSQ(B6:B11)	
25				
26	Std deviation of			
27	x-intercept =	0.62445		
28	B27 =(C18/ABS(B16))*SQRT((1/B22) + B23^2/(B16^2*B24))			

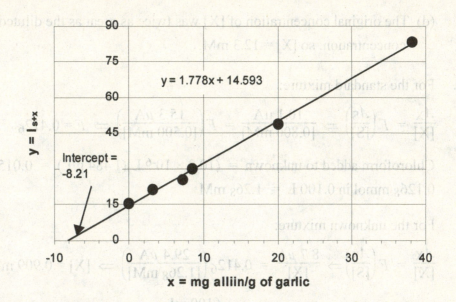

y = 1.778x + 14.593

Intercept = -8.21

$y = I_{s+x}$

x = mg alliin/g of garlic

(a) In cells B20 and B27 of the spreadsheet, the negative x-intercept of the standard addition graph is 8.21 ± 0.62 mg alliin/g garlic.

(b) Two moles of alliin (FM 177.2) produce one mole of allicin (FM 162.3) in the assay. Therefore, the quantity of allicin in garlic is ½(162.3/177.2)(8.21 $\pm$ 0.62 mg/g) = 3.76 ± 0.28 mg allicin/g garlic or 3.8 ± 0.3 mg allicin/g garlic.

5-29. Standard addition is appropriate when the sample matrix is unknown or complex and hard to duplicate, and unknown matrix effects are anticipated. An internal standard can be added to an unknown at the start of a procedure in which uncontrolled losses of sample will occur. The relative amounts of unknown and standard remain constant. The internal standard is excellent if instrument conditions vary from run to run. Variations affect the analyte and standard equally, so the relative signal remains constant. In chromatography the amount of sample injected into the instrument is very small and not very reproducible. However, the relative quantities of standard and analyte remain constant regardless of the sample size.

5-30. (a) $\dfrac{A_X}{[X]} = F\left(\dfrac{A_S}{[S]}\right) \Rightarrow \dfrac{3\,473}{[3.47 \text{ mM}]} = F\left(\dfrac{10\,222}{[1.72 \text{ mM}]}\right) \Rightarrow F = 0.168_4$

(b) $[S] = (8.47 \text{ mM})\left(\dfrac{1.00 \text{ mL}}{10.0 \text{ mL}}\right) = 0.847 \text{ mM}$

(c) $\dfrac{A_X}{[X]} = F\left(\dfrac{A_S}{[S]}\right) \Rightarrow \dfrac{5\,428}{[X]} = 0.168_4\left(\dfrac{4\,431}{[0.847 \text{ mM}]}\right) \Rightarrow [X] = 6.16 \text{ mM}$

(d) The original concentration of [X] was twice as great as the diluted
     concentration, so [X] = 12.3 mM.

5-31. For the standard mixture:

$$\frac{A_X}{[X]} = F\left(\frac{A_S}{[S]}\right) \Rightarrow \frac{10.1\ \mu A}{[0.800\ mM]} = F\left(\frac{15.3\ \mu A}{[0.500\ mM]}\right) \Rightarrow F = 0.412_6$$

Chloroform added to unknown = $(10.2 \times 10^{-6}\ L)(1\ 484\ g/L) = 0.015\ 1_4\ g =$
0.126_8 mmol in 0.100 L = 1.26_8 mM

For the unknown mixture:

$$\frac{A_X}{[X]} = F\left(\frac{A_S}{[S]}\right) \Rightarrow \frac{8.7\ \mu A}{[X]} = 0.412_6\left(\frac{29.4\ \mu A}{[1.26_8\ mM]}\right) \Rightarrow [X] = 0.909\ mM$$

$$[DDT]\ in\ unknown = (0.909\ mM)\left(\frac{100\ mL}{10.0\ mL}\right) = 9.09\ mM$$

5-32. Data in the following table are plotted in the accompanying graph. If the
equation

$$\frac{\text{area of analyte signal}}{\text{area of standard signal}} = F\left(\frac{\text{concentration of analyte}}{\text{concentration of standard}}\right)$$

is obeyed, the graph should be a straight line going through the origin, which it
is. The slope, 1.07_{57}, is the response factor. Over the concentration ratio
analyte/standard = 0.10 to 1.00, the standard deviation of the response factor in
the table is $0.06_{68} = 6.2\%$.

Sample	Concentration ratio $C_{10}H_8/C_{10}D_8$	Area ratio $C_{10}H_8/C_{10}D_8$	$F =$ area ratio/conc. ratio
1	0.10	0.101	1.01_{27}
2	0.50	0.573	1.14_{61}
3	1.00	1.072	1.07_{24}
			mean = 1.07_{57}
			standard deviation 0.06_{68}
		relative standard deviation 6.2%	

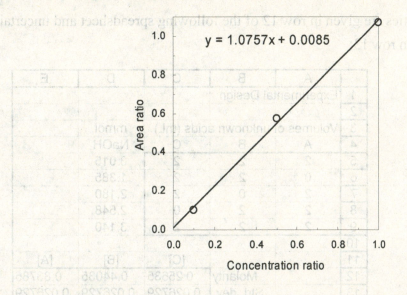

$$y = 1.0757x + 0.0085$$

5-33. To use the internal standard, a known quantity of internal standard is added to a sample of sewage. Analyte signal is measured relative to the signal from internal standard. Since the internal standard is so similar to analyte (differing only by the substitution of D for H), matrix effects are likely to be essentially identical for both compounds. If matrix enhances the response to analyte, it enhances the response to internal standard to the same extent. By measuring unknown relative to the known injection of standard, an accurate measurement of unknown can be made. The key is to use a standard that is almost identical to the analyte in physical properties.

5-34. Molarities are given in row 12 of the following spreadsheet and uncertainties are given in row 13:

	A	B	C	D	E
1	Experimental Design				
2					
3	Volumes of unknown acids (mL)			mmol	
4	A	B	C	NaOH	
5	2	2	2	3.015	
6	0	2	2	1.385	
7	2	0	2	2.180	
8	2	2	0	2.548	
9	2	2	2	3.140	
10					
11			[C]	[B]	[A]
12		Molarity	0.25635	0.44035	0.83785
13		Std. dev.	0.026729	0.026729	0.026729
14			0.995942	0.063896	#N/A
15			R^2	s_y	
16	Highlight cells C12:E14				
17	Type "=LINEST(D5:D9,A5:C9,FALSE,TRUE)"				
18	Press CTRL+SHIFT+ENTER (on PC)				
19	Press COMMAND+RETURN (on Mac)				

CHAPTER 6
CHEMICAL EQUILIBRIUM

6-1. Concentrations in an equilibrium constant are really dimensionless <u>ratios</u> of actual concentrations divided by standard state concentrations. Since standard states are 1 M for solutes, 1 bar for gases, and pure substances for solids and liquids, these are the units we must use. A solvent is approximated as a pure liquid.

6-2. All concentrations in equilibrium constants are expressed as dimensionless ratios of actual concentrations divided by standard-state concentrations.

6-3. Predictions based on free energy or Le Châtelier's principle tell us which way a reaction will go (thermodynamics), but not how long it will take (kinetics). A reaction could be over instantly or it could take forever.

6-4. (a) $K = 1/[Ag^+]^3 [PO_4^{3-}]$ (b) $K = P_{CO_2}^6 / P_{O_2}^{15/2}$

6-5. $K = \dfrac{P_E^3}{P_A^2 [B]} = \dfrac{\left(\dfrac{3.6 \times 10^4 \text{ Torr}}{760 \text{ Torr/atm}} \times 1.013 \dfrac{\text{bar}}{\text{atm}}\right)^3}{\left(\dfrac{2.8 \times 10^3 \text{ Pa}}{10^5 \text{ Pa/bar}}\right)^2 (1.2 \times 10^{-2} \text{ M})} = 1.2 \times 10^{10}$

6-6.

HOBr + OCl⁻	$\rightleftharpoons$ HOCl + OBr⁻	$K_1 = 1/15$
HOCl	$\rightleftharpoons$ H⁺ + OCl⁻	$K_2 = 3.0 \times 10^{-8}$
HOBr	$\rightleftharpoons$ H⁺ + OBr⁻	$K = K_1 K_2 = 2.0 \times 10^{-9}$

6-7. (a) Decrease   (b) give off   (c) negative

6-8. $K = e^{-(59.0 \times 10^3 \text{ J/mol})/(8.314\,472 \text{ J/(K·mol)})(298.15 \text{ K})} = 5 \times 10^{-11}$

6-9. (a) Right because reactant is added

     (b) Right because product is removed

     (c) Neither because solid graphite does not appear in the reaction quotient

     (d) Right. The pressure of reactant and product both increase by a factor of 8. However, reactant appears to the second power in the reaction quotient and product only appears to the first power. Increasing each pressure by the same factor decreases the reaction quotient.

     (e) Smaller. An exothermic reaction liberates heat. Adding heat is like adding a product.

53

6-10. (a) $K = P_{H_2O} = e^{-\Delta G°/RT} = e^{-(\Delta H° - T\Delta S°)/RT}$

$= e^{-\{[(63.11 \times 10^3 \text{ J/mol}) - (298.15\text{K})(148 \text{ J K}^{-1} \text{ mol}^{-1})]/(8.314 \text{ J K}^{-1} \text{ mol}^{-1})(298.15 \text{ K})\}}$

$= 4.7 \times 10^{-4}$ bar

(b) $P_{H_2O} = 1 = e^{-(\Delta H° - T\Delta S°)/RT} \Rightarrow \Delta H° - T\Delta S°$ must be zero.

$\Delta H° - T\Delta S° = 0 \Rightarrow T = \dfrac{\Delta H°}{\Delta S°} = 426 \text{ K} = 153°C$

6-11. (a) Let's designate the equilibrium constant at temperature T_1 as K_1 and the equilibrium constant at temperature T_2 as K_2.

$K_1 = e^{-\Delta G°/RT_1} = e^{-(\Delta H° - T_1\Delta S°)/RT_1} = e^{-\Delta H°/RT_1} \bullet e^{\Delta S°/R}$

Similarly, $K_2 = e^{-\Delta H°/RT_2} \bullet e^{\Delta S°/R}$

Dividing K_1 by K_2 gives $\dfrac{K_1}{K_2} = e^{-(\Delta H°/R)(1/T_1 - 1/T_2)}$

$\Rightarrow \Delta H° = \left(\dfrac{1}{T_2} - \dfrac{1}{T_1}\right)^{-1} R \ln \dfrac{K_1}{K_2}$

Putting in $K_1 = 1.479 \times 10^{-5}$ at $T_1 = 278.15$ K and
$K_2 = 1.570 \times 10^{-5}$ at $T_2 = 283.15$ K gives $\Delta H° = +7.82$ kJ / mol.

(b) $K = e^{-\Delta H°/RT} \bullet e^{\Delta S°/R}$

$\underset{y}{\ln K} = \underset{m}{-\dfrac{\Delta H°}{R}}\underset{x}{\left(\dfrac{1}{T}\right)} + \underset{b}{\dfrac{\Delta S°}{R}}$

$\boxed{\text{A graph of } \ln K \text{ vs. } 1/T \text{ will have a slope of } -\Delta H°/R}$

6-12. (a) $Q = \left(\dfrac{48.0 \text{ Pa}}{10^5 \text{ Pa/bar}}\right)^2 / \left(\dfrac{1\,370 \text{ Pa}}{10^5 \text{ Pa/bar}}\right)\left(\dfrac{3\,310 \text{ Pa}}{10^5 \text{ Pa/bar}}\right)$

$= 5.08 \times 10^{-4} < K$ The reaction will go to the right.

It was not really necessary to convert Pa to bar, since the units cancel.

(b)
$$H_2 \quad + \quad Br_2 \quad \rightleftharpoons \quad 2HBr$$

Initial pressure: $1\,370$ $3\,310$ 48.0

Final pressure: $1\,370 - x$ $3\,310 - x$ $48.0 + 2x$

Note that $2x$ Pa of HBr are formed when x Pa of H_2 are consumed.

$$\dfrac{(48.0 + 2x)^2}{(1\,370 - x)(3\,310 - x)} = 7.2 \times 10^{-4} \Rightarrow x = 4.50 \text{ Pa}$$

$P_{H_2} = 1\,366$ Pa, $P_{Br_2} = 3\,306$ Pa, $P_{HBr} = 57.0$ Pa

(c) Neither, since Q is unchanged.

(d) HBr will be formed, since $\Delta H°$ is positive.

6-13. The concentration of MTBE in solution is 100 µg/mL = 100 mg/L. The molarity is $[\text{MTBE}] = \dfrac{0.100\ \text{g/L}}{88.15\ \text{g/mol}} = 1.13_4 \times 10^{-3}$ M. The pressure in the gas phase is P = $[\text{MTBE}]/K_h = (1.13_4 \times 10^{-3}\ \text{M})/(1.71\ \text{M/bar}) = 0.663$ mbar.

6-14. $[\text{Cu}^+][\text{Br}^-] = K_{sp}$

$[\text{Cu}^+][0.10]\ \ = 5 \times 10^{-9} \ \Rightarrow \ [\text{Cu}^+] = 5 \times 10^{-8}$ M

6-15. $[\text{Ag}^+]^4[\text{Fe(CN)}_6^{4-}] = K_{sp}$

$[1.0 \times 10^{-6}]^4[\text{Fe(CN)}_6^{4-}] = 8.5 \times 10^{-45} \ \Rightarrow \ [\text{Fe(CN)}_6^{4-}] = 8.5 \times 10^{-21}$ M = 8.5 zM

6-16. If we let $x = [\text{Cu}^{2+}]$, then $[\text{SO}_4^{2-}] = \frac{1}{4}x$.

$$K = [\text{Cu}^{2+}]^4\ [\text{OH}^-]^6\ [\text{SO}_4^{2-}] = (x)^4\ (1.0 \times 10^{-6})^6\ (\tfrac{1}{4}x) = 2.3 \times 10^{-69}$$

$$\Rightarrow x = [\text{Cu}^{2+}]\ \ = \left(\frac{(4)(2.3 \times 10^{-69})}{(1.0 \times 10^{-6})^6} \right)^{1/5} = 3.9 \times 10^{-7}\ \text{M}$$

6-17. (a) $[\text{Zn}^{2+}]^2[\text{Fe(CN)}_6^{4-}] = (0.000\ 10)^2[\text{Fe(CN)}_6^{4-}] = 2.1 \times 10^{-16}$

$\Rightarrow [\text{Fe(CN)}_6^{4-}] = 2.1 \times 10^{-8}$ M

(b) $[\text{Zn}^{2+}]^2[\text{Fe(CN)}_6^{4-}] = (5.0 \times 10^{-7})^2[\text{Fe(CN)}_6^{4-}] = 2.1 \times 10^{-16}$

$\Rightarrow [\text{Fe(CN)}_6^{4-}] = 8.4 \times 10^{-4}$ M

6-18. BX_2 coprecipitates with AX_3. This means that some BX_2 is trapped in the AX_3 during precipitation of AX_3.

6-19. For CaSO_4, $K_{sp} = 2.4 \times 10^{-5}$. For Ag_2SO_4, $K_{sp} = 1.5 \times 10^{-5}$.

Removing 99% of the Ca^{2+} reduces $[\text{Ca}^{2+}]$ to 0.000 500 M. The concentration of SO_4^{2-} in equilibrium with 0.000 500 M Ca^{2+} is $2.4 \times 10^{-5}/0.000\ 500 = 0.048$ M. This much $\text{SO}_4{}^{2-}$ <u>will</u> precipitate Ag_2SO_4, because $Q = [\text{Ag}^+]^2\ [\text{SO}_4^{2-}] =$ $(0.030\ 0)^2\ (0.048)\ = 4.3 \times 10^{-5} > K_{sp}$. The separation is not feasible.

When Ag^+ first precipitates, $[\text{SO}_4^{2-}] = 1.5 \times 10^{-5}/(0.030\ 0)^2 = 1.67 \times 10^{-2}$ M.

$[\text{Ca}^{2+}] = 2.4 \times 10^{-5}/1.67 \times 10^{-2} = 0.001\ 4$ M.   97% of the Ca^{2+} has precipitated.

6-20. $\text{BaCrO}_4(s) \rightleftharpoons \text{Ba}^{2+} + \text{CrO}_4^{2-} \qquad K_{sp} = 2.1 \times 10^{-10}$

$\text{Ag}_2\text{CrO}_4(s) \rightleftharpoons 2\text{Ag}^+ + \text{CrO}_4^{2-} \qquad K_{sp} = 1.2 \times 10^{-12}$

The stoichiometries are not identical, so it is not clear that the salt with lower K_{sp} will precipitate first. Let's try each possibility. Suppose that $BaCrO_4$ precipitates first. The concentration of CrO_4^{2-} that will reduce Ba^{2+} to 0.1% of its initial concentration is

$$[Ba^{2+}][CrO_4^{2-}] = [1.0 \times 10^{-5}][CrO_4^{2-}] = 2.1 \times 10^{-10} \Rightarrow [CrO_4^{2-}] = 2.1 \times 10^{-5} \text{ M}.$$

Will this much chromate precipitate 0.010 M Ag^+? We test by evaluating the reaction quotient for Ag_2CrO_4:

$$Q = [Ag^+]^2[CrO_4^{2-}] = (0.010)^2(2.1 \times 10^{-5}) = 2.1 \times 10^{-9} > K_{sp} \text{ for } Ag_2CrO_4$$

Since $Q > K_{sp}$ for Ag_2CrO_4, Ag^+ will precipitate.

Let's try the reverse calculation. If Ag_2CrO_4 precipitates first, the concentration of CrO_4^{2-} that will reduce Ag^+ to 1.0×10^{-5} M is

$$[Ag^+]^2[CrO_4^{2-}] = [1.0 \times 10^{-5}]^2[CrO_4^{2-}] = 1.2 \times 10^{-12} \Rightarrow [CrO_4^{2-}] = 0.012 \text{ M}.$$

This concentration of CrO_4^{2-} exceeds the concentration required to precipitate 99.90% of Ba^{2+}. Neither Ag^+ nor Ba^{2+} can be 99.90% precipitated without precipitating the other ion.

6-21.

Salt	K_{sp}		[Ag$^+$]
			(M, in equilibrium with 0.1 M anion)
AgCl	1.8×10^{-10}	$K_{sp}/0.10$	$= 1.8 \times 10^{-9}$
AgBr	5.0×10^{-13}	$K_{sp}/0.10$	$= 5.0 \times 10^{-12}$
AgI	8.3×10^{-17}	$K_{sp}/0.10$	$= 8.3 \times 10^{-16}$
Ag_2CrO_4	1.2×10^{-12}	$\sqrt{K_{sp}/0.10}$	$= 3.5 \times 10^{-6}$

I^- requires the lowest concentration of Ag^+ to begin precipitating, so I^- precipitates first. The order of precipitation is: I^- before Br^- before Cl^- before CrO_4^{2-}.

6-22.   At low I^- concentration, $[Pb^{2+}]$ decreases with increasing $[I^-]$ because of the reaction $Pb^{2+} + 2I^- \rightarrow PbI_2(s)$. Concentrations of other $Pb^{2+}-I^-$ species are negligible. At high I^- concentration, complex ions form by reactions such as $PbI_2(s) + I^- \rightarrow PbI_3^-$.

6-23.   (a) BF_3 (b) AsF_5

6-24. $\dfrac{[SnCl_2(aq)]}{[Sn^{2+}][Cl^-]^2} = \beta_2 \Rightarrow [SnCl_2(aq)] = \beta_2[Sn^{2+}][Cl^-]^2 = (12)(0.20)(0.20)^2 = 0.096 \text{ M}$

6-25. $[Zn^{2+}] = K_{sp}/[OH^-]^2 = 2._{93} \times 10^{-3} \text{ M}$

$[ZnOH^+] = \beta_1[Zn^{2+}][OH^-] = \beta_1 K_{sp}/[OH^-] = 9 \times 10^{-6} \text{ M}$

$[Zn(OH)_2(aq)] = \beta_2[Zn^{2+}][OH^-]^2 = \beta_3 K_{sp}[OH^-] = 6 \times 10^{-6} \text{ M}$

$[Zn(OH)_3^-] = \beta_3[Zn^{2+}][OH^-]^3 = \beta_3 K_{sp}[OH^-] = 8 \times 10^{-9} \text{ M}$

$[Zn(OH)_4^{2-}] = \beta_4[Zn^{2+}][OH^-]^4 = \beta_4 K_{sp}[OH^-]^2 = 9 \times 10^{-14} \text{ M}$

6-26.

	Na^+	+	OH^-	$\rightleftharpoons$	$NaOH(aq)$
Initial concentration:	1		1		0
Final concentration:	$1-x$		$1-x$		x

$$\frac{x}{(1-x)^2} = 0.2 \Rightarrow x = 0.15 \text{ M}$$

$x = 0.2 - 0.4x + 0.2x^2$

$\underset{a}{0.2x^2} - \underset{b}{1.4x} + \underset{c}{0.2} = 0$

$$x = \frac{-b \pm \sqrt{b^2 - 4ac}}{2a} = \frac{1.4 \pm \sqrt{1.4^2 - 4(0.2)(0.2)}}{2(0.2)} = 6.85 \text{ or } 0.15$$

x cannot be greater than 1 (the initial or formal concentration of NaOH), so the correct answer must be 0.15. That is, 15% of the sodium is in the form $NaOH(aq)$.

6-27. $PbI_2(s) \overset{K_{sp}}{\rightleftharpoons} Pb^{2+} + 2I^- \qquad K_{sp} = [Pb^{2+}][I^-]^2 = 7.9 \times 10^{-9}$

$+$

$ Pb^{2+} + 2I^- \overset{\beta_2}{\rightleftharpoons} PbI_2(aq) \quad \beta_2 = [PbI_2(aq)]/[Pb^{2+}][I^-]^2 = 1.4 \times 10^3$

$PbI_2(s) \rightleftharpoons PbI_2(aq) \qquad K = K_{sp}\beta_2 = 1.1 \times 10^{-5} = [PbI_2(aq)]$

6-28.   Lewis acids and bases are electron pair acceptors and donors, respectively:

$$F_3B + :\overset{..}{\underset{..}{O}}(CH_3)_2 \rightarrow F_3\overset{-}{B} - \overset{+}{\underset{..}{O}}(CH_3)_2$$

$$\underset{\substack{\text{Lewis} \\ \text{acid}}}{} \quad \underset{\substack{\text{Lewis} \\ \text{base}}}{} \qquad\qquad \underset{\text{Adduct}}{}$$

Brønsted acids and bases are proton donors and acceptors, respectively:

$$H_2S + \langle\bigcirc\rangle N: \rightarrow \langle\bigcirc\rangle NH^+ + HS^-$$

$$\underset{\substack{\text{Brønsted} \\ \text{acid}}}{} \underset{\substack{\text{Brønsted} \\ \text{base}}}{}$$

6-29.   (a) An adduct   (b) dative or coordinate covalent

(c) conjugate   (d) $[H^+] > [OH^-]$; $[OH^-] > [H^+]$

6-30. Dissolved CO_2 from the atmosphere lowers the pH by reacting with water to form carbonic acid. Water can be distilled under an inert atmosphere to exclude CO_2, or most CO_2 can be removed by boiling the distilled water.

6-31. SO_2 in the atmosphere reacts with moisture to make H_2SO_3, which is a weak acid. H_2SO_3 can be oxidized to H_2SO_4, which is a strong acid.

6-32.

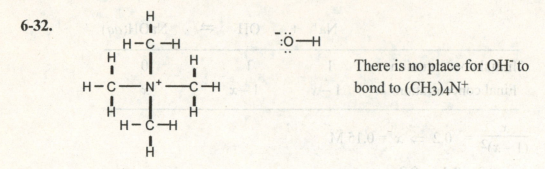

There is no place for OH^- to bond to $(CH_3)_4N^+$.

6-33. (a) HI   (b) H_2O

6-34. $2H_2SO_4 \rightleftharpoons HSO_4^- + H_3SO_4^+$

6-35.

acid	base
(a) H_3O^+	H_2O
(a) $\overset{+}{H_3N}CH_2CH_2\overset{+}{N}H_3$	$\overset{+}{H_3N}CH_2CH_2NH_2$
(b) $C_6H_5CO_2H$	$C_6H_5CO_2^-$
(b) $C_5H_5NH^+$	C_5H_5N

6-36. (a) $[H^+] = 0.010\,M \Rightarrow pH = -\log[H^+] = 2.00$

(b) $[OH^-] = 0.035\,M \Rightarrow [H^+] = K_w/[OH^-] = 2.8_6 \times 10^{-13}\,M \Rightarrow pH = 12.54$

(c) $[H^+] = 0.030\,M \Rightarrow pH = 1.52$

(d) $[H^+] = 3.0\,M \Rightarrow pH = -0.48$

(e) $[OH^-] = 0.010\,M \Rightarrow [H^+] = 1.0 \times 10^{-12}\,M \Rightarrow pH = 12.00$

6-37. (a) $K_w = [H^+][OH^-] = 1.01 \times 10^{-14}$ at $25°C$

$$\underset{x}{}\quad\underset{x}{}$$

$$x^2 = 1.01 \times 10^{-14} \Rightarrow x = [H^+] = 1.00_5 \times 10^{-7}\,M \Rightarrow pH = -\log[H^+] = 6.998$$

(b) At $100°C$, $pH = 6.132$

6-38. Since $[H^+][OH^-] = 1.0 \times 10^{-14}$, $K = [H^+]^4[OH^-]^4 = 1.0 \times 10^{-56}$

6-39. $[La^{3+}][OH^-]^3 = K_{sp} = 2 \times 10^{-21}$

$[OH^-]^3 = K_{sp}/(0.010) \Rightarrow [OH^-] = 5.8 \times 10^{-7}$ M $\Rightarrow$ pH = 7.8

6-40. (a) At 25°C, K_w increases as temperature increases $\Rightarrow$ endothermic

(b) At 100°C, K_w increases as temperature increases $\Rightarrow$ endothermic

(c) At 300°C, K_w decreases as temperature increases $\Rightarrow$ exothermic

6-41. See Table 6-2.

6-42. Weak acids:

RCO_2H Carboxylic acids	$R_3NH^+X^-$ Ammonium ions	M^{n+} Metal ions

Weak bases:

$R_3N:$ Amines	$RCO_2^-M^+$ Carboxylate ions

6-43. $Cl_3CCO_2H \rightleftharpoons Cl_3CCO_2^- + H^+$

$La^{3+} + H_2O \rightleftharpoons LaOH^{2+} + H^+$

6-44.

$HOCH_2CH_2S^- + H_2O \rightleftharpoons HOCH_2CH_2SH + OH^-$

6-45. K_a: $HCO_3^- \rightleftharpoons H^+ + CO_3^{2-}$ K_b: $HCO_3^- + H_2O \rightleftharpoons H_2CO_3 + OH^-$

6-46. (a) $H_3\overset{+}{N}CH_2CH_2\overset{+}{N}H_3 \overset{K_{a1}}{\rightleftharpoons} H_2NCH_2CH_2\overset{+}{N}H_3 + H^+$

$H_2NCH_2CH_2\overset{+}{N}H_3 \overset{K_{a2}}{\rightleftharpoons} H_2NCH_2CH_2NH_2 + H^+$

(b) $^-O_2CCH_2CO_2^- + H_2O \overset{K_{b1}}{\rightleftharpoons} HO_2CCH_2CO_2^- + OH^-$

$HO_2CCH_2CO_2^- + H_2O \overset{K_{b2}}{\rightleftharpoons} HO_2CCH_2CO_2H + OH^-$

6-47. (a), (c)

6-48. $CN^- + H_2O \rightleftharpoons HCN + OH^-$ $K_b = \dfrac{K_w}{K_a} = 1.6 \times 10^{-5}$

6-49. $H_2PO_4^- \overset{K_{a2}}{\rightleftharpoons} HPO_4^{2-} + H^+$ $\qquad\qquad$ $HC_2O_4^- + H_2O \overset{K_{b2}}{\rightleftharpoons} H_2C_2O_4 + OH^-$

6-50. $K_{a1} = \dfrac{K_w}{K_{b3}} = 7.04 \times 10^{-3}$ $\qquad$ $K_{a2} = \dfrac{K_w}{K_{b2}} = 6.25 \times 10^{-8}$

$\qquad\quad$ $K_{a3} = \dfrac{K_w}{K_{b1}} = 4.3 \times 10^{-13}$

6-51. Add the two reactions and multiply their equilibrium constants to get $K = 2.9 \times 10^{-6}$.

6-52. (a) $Ca(OH)_2\,(s) \rightleftharpoons Ca^{2+} + 2OH^-$
$\qquad\qquad\qquad\qquad\qquad\quad x \qquad\quad 2x$

$\qquad\qquad x(2x)^2 = K_{sp} = 10^{-5.19} \Rightarrow x = 1.2 \times 10^{-2}\ M$

$\qquad$ (b) Since some Ca^{2+} reacts with OH^- to form $CaOH^+$, the K_{sp} reaction will be drawn to the right, and the solubility of $Ca(OH)_2$ will be greater than we would expect just on the basis of K_{sp}.

6-53. Reversing the first reaction and then adding the four reactions gives

$\qquad Ca^{2+} + CO_2(g) + H_2O(l) \rightleftharpoons CaCO_3(s) + 2H^+ \qquad K = K_{CO_2}K_1K_2/K_{sp}$

$\qquad K = (3.4 \times 10^{-2})(4.5 \times 10^{-7})(4.7 \times 10^{-11})/(6.0 \times 10^{-9}) = 1.2 \times 10^{-10}$

$\qquad \dfrac{[H^+]^2}{[Ca^{2+}]P_{CO_2}} = \dfrac{(1.8 \times 10^{-7})^2}{[Ca^{2+}][0.10]} = K = 1.2 \times 10^{-10}$

$\qquad [Ca^{2+}] = 2.7 \times 10^{-3}\ M = 0.22\ g/2.00\ L$

CHAPTER 7
ACTIVITY AND SYSTEMATIC TREATMENT OF EQUILIBRIUM

7-1. As ionic strength increases, the charges of the ionic atmospheres increase and the net ionic attractions decrease. There is less tendency for ions to bind to each other.

7-2. (a) true   (b) true   (c) true

7-3. (a) $\frac{1}{2}[0.008\,7 \cdot 1^2 + 0.008\,7 \cdot (-1)^2] = 0.008\,7$ M

     (b) $\frac{1}{2}[0.000\,2 \cdot 3^2 + 0.000\,6 \cdot (-1)^2] = 0.001\,2$ M

7-4. (a) 0.660   (b) 0.54   (c) 0.18   (Eu^{3+} is a lanthanide ion) (d) 0.83

7-5. The ionic strength 0.030 M is halfway between the values 0.01 and 0.05 M. Therefore, the activity coefficient will be halfway between the tabulated values:
$$\gamma = \frac{1}{2}(0.914 + 0.86) = 0.88_7.$$

7-6. (a) $\log \gamma = \dfrac{-0.51 \cdot 2^2 \cdot \sqrt{0.083}}{1 + (600\sqrt{0.083\,/\,305})} = -0.375 \Rightarrow \gamma = 10^{-0.375} = 0.42_2$

     (b) $\gamma = \left(\dfrac{0.083 - 0.05}{0.1 - 0.05}\right)(0.405 - 0.485) + 0.485 = 0.43_2$

7-7. $\gamma = \left(\dfrac{0.083 - 0.05}{0.1 - 0.05}\right)(0.18 - 0.245) + 0.245 = 0.20_2$

7-8. If [ether(aq)] becomes smaller, γ_{ether} must become larger, since
$K (= [\text{ether}(aq)]\,\gamma_{ether})$ is a constant.

7-9. The solubility of Hg_2Br_2 is small, so we assume that Hg_2Br_2 contributes negligible Br^- to 0.001 00 M KBr.

     $\mu = 0.001\,00$ M, $[Br^-] = 0.001\,00$ M, $\gamma_{Hg_2^{2+}} = 0.867$, $\gamma_{Br^-} = 0.964$

     $K_{sp} = 5.6 \times 10^{-23} = [Hg_2^{2+}]\,\gamma_{Hg_2^{2+}}[Br^-]^2\gamma_{Br^-}^2$

          $= [Hg_2^{2+}](0.867)(0.001\,00)^2(0.964)^2 \Rightarrow [Hg_2^{2+}] = 7.0 \times 10^{-17}$ M

Check our assumption: Yes, Br^- from Hg_2Br_2 is negligible.

7-10. The solubility of $Ba(IO_3)_2$ is small, so we assume that $Ba(IO_3)_2$ contributes negligible IO_3^- to 0.100 M $(CH_3)_4NIO_3$.

$\mu = 0.100$ M, $[IO_3^-] = 0.100$ M, $\gamma_{Ba^{2+}} = 0.38$, $\gamma_{IO_3^-} = 0.775$

$$K_{sp} = 1.5 \times 10^{-9} = [Ba^{2+}] \gamma_{Ba^{2+}} [IO_3^-]^2 \gamma_{IO_3^-}^2$$

$$= [Ba^{2+}](0.38)(0.100)^2(0.775)^2 \Rightarrow [Ba^{2+}] = 6.6 \times 10^{-7} \text{ M}$$

7-11. Ionic strength $= 0.010$ M (from HCl) $+ 0.040$ M (from KClO$_4$ that gives

K$^+$ + ClO$_4^-$) $= 0.050$ M.   Using Table 7-1, $\gamma_{H^+} = 0.86$.

$$pH = -\log [H^+]\gamma_{H^+} = -\log[(0.010)(0.86)] = 2.07.$$

7-12. Ionic strength $= 0.010$ M from NaOH $+ 0.012$ M from LiNO$_3$ $= 0.022$ M.

Interpolating in Table 7-1 gives $\gamma_{OH^-} = 0.873$.

$$[H^+]\gamma_{H^+} = \frac{K_w}{[OH^-]\gamma_{OH^-}} = \frac{1.0 \times 10^{-14}}{(0.010)(0.873)} = 1.15 \times 10^{-12}$$

$$pH = -\log(1.15 \times 10^{-12}) = 11.94$$

If we had neglected activities, $pH \approx -\log[H^+] = -\log\dfrac{K_w}{[OH^-]} = 12.00$

7-13. $\varepsilon = 79.755\, e^{-4.6 \times 10^{-3}(323.15 - 293.15)} = 69.474$

$$\log \gamma = \frac{(-1.825 \times 10^6)[(69.474)(323.15)]^{-3/2}(-2)^2 \sqrt{0.100}}{1 + \dfrac{400\sqrt{0.100}}{2.00\sqrt{(69.474)(323.15)}}}$$

$$= -0.4826 \Rightarrow \gamma = 0.329 \quad \text{(In the table, } \gamma = 0.355 \text{ at } 25°C.)$$

7-14. Spreadsheet for Debye-Hückel calculations

	A	B	C	D	E
1	Ionic strength	Gamma $(z = \pm1)$	Gamma $(z = \pm2)$	Gamma $(z = \pm3)$	Gamma $(z = \pm4)$
2	0.0001	0.988	0.955	0.901	0.831
3	0.0003	0.980	0.924	0.836	0.727
4	0.001	0.965	0.867	0.725	0.565
5	0.003	0.942	0.787	0.583	0.383
6	0.01	0.901	0.660	0.393	0.190
7	0.03	0.847	0.515	0.225	0.071
8	0.1	0.769	0.350	0.094	0.015
9					
10	B2 = 10^(-0.51*Sqrt(A2)/(1+400*Sqrt(A2)/305))				
11	C2 = 10^(-0.51*4*Sqrt(A2)/(1+400*Sqrt(A2)/305))				
12	D2 = 10^(-0.51*9*Sqrt(A2)/(1+400*Sqrt(A2)/305))				
13	E2 = 10^(-0.51*16*Sqrt(A2)/(1+400*Sqrt(A2)/305))				

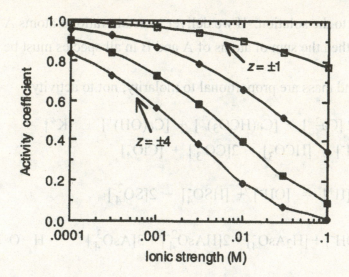

7-15. The equilibrium constant is

$$HA \rightleftharpoons H^+ + A^- \qquad K_a = \frac{[H^+]\gamma_{H^+}\ [A^-]\gamma_{A^-}}{[HA]\gamma_{HA}}.$$

From the table of activity coefficients, $\gamma_{H^+} = 0.83$ and $\gamma_{A^-} = 0.80$ at $\mu = 0.1$ M.

For HA, we estimate

$$\log \gamma_{HA} = k\mu = (0.2)(0.1) = 0.02, \text{ or } \gamma_{HA} = 10^{0.02} = 1.05.$$

Putting activity coefficients for $\mu = 0.1$ M into the equilibrium expression gives

$$K_a = \frac{[H^+]\gamma_{H^+}\ [A^-]\gamma_{A^-}}{[HA]\gamma_{HA}} = \frac{[H^+](0.83)[A^-](0.80)}{[HA](1.05)} = 0.63\frac{[H^+][A^-]}{[HA]}$$

or $\dfrac{[H^+][A^-]}{[HA]} (\mu = 0.1 \text{ M}) = K_a/0.63.$

The activity coefficients at $\mu = 0$ are all 1, so the equilibrium expression is

$$K_a = \frac{[H^+]\gamma_{H^+}\ [A^-]\gamma_{A^-}}{[HA]\gamma_{HA}} = \frac{[H^+](1)[A^-](1)}{[HA](1)} = \frac{[H^+][A^-]}{[HA]} (\mu = 0).$$

The concentration quotient at the two different ionic strengths is

$$\text{Concentration quotient} = \frac{\dfrac{[H^+][A^-]}{[HA]} (\mu = 0)}{\dfrac{[H^+][A^-]}{[HA]} (\mu = 0.1 \text{ M})} = \frac{K_a}{K_a/0.63} = 0.63$$

in agreement with the observed value of 0.63 ± 0.03.

7-16. The charge balance states the magnitude of positive charge equals the magnitude of negative charge in a solution. That is, the solution must be neutral. The mass balance states that atoms are conserved. If we deliver a certain number of atom A into a solution, then the sum of atom A in all species must equal the atoms of A delivered to the solution. If we deliver a certain ratio of atoms A and B into the solution, then the sum of atoms of A and B in all species must be in that same ratio.

7-17. Charge and mass are proportional to molarity, not to activity.

7-18. $[H^+] + 2[Ca^{2+}] + [Ca(HCO_3)^+] + [Ca(OH)^+] + [K^+] =$
$[OH^-] + [HCO_3^-] + 2[CO_3^{2-}] + [ClO_4^-]$

7-19. Charge: $[H^+] = [OH^-] + [HSO_4^-] + 2[SO_4^{2-}]$

7-20. $[H^+] = [OH^-] + [H_2AsO_4^-] + 2[HAsO_4^{2-}] + 3[AsO_4^{3-}]$

$$H-O-\overset{\displaystyle O}{\underset{\displaystyle O^-}{\overset{\|}{\underset{|}{As}}}}-O^-$$

7-21. (a) Charge balance: $2[Mg^{2+}] + [H^+] + [MgBr^+] + [MgOH^+] = [Br^-] + [OH^-]$
Mass balance: total Br = 2(total Mg)
$[MgBr^+] + [Br^-] = 2\{[Mg^{2+}] + [MgBr^+] + [MgOH^+]\}$

(b) $[Mg^{2+}] + [MgBr^+] + [MgOH^+] = 0.2$ M
$[MgBr^+] + [Br^-] = 0.4$ M

7-22. 250 mL of 1.0×10^{-6} M charge $= 0.25 \times 10^{-6}$ moles of charge.
$(0.25 \times 10^{-6} \text{ moles of charge})\left(9.648 \times 10^4 \dfrac{\text{coulombs}}{\text{mole of charge}}\right) = 0.024\,12$ C.
The dielectric constant of air is $\varepsilon = 1$ and the separation is 1.5 m.
Force $= -(8.988 \times 10^9) \dfrac{(0.024\,12)(-0.024\,12)}{(1)(1.5^2)} = 2.3 \times 10^6$ N.
$(2.3 \times 10^6 \text{ N})(0.224\,8 \text{ pounds/N}) = 5.2 \times 10^5$ pounds.
Two elephants do not weigh enough to keep the beakers apart.

7-23. $[CH_3CO_2^-] + [CH_3CO_2H] = 0.1$ M

7-24. $Y_{total} = \frac{3}{2} X_{total}$
$2[X_2Y_2^{2+}] + [X_2Y^{4+}] + 3[X_2Y_3] + [Y^{2-}] = \frac{3}{2}\{2[X_2Y_2^{2+}] + 2[X_2Y^{4+}] + 2[X_2Y_3]\}$
$[Y^{2-}] = [X_2Y_2^{2+}] + 2[X_2Y^{4+}]$

7-25. 3 (total Fe) = 2 (total sulfur)
$3\{[Fe^{3+}] + [Fe(OH)^{2+}] + [Fe(OH)_2^+] + 2[Fe_2(OH)_2^{4+}] + [FeSO_4^+]\}$
$= 2\{[FeSO_4^+] + [SO_4^{2-}] + [HSO_4^-]\}$
We write 2 in front of $[Fe_2(OH)_2^{4+}]$ because $Fe_2(OH)_2^{4+}$ contains 2 Fe.

7-26.   (a)   Pertinent reactions:

$$A^- + H_2O \overset{K_b}{\rightleftharpoons} HA + OH^- \qquad K_b = \frac{[HA]\gamma_{HA}[OH^-]\gamma_{OH^-}}{[A^-]\gamma_{A^-}} = 5.7 \times 10^{-10} \qquad (A)$$

$$H_2O \overset{K_w}{\rightleftharpoons} H^+ + OH^- \qquad K_w = [H^+]\gamma_{H^+}[OH^-]\gamma_{OH^-} = 1.0 \times 10^{-14} \qquad (B)$$

Charge balance: $[H^+] + [Na^+] = [OH^-] + [A^-]$ \hfill (C)

Mass balance: $[Na^+] = 0.01\ M \equiv F$ \hfill (D)

Mass balance: $[HA] + [A^-] = 0.01\ M \equiv F$ \hfill (E)

(b)   Now we neglect activity coefficients. We will make the following substitutions in the charge balance:

$[OH^-] = K_w/[H^+]$

From Equation A: $[HA] = \dfrac{K_b[A^-]}{[OH^-]}$ \hfill (F)

From Equation E: $[HA] = F - [A^-]$ \hfill (G)

Now equate the expressions for [HA] from Equations F and G to solve for [A⁻]

$$\frac{K_b[A^-]}{[OH^-]} = F - [A^-]$$

$$[A^-]\left(\frac{K_b}{[OH^-]} + 1\right) = F \quad \Rightarrow \quad [A^-] = \frac{F[OH^-]}{K_b + [OH^-]}$$

Now substitute $K_w/[H^+]$ for each $[OH^-]$ in the equation above to get

$$[A^-] = \frac{FK_w/[H^+]}{K_b + K_w/[H^+]} = \frac{FK_w}{K_b[H^+] + K_w} \qquad (H)$$

We can now substitute for all terms in the charge balance using [A⁻] from Equation H, $[OH^-] = K_w/[H^+]$, and $[Na^+] = F$:

Charge balance: $[H^+] + [Na^+] = [OH^-] + [A^-]$

$$[H^+] + F = \frac{K_w}{[H^+]} + \frac{FK_w}{K_b[H^+] + K_w} \qquad (I)$$

Equation I has the form we were looking for. The only unknown is [H⁺]. For convenience in finding a numerical solution, I will rearrange as follows:

$$0 = \frac{K_w}{[H^+]} + \frac{FK_w}{K_b[H^+] + K_w} - [H^+] - F \qquad (J)$$

The following spreadsheet evaluates the right side of Equation J in cell D9. Guess a value for [H⁺] in cell B7. Before using Goal Seek in Excel 2007, click the Microsoft Office button at the top left of the spreadsheet, click on Excel Options, and then on Formulas. Set Maximum Change to 1e-14. In earlier versions of Excel, go to Tools and Options and select the Calculations tab and set Maximum change to 1e-14. Then execute Goal Seek to vary [H⁺] in cell B7 until the sum in cell D9 is close to 0 (within 1e-14). The answer in cell B7 is $[H^+] = 4.19 \times 10^{-9}$ M or pH = 8.38.

Cells C15:C17 compute other concentrations from the relations

$$[OH^-] = \frac{K_w}{[H^+]} \qquad [A^-] = \frac{FK_w}{K_b[H^+] + K_w} \qquad [HA] = \frac{K_b[A^-]}{[OH^-]}$$

	A	B	C	D	E
1	Finding the species in NaOAc solution				
2	Using GOAL SEEK to Solve for [H$^+$]				
3	K$_w$ =	1.00E-14			
4	K$_b$ =	5.70E-10			
5	F =	0.01			
6					
7	Guess for [H$^+$] =	4.19E-09		pH = -log[H$^+$] =	8.38
8					
9	K$_w$/[H$^+$] + FK$_w$/(K$_b$[H$^+$] + K$_w$) -[H$^+$] - F =			1.41E-16	
10					
11	Tools --> Options --> Calculation --> Maximum change = 1e-14				
12	Use GOAL SEEK to vary [H$^+$] in cell B7 until the sum in				
13	cell D9 is equal to zero				
14					
15	[OH$^-$] = K$_w$/[H$^+$] =	2.39E-06			
16	[A] = FK$_w$/(K$_b$[H$^+$] + K$_w$) =	1.00E-02			
17	[HA] = K$_b$[A]/[OH] =	2.39E-06			

7-27. (a) Pertinent reactions:

$$Ca(OH)_2(s) \overset{K_{sp}}{\rightleftharpoons} Ca^{2+} + 2OH^- \qquad K_{sp} = [Ca^{2+}]\gamma_{Ca^{2+}}[OH^-]^2\gamma_{OH^-}^2 = 6.5 \times 10^{-6}$$

$$Ca^{2+} + OH^- \overset{K_1}{\rightleftharpoons} CaOH^+ \qquad K_1 = \frac{[CaOH^+]\gamma_{CaOH^+}}{[Ca^{2+}]\gamma_{Ca^{2+}}[OH^-]\gamma_{OH^-}} = 2.0 \times 10^1$$

$$H_2O \overset{K_w}{\rightleftharpoons} H^+ + OH^- \qquad K_w = [H^+]\gamma_{H^+}[OH^-]\gamma_{OH^-} = 1.0 \times 10^{-14}$$

Charge balance: $2[Ca^{2+}] + [CaOH^+] + [H^+] = [OH^-]$

Mass balance: $\underbrace{[OH^-] + [CaOH^+]}_{\text{species containing OH}^-} = \underbrace{2\{[Ca^{2+}] + [CaOH^+]\}}_{\text{species containing Ca}^{2+}} + [H^+]$

(Mass balance gives the same result as charge balance.)

There are 4 equations (3 equilibria and charge balance) and 4 unknowns:

$[Ca^{2+}]$, $[CaOH^+]$, $[H^+]$, and $[OH^-]$.

(b) The approximations we make are to disregard the activity coefficients and to neglect [H$^+$] in the charge balance because [H$^+$] $\ll$ [OH$^-$] in basic solution. The charge balance becomes

$$2[Ca^{2+}] + [CaOH^+] = [OH^-] \qquad (A)$$

Substituting $[CaOH^+] = K_1[Ca^{2+}][OH^-]$ into (A) gives

$$2[Ca^{2+}] + K_1[Ca^{2+}][OH^-] = [OH^-] \quad \Rightarrow \quad [Ca^{2+}] = \frac{[OH^-]}{2 + K_1[OH^-]}$$

Substitute this expression for $[Ca^{2+}]$ into the solubility product:

$$K_{sp} = [Ca^{2+}][OH^-]^2 = \frac{[OH^-]^3}{2 + K_1[OH^-]} \qquad (B)$$

We solve Equation B in the spreadsheet by guessing $[OH^-]$ in cell C4 until the expression in cell D4 is equal to K_{sp}. We used Goal Seek for this purpose.

	A	B	C	D
1	Ca(OH)$_2$ solubility			
2				
3	K$_{sp}$ =		[OH$^-$]$_{guess}$ =	[OH$^-$]^3/(2 + K$_1$[OH$^-$]) =
4	6.5E-06		0.0253528	6.5000E-06
5	K$_1$ =			
6	2.0E+01		[Ca^{2+}] =	[CaOH$^+$] =
7			0.0101126	0.0051276
8				
9	D4 = C4^3/(2 + A6*C4)		[H$^+$] = K$_w$/[OH$^-$] =	
10	C7 = A4/C4^2			3.94E-13
11	D7 =A6*C7*C4			

Results: $[Ca^{2+}]$ = 0.010 1 M $[CaOH^+]$ = 0.005 1 M

$[OH^-]$ = 0.025 4 M $[H^+] = K_w/[OH^-] = 3.9 \times 10^{-13}$ M

Total dissolved Ca = 0.010 1 + 0.005 1 = 0.015 2 M

The formula mass of Ca(OH)$_2$ is 74.09 g/mol, so 0.015 2 M is 1.1$_3$ g/L. The *Handbook of Chemistry and Physics* lists the solubility of Ca(OH)$_2$ as 1.85 g/L at 0°C and 0.77 g/L at 100°C.

7-28. (a) Pertinent reactions:

$$Zn^{2+} + SO_4^{2-} \rightleftharpoons ZnSO_4(aq) \quad K_{ion\ pair} = \frac{[ZnSO_4(aq)]}{[Zn^{2+}]\gamma_{Zn^{2+}}[SO_4^{2-}]\gamma_{SO_4^{2-}}} = 2.2 \times 10^2$$

$$(\gamma_{ZnSO_4} = 1 \text{ because the ion pair is neutral})$$

$$H_2O \overset{K_w}{\rightleftharpoons} H^+ + OH^- \qquad K_w = [H^+]\gamma_{H^+}[OH^-]\gamma_{OH^-} = 1.0 \times 10^{-14}$$

Charge balance: $2[Zn^{2+}] + [H^+] = 2[SO_4^{2-}] + [OH^-]$

Mass balance: 0.010 M = $[Zn^{2+}] + [ZnSO_4(aq)]$

0.010 M = $[SO_4^{2-}] + [ZnSO_4(aq)]$

5 equations and 5 unknowns: $[Zn^{2+}]$, $[SO_4^{2-}]$, $[ZnSO_4(aq)]$, $[H^+]$, $[OH^-]$

Since there is no coupling of the zinc sulfate and water equilibria, $[H^+] = [OH^-]$
$= 1.0 \times 10^{-7}$ M and we can disregard H^+ and OH^- in this problem.

Substituting $[ZnSO_4(aq)] = 0.010 - [Zn^{2+}]$ and $[SO_4^{2-}] = [Zn^{2+}]$ into the ion
pair equilibrium expression gives

$$K_{\text{ion pair}} = \frac{[ZnSO_4(aq)]}{[Zn^{2+}]\gamma_{Zn^{2+}}[SO_4^{2-}]\gamma_{SO_4^{2-}}} = 2.2 \times 10^2$$

$$K_{\text{ion pair}} = \frac{0.010 - [Zn^{2+}]}{[Zn^{2+}]\gamma_{Zn^{2+}}[Zn^{2+}]\gamma_{SO_4^{2-}}} = 2.2 \times 10^2$$

Rearranging gives

$$(220)[Zn^{2+}]^2\gamma_{Zn^{2+}}\gamma_{SO_4^{2-}} + [Zn^{2+}] - 0.010 = 0. \tag{A}$$

Setting $\gamma_{Zn^{2+}} = \gamma_{SO_4^{2-}} = 1$ allows us to solve quadratic Equation (A) to find
$[Zn^{2+}] = 4.8_4 \times 10^{-3}$ M.

(b) Ionic strength $= \frac{1}{2}([Zn^{2+}] \cdot 4 + [SO_4^{2-}] \cdot 4) = 0.019_4$ M

Interpolation in Table 7-1 gives $\gamma_{Zn^{2+}} = 0.630$ and $\gamma_{SO_4^{2-}} = 0.609$.

Putting these values of γ into Equation (A) gives $[Zn^{2+}] = 6.4_7 \times 10^{-3}$ M

3rd iteration: $\mu = 4 (6.4_7 \times 10^{-3}) = 0.025_9$ M
$\gamma_{Zn^{2+}} = 0.599$ $\gamma_{SO_4^{2-}} = 0.574$ $[Zn^{2+}] = 6.6_5 \times 10^{-3}$ M

4th iteration: $\mu = 0.026_6$ M
$\gamma_{Zn^{2+}} = 0.596$ $\gamma_{SO_4^{2-}} = 0.571$ $[Zn^{2+}] = 6.6_7 \times 10^{-3}$ M

$$\text{Ion-paired percent} = \frac{0.010 - 6.6_7 \times 10^{-3}}{0.010} \times 100 = 33\%$$

Ionic strength $= 4 (6.6_7 \times 10^{-3}) = 0.027$ M

(c) Acid hydrolysis: $Zn^{2+} + H_2O \overset{K_a}{\rightleftharpoons} ZnOH^+ + H^+$
Base hydrolysis: $SO_4^{2-} + H_2O \overset{K_b}{\rightleftharpoons} HSO_4^- + OH^-$

Appendix I gives the formation constant (β_1) for $ZnOH^+$.

We can find K_a for Zn^{2+} by adding the β_1 reaction to the K_w reaction:

$Zn^{2+} + OH^- \overset{\beta_1}{\rightleftharpoons} ZnOH^+$ $\log \beta_1 = 5.0 \Rightarrow \beta_1 = 10^{5.0}$

$\phantom{Zn^{2+} +} H_2O \overset{K_w}{\rightleftharpoons} H^+ + OH^-$ $K_w = 10^{-14.00}$

$\overline{Zn^{2+} + H_2O \overset{K_a}{\rightleftharpoons} ZnOH^+ + H^+ \quad K_a = \beta_1 K_w = 10^{-9.0}}$

The base hydrolysis constant for SO_4^{2-} is K_w/K_{a2} for H_2SO_4:

$$SO_4^{2-} + H^+ \overset{1/K_{a2}}{\rightleftharpoons} HSO_4^- \qquad 1/K_{a2} = 1/10^{-1.99}$$

$$H_2O \overset{K_w}{\rightleftharpoons} H^+ + OH^- \qquad K_w = 10^{-14.00}$$

$$SO_4^{2-} + H_2O \overset{K_b}{\rightleftharpoons} HSO_4^- + OH^- \qquad K_b = K_w/K_{a2} = 10^{-12.01}$$

7-29. Pertinent reactions:

$$LiF(s) \rightleftharpoons Li^+ + F^- \qquad K_{sp} = [Li^+]\gamma_{Li^+}[F^-]\gamma_{F^-} = 0.0017$$

$$LiF(s) \rightleftharpoons LiF(aq) \qquad K_{ion\ pair} = [LiF(aq)]\gamma_{LiF} = 0.0029$$

$$H_2O \overset{K_w}{\rightleftharpoons} H^+ + OH^- \qquad K_w = [H^+]\gamma_{H^+}[OH^-]\gamma_{OH^-} = 1.0 \times 10^{-14}$$

Charge balance: $[Li^+] + [H^+] = [F^-] + [OH^-]$

Mass balance: $[Li^+] + [LiF(aq)] = [F^-] + [LiF(aq)]$

There are five equations and five unknowns: $[Li^+]$, $[F^-]$, $[LiF(aq)]$, $[H^+]$, and $[OH^-]$. There is no coupling between the LiF reactions and the dissociation of water, so $[H^+] = [OH^-] = 1 \times 10^{-7}$ M. $[H^+]$ and $[OH^-]$ cancel in the charge balance, leaving $[Li^+] = [F^-]$. Also, the ion-pairing equilibrium constant tells us that $[LiF(aq)] = 0.0029$ M with the good approximation that $\gamma_{LiF} = 1$. All that is left is to find $[Li^+]$ and $[F^-]$ from the solubility product with the condition $[Li^+] = [F^-]$.

$$[Li^+]\gamma_{Li^+}[F^-]\gamma_{F^-} = K_{sp}$$

$$[Li^+]^2\gamma_{Li^+}\gamma_{F^-} = K_{sp} \quad \Rightarrow \quad [Li^+] = [F^-] = \sqrt{K_{sp}/\gamma_{Li^+}\gamma_{F^-}}$$

1st iteration: Let $\gamma_{Li^+} = \gamma_F = 1 \quad \Rightarrow \quad [Li^+] = [F^-] = \sqrt{K_{sp}} = 0.041$ M

2nd iteration: $\mu = 0.041$ M $\quad \Rightarrow \quad \gamma_{Li^+} = 0.843$ and $\gamma_F = 0.824$

$$\Rightarrow \quad [Li^+] = [F^-] = \sqrt{K_{sp}/\gamma_{Li^+}\gamma_{F^-}} = 0.049 \text{ M}$$

3rd iteration: $\mu = 0.049$ M $\quad \Rightarrow \quad \gamma_{Li^+} = 0.834$ and $\gamma_F = 0.812$

$$\Rightarrow \quad [Li^+] = [F^-] = \sqrt{K_{sp}/\gamma_{Li^+}\gamma_{F^-}} = 0.050 \text{ M}$$

4th iteration: $\mu = 0.049$ M $\quad \Rightarrow \quad \gamma_{Li^+} = 0.833$ and $\gamma_F = 0.811$

$$\Rightarrow \quad [Li^+] = [F^-] = \sqrt{K_{sp}/\gamma_{Li^+}\gamma_{F^-}} = 0.050 \text{ M}$$

7-30.   (a)

$$CaCO_3(s) \rightleftharpoons Ca^{2+} + CO_3^{2-} \qquad K_{sp} = 4.5 \times 10^{-9}$$

$$CO_2(aq) + H_2O \rightleftharpoons HCO_3^- + H^+ \qquad K_1 = 4.46 \times 10^{-7}$$

$$CO_3^{2-} + H^+ \rightleftharpoons HCO_3^- \qquad 1/K_2 = 1/(4.69 \times 10^{-11})$$

$$\overline{CaCO_3(s) + CO_2(aq) + H_2O \rightleftharpoons Ca^{2+} + 2HCO_3^-} \qquad \overline{K = K_{sp}K_1/K_2}$$

$$= 4.2_8 \times 10^{-5}$$

(b)   The equilibrium constant for the net reaction is

$$\frac{[Ca^{2+}][HCO_3^-]^2}{[CO_2(aq)]} = K = 4.2_8 \times 10^{-5}$$

We can substitute into this equation $[HCO_3^-] = 2[Ca^{2+}]$ and $[CO_2(aq)] = K_{CO_2}P_{CO_2}$ (where $K_{CO_2} = 0.032$ and $P_{CO_2} = 3.8 \times 10^{-4}$ bar) to get

$$\frac{[Ca^{2+}](2[Ca^{2+}])^2}{K_{CO_2}P_{CO_2}} = K \Rightarrow [Ca^{2+}] = 5.0_7 \times 10^{-4} \text{ M} = 20 \text{ mg/L}$$

(c)   If $[Ca^{2+}] = 80$ mg/L $= 2.0 \times 10^{-3}$ M, then

$$P_{CO_2} = \frac{[Ca^{2+}](2[Ca^{2+}])^2}{K_{CO_2}K} = 0.023 \text{ bar}$$

The partial pressure of CO_2 in the river is about (0.023 bar)/(3.8×10^{-4} bar) = 60 times higher than the atmospheric pressure of CO_2. There must be a source of extra CO_2 such as respiration in the river or inflow of ground water that is very rich in CO_2 and not in equilibrium with the atmosphere.

CHAPTER 8
MONOPROTIC ACID-BASE EQUILIBRIA

8-1. HBr (or any other acid or base) drives the reaction $H_2O \rightleftharpoons H^+ + OH^-$ to the left, according to Le Châtelier's principle. If, for example, the solution contains 10^{-4} M HBr, the concentration of OH^- from H_2O is $K_w/[H^+] = 10^{-10}$ M. The concentration of H^+ from H_2O must also be 10^{-10} M, since H^+ and OH^- are created in equimolar quantities.

8-2. (a) $pH = -\log [H^+] = -\log (1.0 \times 10^{-3}) = 3.00$

(b) $[H^+] = K_w /[OH^-] = (1.0 \times 10^{-14})/(1.0 \times 10^{-2}) = 1.0 \times 10^{-12}$ M

$pH = -\log [H^+] = 12.00$

8-3. Charge balance: $[H^+] = [OH^-] + [ClO_4^-] \Rightarrow [OH^-] = [H^+] - 5.0 \times 10^{-8}$

Mass balance is the same as charge balance.

Equilibrium: $[H^+] [OH^-] = K_w$

$[H^+] ([H^+] - 5.0 \times 10^{-8}) = 1.0 \times 10^{-14} \Rightarrow [H^+] = 1.28 \times 10^{-7}$ M

$pH = -\log [H^+] = 6.89$

$[OH^-] = K_w/[H^+] = 7.8 \times 10^{-8}$ M $\Rightarrow [H^+]$ from $H_2O = 7.8 \times 10^{-8}$ M

Fraction of total $[H^+]$ from $H_2O = \dfrac{7.8 \times 10^{-8} \text{ M}}{1.28 \times 10^{-7} \text{ M}} = 0.61$

8-4. (a) $pH = -\log [H^+]\gamma_{H^+}$

$1.092 = -\log (0.100) \gamma_{H^+} \Rightarrow \gamma_{H^+} = 0.809$

The tabulated activity coefficient is 0.83.

(b) $2.102 = -\log (0.010\,0)\gamma_{H^+} \Rightarrow \gamma_{H^+} = 0.791$

(c) The activity coefficient depends somewhat on the identity of the counterions in solution.

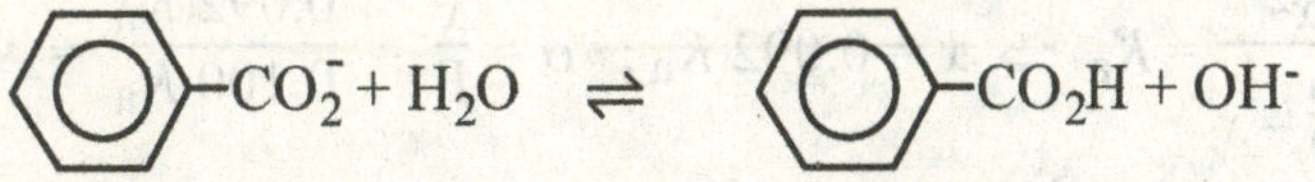

8-5. (a)

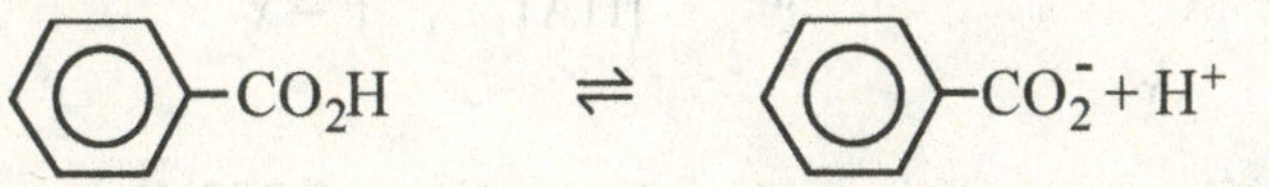

(b)

(c)

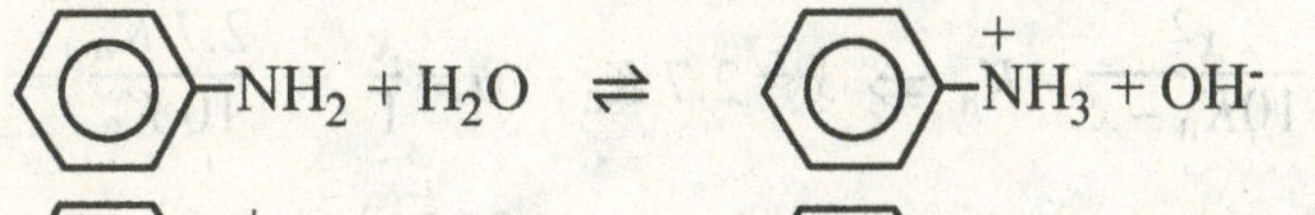

(d)

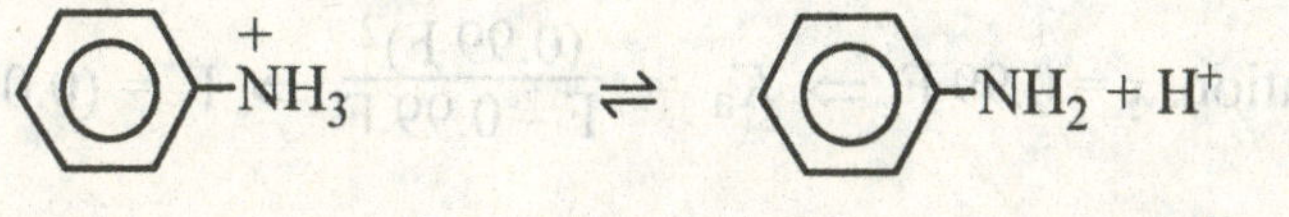

8-6. Let $x = [H^+] = [A^-]$ and $0.100 - x = [HA]$.

$$\frac{x^2}{0.100 - x} = 1.00 \times 10^{-5} \Rightarrow x = 9.95 \times 10^{-4} \text{ M} \Rightarrow pH = -\log x = 3.00$$

$$\alpha = \frac{[A^-]}{[A^-] + [HA]} = \frac{9.95 \times 10^{-4}}{0.100} = 9.95 \times 10^{-3}$$

8-7.

$$BH^+ \overset{K_a}{\rightleftharpoons} B + H^+ \qquad K_a = K_w/K_b = 1.00 \times 10^{-10}$$

$$\underset{0.100 - x}{} \qquad \underset{x}{} \quad \underset{x}{}$$

$$\frac{x^2}{0.100 - x} = 1.00 \times 10^{-10} \Rightarrow x = [B] = [H^+] = 3.16 \times 10^{-6} \text{ M} \Rightarrow pH = 5.50$$

8-8.

$$(CH_3)_3NH^+ \rightleftharpoons (CH_3)_3N + H^+ \qquad K_a = 1.59 \times 10^{-10}$$

$$\underset{F - x}{} \qquad \underset{x}{} \quad \underset{x}{}$$

$$\frac{x^2}{0.060 - x} = K_a \Rightarrow x = 3.0_9 \times 10^{-6} \Rightarrow pH = 5.51$$

$$[(CH_3)_3N] = x = 3.1 \times 10^{-6} \text{ M}, \quad [(CH_3)_3NH^+] = F - x = 0.060 \text{ M}$$

8-9. $HA \overset{K_a}{\rightleftharpoons} H^+ + A^-$. $Q = \frac{[A^-][H^+]}{[HA]}$. When the system is at equilibrium, $Q = K_a$.

Let's call the concentrations at equilibrium $[A^-]_e$, $[H^+]_e$, and $[HA]_e$. If the solution is diluted by a factor of 2, the concentrations become $\frac{1}{2}[A^-]_e$, $\frac{1}{2}[H^+]_e$, and $\frac{1}{2}[HA]_e$.

The reaction quotient becomes $Q = \dfrac{\frac{1}{2}[A^-]_e \frac{1}{2}[H^+]_e}{\frac{1}{2}[HA]_e} = \frac{1}{2}\dfrac{[A^-]_e[H^+]_e}{[HA]_e} = \frac{1}{2}K_a$.

Since $Q < K_a$, the concentrations of products must increase and the concentration of reactant must decrease to attain equilibrium. That means that the weak acid dissociates further as it is diluted in order to stay in equilibrium.

8-10.

$$HA \overset{K_a}{\rightleftharpoons} H^+ + A^- \qquad K_a = \frac{[H^+][A^-]}{[HA]} = \frac{x^2}{F - x}$$

$$\underset{F - x}{} \qquad \underset{x}{} \quad \underset{x}{}$$

For $F = \dfrac{K_a}{10}$, $\dfrac{x^2}{\frac{K_a}{10} - x} = K_a \Rightarrow x = 0.092\, K_a$; $\alpha = \dfrac{x}{F} = \dfrac{0.092\, K_a}{0.100\, K_a} = 92\%$

For $F = 10\, K_a$, $\dfrac{x^2}{10K_a - x} = K_a \Rightarrow x = 2.7\, K_a$; $\alpha = \dfrac{x}{F} = \dfrac{2.7\, K_a}{10\, K_a} = 27\%$

For 99% dissociation, $x = 0.99\, F \Rightarrow K_a = \dfrac{(0.99\, F)^2}{F - 0.99\, F} \Rightarrow F = (0.010\,2)K_a$

8-11.

$$K_a = \frac{(10^{-2.78})^2}{0.045\,0 - 10^{-2.78}} = 6.35 \times 10^{-5} = pK_a = 4.20$$

8-12. $HA \rightleftharpoons H^+ + A^- \qquad \alpha = 0.006\,0 = \dfrac{x}{F}$

 $F-x \qquad\ x \qquad\ x$

$F = 0.045\,0$ M and $x = (0.006\,0)(0.045\,0$ M$) = 2.7 \times 10^{-4}$ M

$$\Rightarrow K_a = \frac{x^2}{F-x} = 1.6 \times 10^{-6} \Rightarrow pK_a = 5.79$$

8-13. (a) $\quad HA \rightleftharpoons H^+ + A^-$

 $F-x \quad\ x \quad\ x$

$$\frac{x^2}{F-x} = K_a \Rightarrow \frac{x^2}{0.010-x} = 9.8 \times 10^{-5} \Rightarrow x = 9.4 \times 10^{-4}$$

$$\Rightarrow pH = 3.03$$

$$\alpha = \frac{[A^-]}{[HA]+[A^-]} = \frac{x}{F} = 9.4\%$$

(b) pH = 7.00 because the acid is so dilute. From the K_a equilibrium we write

$$[A^-] = \frac{K_a}{[H^+]}[HA] = \frac{9.8 \times 10^{-5}}{1.0 \times 10^{-7}}[HA] = 980\,[HA]$$

$$\alpha = \frac{[A^-]}{[HA]+[A^-]} = \frac{980\,[HA]}{[HA]+980\,[HA]} = \frac{980}{981} = 99.9\%$$

8-14. Phenol is a weak acid, so it will contribute negligible ionic strength. The ionic strength of the solution is 0.050 M.

 $HA \rightleftharpoons H^+ + A^- \qquad\qquad K_a = 1.01 \times 10^{-10}$

 $F-x \qquad\ x \qquad\ x$

$$\frac{[H^+]\gamma_{H^+}[A^-]\gamma_{A^-}}{[HA]\gamma_{HA}} = K_a \Rightarrow \frac{(x)(0.86)(x)(0.835)}{(0.050\,0-x)(1.00)} = 1.01 \times 10^{-10} \Rightarrow x = 2.65 \times 10^{-6}$$

$$pH = -\log[H^+]\gamma_{H^+} = -\log[2.65 \times 10^{-6}](0.86) = 5.64$$

$$\alpha = \frac{[A^-]}{[HA]+[A^-]} = \frac{2.65 \times 10^{-6}}{0.050\,0} = 5.30 \times 10^{-5} = 0.005\,3\%$$

8-15.
$$Cr^{3+} + H_2O \overset{K_{a1}}{\rightleftharpoons} Cr(OH)^{2+} + H^+$$

$$0.010 - x \qquad\qquad\qquad\qquad x \qquad\quad x$$

$$\frac{x^2}{0.010 - x} = 10^{-3.66} \Rightarrow x = 1.3_7 \times 10^{-3} \text{ M}$$

$$pH = -\log x = 2.86 \qquad\qquad \alpha = \frac{x}{0.010} = 0.14$$

8-16.
$$HNO_3 \rightleftharpoons H^+ + NO_3^-$$

$$F - x \qquad\quad x \qquad x$$

$$\frac{x^2}{F - x} = 26.8 \quad \Rightarrow x = 0.099\,6 \text{ M when } F = 0.100 \text{ M} \Rightarrow \alpha = \frac{x}{F} = 99.6\%$$

$$\Rightarrow x = 0.965 \text{ M when } F = 1.00 \text{ M} \Rightarrow \alpha = \frac{x}{F} = 96.5\%$$

8-17. The initial spreadsheet follows (on the left). Guess a value for x in cell A4. The formula in cell B4 is "=A4^2/(A6-A4)". In Excel 2007, click the Microsoft Office button at the top left of the spreadsheet, click on Excel Options, and then on Formulas. Set Maximum Change to a small number such as 1e-20 to find an answer with high precision. In earlier versions of Excel, in the Tools menu, go to Options and then Calculation. Check Iteration and set Maximum change to 1e-20. Highlight cell B4 and select Goal Seek. Set cell <u>B4</u> To value <u>1e-5</u> By changing cell <u>A4</u>. Click OK and Goal Seek finds the solution in the second spreadsheet (on the right). The value $x = 9.95 \times 10^{-5}$ makes the quotient $x^2/(F\text{-}x)$ equal to 1.00×10^{-5}.

	A	B
1	Using Excel GOAL SEEK	
2		
3	x =	x^2/(F-x) =
4	0.01	1.1111E-03
5	F =	
6	0.1	

	A	B
1	Using Excel GOAL SEEK	
2		
3	x =	x^2/(F-x) =
4	0.00099501	1.0000E-05
5	F =	
6	0.1	

Before executing Goal Seek   After executing Goal Seek

8-18. The "fishy" smell comes from volatile amines (RNH_2). Lemon juice protonates the amines, giving much less volatile ammonium ions (RNH_3^+).

8-19. Let $x = [OH^-] = [BH^+]$ and $0.100 - x = [B]$. $\dfrac{x^2}{0.100 - x} = 1.00 \times 10^{-5}$

$$\Rightarrow x = 9.95 \times 10^{-4} \text{ M} \Rightarrow [H^+] = \frac{K_w}{x} = 1.005 \times 10^{-11} \Rightarrow pH = 11.00$$

$$\alpha = \frac{[BH^+]}{[B] + [BH^+]} = \frac{9.95 \times 10^{-4}}{0.100} = 9.95 \times 10^{-3}$$

8-20. $(CH_3)_3N + H_2O \rightleftharpoons (CH_3)_3NH^+ + OH^-$ $K_b = K_w/K_a = 6.3 \times 10^{-5}$
    $F-x$ x x

$$\frac{x^2}{0.060-x} = K_b \Rightarrow x = 1.9_1 \times 10^{-3} \Rightarrow pH = -\log\frac{K_w}{x} = 11.28$$

$[(CH_3)_3NH^+] = x = 1.9_1 \times 10^{-3}\ M$, $[(CH_3)_3N] = F-x = 0.058\ M$

8-21. $CN^- + H_2O \rightleftharpoons HCN + OH^-$ $K_b = K_w/K_a = 1.6 \times 10^{-5}$
    $F-x$ x x

$$\frac{x^2}{0.050-x} = K_b \Rightarrow x = 8.9 \times 10^{-4} \Rightarrow pH = -\log\frac{K_w}{x} = 10.95$$

8-22. $CH_3CO_2^- + H_2O \rightleftharpoons CH_3CO_2H + OH^-$ $K_b = K_w/K_a = 5.7 \times 10^{-10}$
    $F-x$ x x

$$\frac{x^2}{(1.00 \times 10^{-1})-x} = K_b \Rightarrow x = 7.6 \times 10^{-6} \Rightarrow \alpha = \frac{x}{F} = 0.0076\%$$

$$\frac{x^2}{(1.00 \times 10^{-2})-x} = K_b \Rightarrow x = 2.4 \times 10^{-6} \Rightarrow \alpha = \frac{x}{F} = 0.024\%$$

For 1.00×10^{-12} M sodium acetate, pH = 7.00 and we can say

$$[HA] = \frac{K_b[A^-]}{[OH^-]} = (5.7 \times 10^{-3})[A^-]$$

$$\alpha = \frac{[HA]}{[HA]+[A^-]} = \frac{(5.7 \times 10^{-3})[A^-]}{(5.7 \times 10^{-3})[A^-]+[A^-]} = 0.57\%$$

The more dilute the solution, the greater is α.

8-23.   B   + H_2O $\rightleftharpoons$ BH^+ + OH^-
    $F-(K_w/10^{-9.28})$ $K_w/10^{-9.28}$ $K_w/10^{-9.28}$

$$K_b = \frac{(K_w/10^{-9.28})^2}{F-(K_w/10^{-9.28})} = \frac{(K_w/10^{-9.28})^2}{0.10-(K_w/10^{-9.28})} = 3.6 \times 10^{-9}$$

8-24. $B + H_2O \rightleftharpoons BH^+ + OH^-$ $\alpha = 0.020 = \frac{x}{F} \Rightarrow x = 2.0 \times 10^{-3}\ M$
    $0.10-x$ x x

$$K_b = \frac{x^2}{0.10-x} = \frac{(2.0 \times 10^{-3})^2}{0.10-(2.0 \times 10^{-3})} = 4.1 \times 10^{-5}$$

8-25. As $[B] \rightarrow 0$, pH $\rightarrow 7$ and $[OH^-] \rightarrow 10^{-7}\ M$.

$$K_b = \frac{[BH^+][OH^-]}{[B]} = 10^{-7}\frac{[BH^+]}{[B]} \Rightarrow [BH^+] = 10^7 K_b[B]$$

$$\alpha = \frac{[BH^+]}{[B]+[BH^+]} = \frac{10^7 K_b[B]}{[B]+10^7 K_b[B]} = \frac{10^7 K_b}{1+10^7 K_b}$$

For $K_b = 10^{-4}$, we have $\alpha = \dfrac{10^7 K_b}{1 + 10^7 K_b} = \dfrac{10^7 \, 10^{-4}}{1 + 10^7 \, 10^{-4}} = 0.999$

For $K_b = 10^{-10}$, we have $\alpha = \dfrac{10^7 K_b}{1 + 10^7 K_b} = \dfrac{10^7 \, 10^{-10}}{1 + 10^7 \, 10^{-10}} = 0.000\ 999$

8-26. I would weigh out 0.020 0 mol of acetic acid (= 1.201 g) and place it in a beaker with ~75 mL of water. While monitoring the pH with a pH electrode, I would add 3 M NaOH (~4 mL is required) until the pH is exactly 5.00. I would then pour the solution into a 100 mL volumetric flask and wash the beaker several times with a few milliliters of distilled water. Each washing would be added to the volumetric flask, to ensure quantitative transfer from the beaker to the flask. After swirling the volumetric flask to mix the solution, I would carefully add water up to the 100 mL mark, insert the cap, and invert 20 times to ensure complete mixing.

8-27. The pH of a buffer depends on the ratio of the concentrations of HA and A⁻ (pH = pK_a + log [A⁻]/[HA]). When the volume of solution is changed, both concentrations are affected equally and their ratio does not change.

8-28. Buffer capacity measures the ability to maintain the original [A⁻]/[HA] ratio when acid or base is added. A more concentrated buffer has more A⁻ and HA, so a smaller fraction of A⁻ or HA is consumed by added acid or base. Therefore, there is a smaller change in the ratio [A⁻]/[HA].

8-29. At very low or very high pH, there is so much acid or base in the solution already that small additions of acid or base will hardly have any effect. At low pH, the buffer is H_3O^+/H_2O; and at high pH, the buffer is H_2O/OH^-.

8-30. When pH = pK_a, the ratio of concentrations [A⁻]/[HA] is unity. A given increment of added acid or base has the least effect on the ratio [A⁻]/[HA] when the concentrations of A⁻ and HA are initially equal.

8-31. The Henderson-Hasselbalch is just a rearranged form of the K_a equilibrium expression, which is always true. When we make the approximation that [HA] and [A⁻] are unchanged from what we added, we are neglecting acid dissociation and base hydrolysis, which can change the concentrations in dilute solutions of moderately strong acids or bases.

8-32.

acid	pK_a
hydrogen peroxide	11.65
propanoic acid	4.87
cyanoacetic acid	2.47
4-aminobenzenesulfonic acid	3.23 $\leftarrow$

most suitable because pK_a is closest to desired pH

8-33. $pH = pK_a + \log \dfrac{[A^-]}{[HA]} = 5.00 + \log \dfrac{0.050}{0.100} = 4.70$

8-34. $pH = 3.744 + \log \dfrac{[HCO_2^-]}{[HCO_2H]}$

pH:	3.000	3.744	4.000
$[HCO_2^-]/[HCO_2H]$:	0.180	1.00	1.80

8-35. $pH = 3.57 + \log \dfrac{[HCO_2^-]}{[HCO_2H]}$ where 3.57 is pK_a at $\mu = 0.1$ M

Substituting pH = 3.744 gives $[HCO_2^-]/[HCO_2H] = 1.5$

8-36. $pH = pK_a + \log \dfrac{[NO_2^-]}{[HNO_2]}$, where $pK_a = 14.00 - pK_b = 3.15$

(a) If pH = 2.00, $[HNO_2]/[NO_2^-] = 14$

(b) If pH = 10.00, $[HNO_2]/[NO_2^-] = 1.4 \times 10^{-7}$

8-37. (a) HEPES is an acid with $pK_a = 7.56$. When it is dissolved in water, the solution will be acidic and will require NaOH to raise the pH to 7.45.

(b) 1. Weigh out $(0.250 \text{ L})(0.050\,0 \text{ M}) = 0.012\,5$ mol of HEPES and dissolve in ~200 mL.

2. Adjust the pH to 7.45 with NaOH.

3. Dilute to 250 mL.

8-38. 213 mL of 0.006 66 M 2,2'-bipyridine = 1.41_9 mmol base. We will add x mol H$^+$ to get a pH of 4.19.

2,2'-bipyridine	+	H$^+$	$\rightarrow$	2,2'-bipyridineH$^+$
Initial mmol: 1.41_9		x		—
Final mmol: $1.41_9 - x$		—		x

$$pH = pK_a + \log \frac{[\text{bipyridine}]}{[\text{bipyridineH}^+]}$$

$$4.19 = 4.34 + \log \frac{1.41_9 - x}{x} \Rightarrow x = 0.831 \text{ mmol}$$

$$\text{volume} = \frac{0.831 \text{ mmol}}{0.246 \text{ mmol/mL}} = 3.38 \text{ mL}$$

8-39. (a)

(b) FM of imidazole = 68.08.   FM of imidazole hydrochloride = 104.54.

$$pH = 6.993 + \log \frac{1.00/68.08}{1.00/104.54} = 7.18$$

(c)

	B	+	H$^+$	→	BH$^+$
Initial mmol:	14.6$_9$		2.46		9.57
Final mmol:	12.2$_3$		—		12.0$_3$

$$pH = 6.993 + \log \frac{12.2_3}{12.0_3} = 7.00$$

(d) The imidazole must be half neutralized to obtain pH = pK_a. Since there are 14.6$_9$ mmol of imidazole, this will require $\frac{1}{2}$(14.6$_9$) = 7.34 mmol of HClO$_4$ = 6.86 mL.

8-40. (a) $pH = 2.865 + \log \dfrac{0.040\,0}{0.080\,0} = 2.56$

(b) Using Eqns. (8-20) and (8-21), and neglecting [OH$^-$], we can write

$$K_a = 1.36 \times 10^{-3} = \frac{[\text{H}^+](0.040\,0 + [\text{H}^+])}{0.080\,0 - [\text{H}^+]} \Rightarrow [\text{H}^+] = 2.48 \times 10^{-3} \text{ M}$$

$$\Rightarrow pH = 2.61$$

(c) 0.080 mol of HNO$_3$ + 0.080 mol of Ca(OH)$_2$ react completely, leaving an excess of 0.080 mol of OH$^-$. This much OH$^-$ converts 0.080 mol of ClCH$_2$CO$_2$H into 0.080 mol of ClCH$_2$CO$_2^-$. The final concentrations are [ClCH$_2$CO$_2^-$] = 0.020 + 0.080 = 0.100 M and [ClCH$_2$CO$_2$H] = 0.180 − 0.080 = 0.100 M. So pH = pK_a = 2.86.

8-41.

$$HA \quad + \quad OH^- \quad \rightarrow \quad A^- \quad + \quad H_2O$$

Initial moles: 0.022 4 x —

Final moles: 0.022 4 − x — x

$$pH = 7.40 = pK_a + \log \frac{[A^-]}{[HA]} = 7.48 + \log \frac{x}{0.022\,4 - x} \Rightarrow x = 0.010\,17 \text{ mol.}$$

$$\text{volume} = \frac{0.010\,17 \text{ mol}}{0.626 \text{ M}} = 16.2 \text{ mL}$$

8-42. (a) Since pK_a for acetic acid is 4.756, we expect the solution to be acidic and will

ignore $[OH^-]$ in comparison to $[H^+]$.

$[HA] = 0.002\,0 - [H^+]$ $[A^-] = 0.004\,00 + [H^+]$

$$K_a = 1.75 \times 10^{-5} = \frac{[H^+](0.004\,00 + [H^+])}{0.002\,00 - [H^+]} \Rightarrow [H^+] = 8.69 \times 10^{-6} \text{ M}$$

$\Rightarrow pH = 5.06$ $[HA] = 0.001\,99 \text{ M}$ $[A^-] = 0.004\,01 \text{ M}$

If you used $K_a = 10^{-4.756}$ instead of rounding to 1.75×10^{-5}, then $[H^+] = 8.71 \times 10^{-6}$ M.

(b) We use Goal Seek to vary cell B5 until cell D4 is equal to K_a.

	A	B	C	D	E
1	Ka = 10^-pKa =	1.75E-05		Reaction quotient	
2	Kw =	1.00E-14		for Ka =	
3	FHA =	0.002000		[H+][A-]/[HA] =	
4	FA =	0.004000		1.75E-05	
5	H =	8.693E-06	<-Goal Seek solution		
6	OH = Kw/H =	1.15E-09		D4 = H*(FA+H-OH)/(FHA-H+OH)	
7	pH = -logH =	5.0608262			
8	[HA] =	0.0019913		B8 = FHA-H+OH	
9	[A-] =	0.0040087		B9 = FA+H-OH	

8-43. (a) If we dissolve B and BH^+Br^- (where Br^- is an inert anion), the mass balance is

$F_{BH^+} + F_B = [BH^+] + [B]$ and the charge balance is $[Br^-] + [OH^-] = [BH^+] +$

$[H^+]$. Noting that $[Br^-] = F_{BH^+}$, the charge balance can be rewritten as

$$[BH^+] = F_{BH^+} + [OH^-] - [H^+] \qquad (A)$$

Substituting this expression into the mass balance gives

$$[B] = F_B - [OH^-] + [H^+] \qquad (B)$$

If we assume that $[B] = 0.010\,0$ M and $[BH^+] = 0.020\,0$ M, we calculate

$$pH = pK_a + \log \frac{[B]}{[BH^+]} = 12.00 + \log \frac{0.010\,0}{0.020\,0} = 11.70$$

If we do not assume that $[B] = 0.010\,0$ M and $[BH^+] = 0.020\,0$ M, we use

Equations A and B. Since the solution is basic, we neglect $[H^+]$ relative to

$[OH^-]$ and write $[B] = 0.010\,0 - x$ and $[BH^+] = 0.020\,0 + x$, where $x =$

$[OH^-]$.

Then we can say $K_b = 10^{-2.00} = \dfrac{[BH^+][OH^-]}{[B]} = \dfrac{(0.020\,0 + x)(x)}{(0.010\,0 - x)} \Rightarrow$

$x = 0.003\,03$ M $\qquad$ pH $= -\log \dfrac{K_w}{x} = 11.48$.

(b) We use Goal Seek to vary cell B5 until cell D4 is equal to K_b.

	A	B	C	D	E
1	Kb =	1.00E-02		Reaction quotient	
2	Kw =	1.00E-14		for Kb =	
3	FBH =	0.02		[OH-][BH+]/[B] =	
4	FB =	0.01		0.01	
5	OH =	3.028E-03	<-Goal Seek solution		
6	H = Kw/OH =	3.303E-12		D4 = OH*(FBH-H+OH)/(FB+H-OH)	
7	pH = -logH =	11.481121			
8	BH =	0.0230278		C8 = FBH-H+OH	
9	B =	0.0069722		C9 = FB+H-OH	

8-44. $K_a = \dfrac{[HPO_4^{2-}][H^+]\,\gamma_{HPO_4^{2-}}\,\gamma_{H^+}}{[H_2PO_4^-]\,\gamma_{H_2PO_4^-}} = 10^{-7.20}$

To find pH, rearrange the K_a expression to solve for the activity of H^+, which is $[H^+]\gamma_{H^+}$:

$[H^+]\gamma_{H^+} = \dfrac{K_a[H_2PO_4^-]\,\gamma_{H_2PO_4^-}}{[HPO_4^{2-}]\,\gamma_{HPO_4^{2-}}}$

At $\mu = 0.1$ M, the activity coefficients are $\gamma_{H^+} = 0.83$, $\gamma_{H_2PO_4^-} = 0.775$, and

$\gamma_{HPO_4^{2-}} = 0.355$, so

$[H^+]\gamma_{H^+} = \dfrac{10^{-7.20}[H_2PO_4^-](0.775)}{[HPO_4^{2-}](0.355)} = 1.38 \times 10^{-7}$ when $[H_2PO_4^-] = [HPO_4^{2-}]$

pH $= -\log \mathcal{A}_{H^+} = -\log[H^+]\gamma_{H^+} = -\log(1.38 \times 10^{-7}) = 6.86$

8-45. Equilibria: $\beta_1 = \dfrac{[AlOH^{2+}][H^+]}{[Al^{3+}]}$ $\qquad$ (a)

$\beta_2 = \dfrac{[Al(OH)_2^+][H^+]^2}{[Al^{3+}]}$ $\qquad$ (b)

$\beta_3 = \dfrac{[Al(OH)_3(aq)][H^+]^3}{[Al^{3+}]}$ $\qquad$ (c)

$\beta_4 = \dfrac{[Al(OH)_4^-][H^+]^4}{[Al^{3+}]}$ $\qquad$ (d)

$K_{22} = \dfrac{[Al_2(OH)_2^{4+}][H^+]^2}{[Al^{3+}]^2}$ $\qquad$ (e)

$$K_{43} = \frac{[Al_3(OH)_4^{5+}][H^+]^4}{[Al^{3+}]^3} \qquad (f)$$

$$K_w = [H^+][OH^-] \qquad (g)$$

Charge balance: $3[Al^{3+}] + 2[AlOH^{2+}] + [Al(OH)_2^+] + 4[Al_2(OH)_2^{4+}] +$

$\qquad\qquad 5[Al_3(OH)_4^{5+}] + [H^+] = [Al(OH)_4^-] + [OH^-] + [ClO_4^-] \quad (h)$

Mass balances: $3F = [ClO_4^-] \qquad\qquad\qquad\qquad\qquad\qquad (i)$

$\qquad F = [Al^{3+}] + [AlOH^{2+}] + [Al(OH)_2^+] + 2[Al_2(OH)_2^{4+}] +$

$\qquad\qquad [Al(OH)_3(aq)] + [Al(OH)_4^-] + 3[Al_3(OH)_4^{5+}] \qquad (j)$

We have 10 equations and 10 unknowns, so the problem can, in principle, be solved.

CHAPTER 9
POLYPROTIC ACID-BASE EQUILIBRIA

9-1. The K_a reaction, with a much greater equilibrium constant than K_b, releases H^+:

$$HA^- \rightleftharpoons H^+ + A^{2-} \qquad K_a$$

Each mole of H^+ reacts with one mole of OH^- from the K_b reaction:

$$HA^- + H_2O \rightleftharpoons H_2A + OH^-.$$

The net result is that the K_b reaction is driven almost as far toward completion as the K_a reaction.

9-2.

$H_3\overset{+}{N}\overset{\underset{|}{R}}{\underset{|}{C}}CO_2^-$ pK values apply to $-NH_3^+$, $-CO_2H$, and, in some cases, R.

9-3.

$$\text{(pyrrolidine-}CO_2^-) + H_2O \rightleftharpoons (\text{pyrrolidine-}\overset{+}{N}H_2\text{-}CO_2^-) + OH^- \qquad K_{b1} = \frac{K_w}{K_2} = 4.37 \times 10^{-4}$$

$$(\text{pyrrolidine-}\overset{+}{N}H_2\text{-}CO_2^-) + H_2O \rightleftharpoons (\text{pyrrolidine-}\overset{+}{N}H_2\text{-}CO_2H) + OH^- \qquad K_{b2} = \frac{K_w}{K_1} = 8.93 \times 10^{-13}$$

9-4. (a) $\dfrac{x^2}{0.100 - x} = K_1 \Rightarrow x = 3.11 \times 10^{-3} = [H^+] = [HA^-] \Rightarrow pH = 2.51$

$[H_2A] = 0.100 - x = 0.0969 \text{ M} \qquad [A^{2-}] = \dfrac{K_2[HA^-]}{[H^+]} = 1.00 \times 10^{-8} \text{ M}$

(b) $[H^+] \approx \sqrt{\dfrac{K_1 K_2 F + K_1 K_w}{K_1 + F}} = 1.00 \times 10^{-6} \Rightarrow pH = 6.00$

$[HA^-] \approx 0.100 \text{ M}$

$[H_2A] = \dfrac{[H^+][HA^-]}{K_1} = 1.00 \times 10^{-3} \text{ M} \qquad [A^{2-}] = \dfrac{K_2[HA^-]}{[H^+]} = 1.00 \times 10^{-3} \text{ M}$

(c) $\dfrac{x^2}{0.100 - x} = \dfrac{K_w}{K_2} \Rightarrow x = [OH^-] = [HA^-] = 3.16 \times 10^{-4} \text{ M} \Rightarrow pH = 10.50$

$[A^{2-}] = 0.100 - x = 9.97 \times 10^{-2} \text{ M} \qquad [H_2A] = \dfrac{[H^+][HA^-]}{K_1} = 1.00 \times 10^{-10} \text{ M}$

	pH	$[H_2A]$	$[HA^-]$	$[A^{2-}]$
0.100 M H_2A	2.51	9.69×10^{-2}	3.11×10^{-3}	1.00×10^{-8}
0.100 M NaHA	6.00	1.00×10^{-3}	1.00×10^{-1}	1.00×10^{-3}
0.100 M Na_2A	10.50	1.00×10^{-10}	3.16×10^{-4}	9.97×10^{-2}

82

9-5. (a) $H_2M = H^+ + HM^-$ $K_1 = 1.42 \times 10^{-3}$

    $F - x$ x x

$$\frac{x^2}{0.100 - x} = K_1 \Rightarrow x = 1.12 \times 10^{-2} \Rightarrow pH = -\log x = 1.95$$

$[H_2M] = 0.100 - x = 0.089 \text{ M}$

$[HM^-] = x = 1.12 \times 10^{-2} \text{ M}$ $[M^{2-}] = \dfrac{[HM^-] K_2}{[H^+]} = 2.01 \times 10^{-6} \text{ M}$

(b) $[H^+] = \sqrt{\dfrac{K_1 K_2 (0.100) + K_1 K_w}{K_1 + 0.100}} = 5.30 \times 10^{-5} \Rightarrow pH = 4.28$

$[HM^-] \approx 0.100 \text{ M}$ $[H_2M] = \dfrac{[HM^-][H^+]}{K_1} = 3.7 \times 10^{-3} \text{ M}$

$[M^{2-}] = \dfrac{K_2[HM^-]}{[H^+]} = 3.8 \times 10^{-3} \text{ M}$

The method of Box 9-2 would give more accurate answers, since $[HM^-]$ is not that much greater than $[H_2M]$ or $[M^{2-}]$ in this case.

(c) $M^{2-} + H_2O \rightleftharpoons HM^- + OH^-$ $K_{b1} = K_w/K_{a2} = 4.98 \times 10^{-9}$

    $F - x$ x x

$$\frac{x^2}{0.100 - x} = K_{b1} \Rightarrow x = 2.23 \times 10^{-5} \Rightarrow pH = -\log \frac{K_w}{x} = 9.35$$

$[M^{2-}] = 0.100 - x = 0.100 \text{ M}$ $[HM^-] = x = 2.23 \times 10^{-5} \text{ M}$

$[H_2M] = \dfrac{[H^+][HM^-]}{K_1} = 7.04 \times 10^{-12} \text{ M}$

9-6. $HN\underset{F-x}{\bigcirc}NH + H_2O \rightleftharpoons HN\underset{x}{\bigcirc}\overset{+}{N}H_2 + OH^-$ $\underset{x}{}$ $K_{b1} = \dfrac{K_w}{K_2} = 5.38 \times 10^{-5}$

$$\frac{x^2}{0.300 - x} = K_{b1} \Rightarrow x = 3.99 \times 10^{-3} \text{ M} \Rightarrow pH = -\log K_w/x = 11.60$$

$[B] = 0.300 - x = 0.296 \text{ M}$ $[BH^+] = x = 3.99 \times 10^{-3} \text{ M}$

$[BH_2^{2+}] = \dfrac{[BH^+][H^+]}{K_1} = 2.15 \times 10^{-9} \text{ M}$

9-7. For H_2A, $K_1 = 5.62 \times 10^{-2}$ and $K_2 = 5.42 \times 10^{-5}$

First approximation ($[HA^-]_1 \approx 0.001\ 00 \text{ M}$):

$$[H^+]_1 = \sqrt{\frac{K_1 K_2 (0.001\ 00) + K_1 K_w}{K_1 + 0.001\ 00}} = 2.31 \times 10^{-4} \text{ M} \Rightarrow pH_1 = 3.64$$

$[H_2A]_1 = \dfrac{[H^+]_1\ [HA^-]_1}{K_1} = 4.10 \times 10^{-6} \text{ M}$

$$[A^{2-}]_1 = \frac{K_2[HA^-]_1}{[H^+]} = 2.35 \times 10^{-4} \text{ M}$$

Second approximation:

$$[HA^-]_2 \approx 0.001\,00 - [H_2A]_1 - [A^{2-}]_1 = 0.000\,761 \text{ M}$$

$$[H^+]_2 = \sqrt{\frac{K_1K_2(0.000\,761) + K_1K_w}{K_1 + 0.000\,761}} = 2.02 \times 10^{-4} \text{ M} \Rightarrow pH_2 = 3.70$$

$$[H_2A]_2 = \frac{[H^+]_2[HA^-]_2}{K_1} = 2.73 \times 10^{-6} \text{ M}$$

$$[A^{2-}]_2 = \frac{K_2[HA^-]_2}{[H^+]_2} = 2.04 \times 10^{-4} \text{ M}$$

Third approximation:

$$[HA^-]_3 \approx 0.001\,00 - [H_2A]_2 - [A^{2-}]_2 = 0.000\,793 \text{ M}$$

$$[H^+]_3 = \sqrt{\frac{K_1K_2(0.000\,793) + K_1K_w}{K_1 + 0.000\,793}} = 2.06 \times 10^{-4} \text{ M} \Rightarrow pH_3 = 3.69$$

$$[H_2A]_3 = \frac{[H^+]_3[HA^-]_3}{K_1} = 2.90 \times 10^{-6} \text{ M}$$

$$[A^{2-}]_3 = \frac{K_2[HA^-]_3}{[H^+]_3} = 2.09 \times 10^{-4} \text{ M}$$

9-8. (a) Charge balance: $[K^+] + [H^+] = [OH^-] + [HP^-] + 2[P^{2-}]$ (1)

Mass balance: $[K^+] = [H_2P] + [HP^-] + [P^{2-}]$ (2)

Equilibria: $K_1 = \dfrac{[H^+]\gamma_{H^+}[HP^-]\,\gamma_{HP^-}}{[H_2P]\,\gamma_{H_2P}}$ (3)

$K_2 = \dfrac{[H^+]\gamma_{H^+}[P^{2-}]\,\gamma_{P^{2-}}}{[HP^-]\,\gamma_{HP^-}}$ (4)

$K_w = [H^+]\,\gamma_{H^+}[OH^-]\,\gamma_{OH^-}$ (5)

Solving for $[K^+]$ in Eqns. (1) and (2) and equating the results gives

$[H_2P] + [H^+] - [P^{2-}] - [OH^-] = 0$

Making substitutions from Eqns. (3), (4), and (5), we can write

$$\frac{[H^+]\gamma_{H^+}[HP^-]\,\gamma_{HP^-}}{K_1\,\gamma_{H_2P}} + [H^+] - \frac{K_2[HP^-]\,\gamma_{HP^-}}{[H^+]\,\gamma_{H^+}\,\gamma_{P^{2-}}} - \frac{K_w}{[H^+]\,\gamma_{H^+}\,\gamma_{OH^-}} = 0$$

which can be rearranged to

$$[H^+] = \sqrt{\dfrac{\dfrac{K_1 K_2 [HP^-] \gamma_{HP^-} \gamma_{H_2P}}{\gamma_{H^+} \gamma_{P^{2-}}} + \dfrac{K_1 K_w \gamma_{H_2P}}{\gamma_{H^+} \gamma_{OH^-}}}{K_1 \gamma_{H_2P} + [HP^-] \gamma_{H^+} \gamma_{HP^-}}}$$

(b) The ionic strength of 0.050 M KHP is 0.050 M, since the only major ions are K^+ and HP^-.

$[HP^-] \approx 0.050$ M, $\gamma_{HP^-} = 0.835$, $\gamma_{P^{2-}} = 0.485$, $\gamma_{H_2P} \approx 1.00$,

$\gamma_{H^+} = 0.86$, $\gamma_{OH^-} = 0.81$. Using these values in the previous equation

gives $[H^+] = 1.09 \times 10^{-4} \Rightarrow pH = -\log [H^+]\gamma_{H^+} = 4.03$.

9-9. Case (a): pH = 6.002, $[HM^-] = 9.80 \times 10^{-3}$ M, $[H_2M] = 9.76 \times 10^{-5}$ M,

$[M^{2-}] = 9.85 \times 10^{-5}$ M

Case (b):

	A	B	C	D	E	F	G	H	I	J
1	Box 9-1 Successive Approximations									
2										
3	pK_{a1} =	4		1st approx.	2nd approx.	3rd approx.	4th approx.	5th approx.		15th approx.
4	pK_{a2} =	5	[HA⁻] =	0.01000	0.003675	0.007675	0.005146	0.006745		0.00613201
5	K_{a1} =	0.0001	[H⁺] =	3.15E-05	3.12E-05	3.14E-05	3.13E-05	3.14E-05		3.14E-05
6	K_{a2} =	0.00001	[H₂A] =	3.15E-03	1.15E-03	2.41E-03	1.61E-03	2.12E-03		1.92E-03
7	F =	0.01	[A²⁻] =	3.18E-03	1.18E-03	2.44E-03	1.64E-03	2.15E-03		1.95E-03
8	K_w =	1.00E-14	pH =	4.50	4.51	4.50	4.50	4.50		4.50
9										
10	Cell D4: [HA⁻] = F									
11	Cell D5: [H⁺] =SQRT((Ka1*Ka2*D4+Ka1*Kw)/(Ka1+D4))									
12	Cell D6: [H₂A] = D4*D5/Ka1									
13	Cell D7: [A²⁻] = Ka2*D4/D5									
14	Cell D8: pH = -log10(D5)									
15	Cell E4: [HA⁻] = F-D6-D7									
16	After computing E4, then highlight cells D5:E8 and FILL RIGHT									
17	After completing column E, highlight cells E4:F:8 and FILL RIGHT									
18	Continue to highlight each column and FILL RIGHT									

9-10. Mass balance: $F = [Na^+] = [H_2A] + [HA^-] + [A^{2-}]$ (A)

Charge balance: $[Na^+] + [H^+] = [HA^-] + 2[A^{2-}] + [OH^-]$ (B)

Equilibria: $[H_2A] = [H+][HA^-]/K_1$ (C)

$[A^{2-}] = K_2[HA^-]/[H^+]$ (D)

$[OH^-] = K_w/[H^+]$ (E)

Substitute $[Na^+] = [H_2A] + [HA^-] + [A^{2-}]$ from Equation A for $[Na^+]$ in
Equation B:

$[H_2A] + \cancel{[HA^-]} + \cancel{[A^{2-}]} + [H^+] = \cancel{[HA^-]} + \cancel{2}[A^{2-}] + [OH^-]$

$[H_2A] + [H^+] = [A^{2-}] + [OH^-]$ (F)

Substitute expressions for $[H_2A]$ from Equation C, $[A^{2-}]$ from Equation D, and $[OH^-]$ from Equation (E) into Equation F:

$$[H^+][HA^-]/K_1 + [H^+] = K_2[HA^-]/[H^+] + K_w/[H^+]$$

Multiply all terms by $[H^+]$ and factor $[H^+]^2$ out on the left side of the equation:

$$[H^+]^2[HA^-]/K_1 + [H^+]^2 = K_2[HA^-] + K_w$$

$$[H^+]^2\{[HA^-]/K_1 + 1\} = K_2[HA^-] + K_w$$

Solve for $[H^+]$:

$$[H^+] = \sqrt{\frac{K_2[HA^-] + K_w}{[HA^-]/K_1 + 1}} = \sqrt{\frac{K_1K_2[HA^-] + K_1K_w}{[HA^-] + K_1}}$$

The last expression is equivalent to Equation 9-10. Substituting $[HA^-] \approx F$ gives Equation 9-11.

9-11. $["H_2CO_3"] = [CO_2(aq)] = KP_{CO_2} = 10^{-1.5} \cdot 10^{-3.4} = 10^{-4.9}$ M

$$H_2CO_3 \rightleftharpoons HCO_3^- + H^+ \qquad K_{a1} = 4.46 \times 10^{-7}$$

$$\begin{array}{ccc} 10^{-4.9} - x & x & x \end{array}$$

$$\frac{x^2}{10^{-4.9} - x} = K_{a1} \Rightarrow x = 2.1_6 \times 10^{-6} \text{ M} \Rightarrow pH = 5.67$$

9-12.   (a)   From Equation C, $[CO_3^{2-}] = \dfrac{K_{a2}[HCO_3^-]}{[H^+]}$ (F)

From Equation B, $[HCO_3^-] = \dfrac{K_{a1}[CO_2(aq)]}{[H^+]}$ (G)

Substituting $[HCO_3^-]$ from Equation G into Equation F gives

$$[CO_3^{2-}] = \frac{K_{a2}K_{a1}[CO_2(aq)]}{[H^+]^2} \qquad (H)$$

Substituting for $[CO_2(aq)]$ from Equation A into Equation H gives

$$[CO_3^{2-}] = \frac{K_{a2}K_{a1}K_H P_{CO_2}}{[H^+]^2} \qquad (I)$$

(b)   For $P_{CO_2} = 800$ µbar, pH = 7.8, 0°C, we find

$$[CO_3^{2-}] = \frac{K_{a2}K_{a1}K_H P_{CO_2}}{[H^+]^2} =$$

$$\frac{10^{-9.3762} \text{ mol kg}^{-1}\; 10^{-6.1004} \text{ mol kg}^{-1}\; 10^{-1.2073} \text{ mol kg}^{-1} \text{ bar}^{-1} (800 \times 10^{-6} \text{ bar})}{[10^{-7.8} \text{ mol kg}^{-1}]^2}$$

$$= 6.6 \times 10^{-5} \text{ mol kg}^{-1}$$

For $P_{CO_2} = 800$ µbar, pH = 7.8, 30°C, we find

$$[CO_3^{2-}] = \frac{K_{a2}K_{a1}K_H P_{CO_2}}{[H^+]^2} =$$

$$\frac{10^{-8.8324} \text{ mol kg}^{-1} \ 10^{-5.8008} \text{ mol kg}^{-1} \ 10^{-1.6048} \text{ mol kg}^{-1} \text{ bar}^{-1} \ (800 \times 10^{-6} \text{ bar})}{[10^{-7.8} \text{ mol kg}^{-1}]^2}$$

$$= 1.8 \times 10^{-4} \text{ mol kg}^{-1}$$

(c) The equilibrium expressions for aragonite and calcite are

$$CaCO_3(s, \textit{aragonite}) \rightleftharpoons Ca^{2+} + CO_3^{2-} \tag{D}$$

$$K_{sp}^{arg} = [Ca^{2+}][CO_3^{2-}] \quad = 10^{-6.1113} \text{ mol}^2 \text{ kg}^{-2} \text{ at } 0°C$$

$$= 10^{-6.1391} \text{ mol}^2 \text{ kg}^{-2} \text{ at } 30°C$$

$$CaCO_3(s, \textit{calcite}) \rightleftharpoons Ca^{2+} + CO_3^{2-} \tag{E}$$

$$K_{sp}^{cal} = [Ca^{2+}][CO_3^{2-}] \quad = 10^{-6.3652} \text{ mol}^2 \text{ kg}^{-2} \text{ at } 0°C$$

$$= 10^{-6.3713} \text{ mol}^2 \text{ kg}^{-2} \text{ at } 30°C$$

The reaction quotient at 0°C is

$$[Ca^{2+}][CO_3^{2-}] = [0.010 \text{ mol kg}^{-1}][6.6 \times 10^{-5} \text{ mol kg}^{-1}]$$

$$= 6.6 \times 10^{-7} \text{ mol}^2 \text{ kg}^{-2} = 10^{-6.18} \text{ mol}^2 \text{ kg}^{-2} < 10^{-6.1113} \text{ mol}^2 \text{ kg}^{-2},$$

so aragonite will dissolve.

But $10^{-6.18} \text{ mol}^2 \text{ kg}^{-2} > 10^{-6.3652} \text{ mol}^2 \text{ kg}^{-2}$, so calcite will not dissolve.

The reaction quotient at 30°C is

$$[Ca^{2+}][CO_3^{2-}] = [0.010 \text{ mol kg}^{-1}][1.84 \times 10^{-4} \text{ mol kg}^{-1}]$$

$$= 1.84 \times 10^{-6} \text{ mol}^2 \text{ kg}^{-2} = 10^{-5.74} \text{ mol}^2 \text{ kg}^{-2} > 10^{-6.1113} \text{ mol}^2 \text{ kg}^{-2},$$

so neither aragonite nor calcite dissolve.

9-13. $pH = pK_a + \log \dfrac{[CO_3^{2-}]}{[HCO_3^-]}$

$$10.00 = 10.329 + \log \frac{(x \text{ g})/(105.99 \text{ g/mol})}{(5.00 \text{ g})/(84.01 \text{ g/mol})} \implies x = 2.96 \text{ g}$$

9-14. We begin with $(25.0 \text{ mL})(0.023\ 3 \text{ M}) = 0.582_5$ mmol salicylic acid (H_2A, $pK_1 = 2.972$, $pK_2 = 13.7$). At pH 3.50, there will be a mixture of H_2A and HA^-.

	H_2A	+	OH^-	$\rightarrow$	HA^-	+	H_2O
Initial mmol:	0.582_5		x		—		
Final mmol:	$0.582_5 - x$		—		x		

$$3.50 = 2.972 + \log \frac{x}{0.582_5 - x} \implies x = 0.449_3 \text{ mmol}$$

$$(0.449_3 \text{ mmol})/(0.202 \text{ M}) = 2.223 \text{ mL NaOH}$$

9-15. Picolinic acid is HA, the intermediate form of a diprotic system with $pK_1 = 1.01$ and $pK_2 = 5.39$. To achieve pH 5.50, we need a mixture of HA + A$^-$.

$$\text{HA} \quad + \quad \text{OH}^- \quad \rightarrow \quad \text{A}^-$$

Initial mmol:	10.0	x	
Final mmol:	$10.0 - x$	—	x

$$5.50 = 5.39 + \log\frac{x}{10.0 - x} \Rightarrow x = 5.63 \text{ mmol} \approx 5.63 \text{ mL NaOH}$$

Procedure: Dissolve 10.0 mmol (1.23 g) picolinic acid in ≈ 75 mL H_2O in a beaker. Add NaOH (≈ 5.63 mL) until the measured pH is 5.50. Transfer to a 100 mL volumetric flask and use small portions of H_2O to rinse the contents of the beaker into the flask. Dilute to 100.0 mL and mix well.

9-16. At pH 2.80, we have a mixture of SO_4^{2-} and HSO_4^-, since pK_a for HSO_4^- is 1.99.

$$2.80 = 1.987 + \log\frac{[SO_4^{2-}]}{[HSO_4^-]} \Rightarrow HSO_4^- = 0.153_8 \,[SO_4^{2-}]$$

The reaction between H_2SO_4 and SO_4^{2-} produces 2 moles of HSO_4^-:

$$\text{H}_2\text{SO}_4 \quad + \quad \text{SO}_4^{2-} \quad \rightarrow \quad 2\text{HSO}_4^-$$

Initial mmol:	x	y	—
Final mmol:	—	$y - x$	$2x$

The Henderson-Hasselbalch equation told us that $[HSO_4^-] = 0.153_8 \,[SO_4^{2-}]$ $\Rightarrow 2x = 0.153_8 \,(y - x)$. Since the total sulfur is 0.200 M, $x + y = 0.200$ mol. Substituting $x = 0.200 - y$ into the equation $2x = 0.153_8 \,(y - x)$ gives $Na_2SO_4 = y = 0.186_7$ mol $= 26.5_2$ g and $H_2SO_4 = x = 0.013\,3$ mol $= 1.31$ g.

9-17. pK_2 for phosphoric acid is 7.20, so it has a high buffer capacity at pH 7.45 (from the buffer pair $H_2PO_4^-/HPO_4^{2-}$). At pH 8.5, the buffer capacity of phosphate would be low and it would not be very useful.

9-18.

$$\overset{+}{\text{N}}\text{H}_3 \qquad\qquad \overset{+}{\text{N}}\text{H}_3 \qquad\qquad \overset{+}{\text{N}}\text{H}_3 \qquad\qquad \text{NH}_2$$
$$| \qquad\qquad\quad | \qquad\qquad\quad | \qquad\qquad\quad |$$
$$\text{CHCH}_2\text{CH}_2\text{CO}_2\text{H} \underset{}{\overset{K_1}{\rightleftharpoons}} \text{CHCH}_2\text{CH}_2\text{CO}_2\text{H} \overset{K_2}{\rightleftharpoons} \text{CHCH}_2\text{CH}_2\text{CO}_2^- \overset{K_3}{\rightleftharpoons} \text{CHCH}_2\text{CH}_2\text{CO}_2^-$$
$$| \qquad\qquad\quad | \qquad\qquad\quad | \qquad\qquad\quad |$$
$$\text{CO}_2\text{H} \qquad\qquad \text{CO}_2^- \qquad\qquad \text{CO}_2^- \qquad\qquad \text{CO}_2^-$$

glutamic acid

9-19. (a) For 0.050 0 M KH$_2$PO$_4$, [H$^+$] $= \sqrt{\dfrac{K_1K_2(0.050\,0) + K_1K_w}{K_1 + 0.050\,0}} = 1.99 \times 10^{-5}$

$\Rightarrow$ pH $= 4.70$

$4.70 = 2.148 + \log\dfrac{[\text{H}_2\text{PO}_4^-]}{[\text{H}_3\text{PO}_4]} \Rightarrow \dfrac{[\text{H}_3\text{PO}_4]}{[\text{H}_2\text{PO}_4^-]} = 2.8 \times 10^{-3}$

(b) For 0.050 0 M K$_2$HPO$_4$, [H$^+$] $= \sqrt{\dfrac{K_2K_3(0.050\,0) + K_2K_w}{K_2 + 0.050\,0}} = 1.99 \times 10^{-10}$

$\Rightarrow$ pH $= 9.70$

$9.70 = 2.148 + \log\dfrac{[\text{H}_2\text{PO}_4^-]}{[\text{H}_3\text{PO}_4]} \Rightarrow \dfrac{[\text{H}_3\text{PO}_4]}{[\text{H}_2\text{PO}_4^-]} = 2.8 \times 10^{-8}$

9-20. (a)
$$\text{H}_3\text{PO}_4 \xrightarrow{\ pK_1 = 2.148\ } \text{H}_2\text{PO}_4^- \xrightarrow{\ pK_2 = 7.198\ } \text{HPO}_4^{2-} \xrightarrow{\ pK_3 = 12.375\ } \text{PO}_4^{3-}$$

pH $\approx$ (2.148+7.198)/2   pH $\approx$ (7.198+12.375)/2

= 4.67   = 9.79

pH 7.45 corresponds to a mixture of NaH$_2$PO$_4$ and Na$_2$HPO$_4$. (You could get the same result by mixing other combinations such as H$_3$PO$_4$ and Na$_3$PO$_4$ or H$_3$PO$_4$ and Na$_2$HPO$_4$.)

(b) pH $= pK_2 + \log\dfrac{[\text{HPO}_4^{2-}]}{[\text{H}_2\text{PO}_4^-]}$

$7.45 = 7.198 + \log\dfrac{[\text{HPO}_4^{2-}]}{[\text{H}_2\text{PO}_4^-]} \Rightarrow \dfrac{[\text{HPO}_4^{2-}]}{[\text{H}_2\text{PO}_4^-]} = 1.78_6$

Combining this last result with [HPO$_4^{2-}$] + [H$_2$PO$_4^-$] = 0.050 0 M gives [HPO$_4^{2-}$] = 0.032 0$_5$ M and [H$_2$PO$_4^-$] = 0.017 9$_5$ M. Use 4.55 g of Na$_2$HPO$_4$ and 2.15 g of NaH$_2$PO$_4$.

(c) Here is one of several ways: Weigh out 0.050 0 mol Na_2HPO_4 and dissolve it in 900 mL of water. Add HCl while monitoring the pH with a pH electrode. When the pH is 7.45, stop adding HCl and dilute up to exactly 1 L with H_2O.

9-21. Lysine hydrochloride (H_2L^+) is

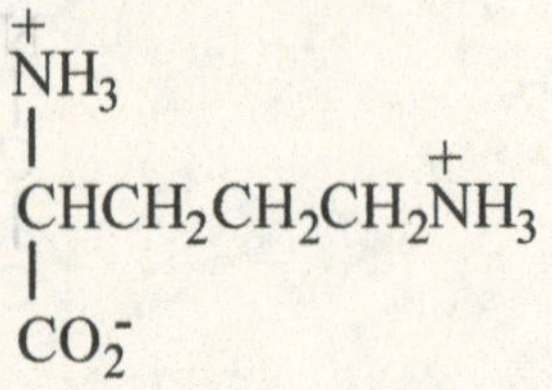

$$\overset{+}{N}H_3$$
$$|$$
$$CHCH_2CH_2CH_2\overset{+}{N}H_3$$
$$|$$
$$CO_2^-$$

for which $[H^+] = \sqrt{\dfrac{K_1K_2(0.010\,0)+K_1K_w}{K_1+0.010\,0}} = 2.32 \times 10^{-6}$ M $\Rightarrow$ pH = 5.64

$[H_2L^+] = 0.010\,0$ M

$[H_3L^{2+}] = \dfrac{[H^+][H_2L^+]}{K_1} = 1.36 \times 10^{-6}$ M

$[HL] = \dfrac{K_2[H_2L^+]}{[H^+]} = 3.68 \times 10^{-6}$ M $[L^-] = \dfrac{K_3[HL]}{[H^+]} = 2.40 \times 10^{-11}$ M

9-22.

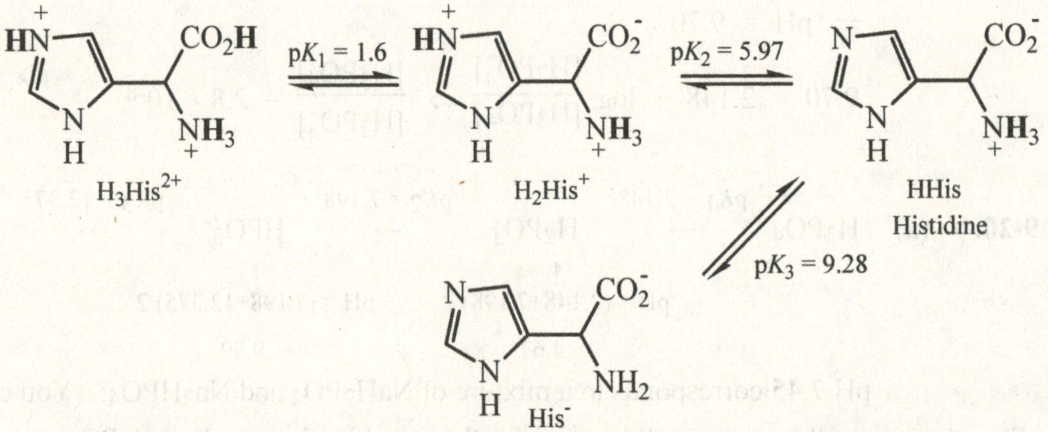

Histidine hydrochloride (FM 191.62) is $H_2His^+Cl^-$, the intermediate form between pK_1 and pK_2. A pH of 9.30 requires neutralizing all of the H_2His^+ to HHis and then adding more KOH to create a mixture of and HHis + His⁻. Therefore, we must add 1 mol KOH for each mol of His·HCl to get to HHis and then add some more KOH to obtain the mixture of HHis and His⁻. Initial mol of His·HCl = 10.0 g/(191.62 g/mol) = 0.052 1$_9$ mol. We require 0.052 1$_9$ mol of KOH plus the amount x in the following table to obtain the correct mixture:

	HHis	+	OH⁻	→	His⁻
Initial mol:	0.052 1$_9$		x		—
Final mol:	0.052 1$_9 - x$		—		x

$$pH = pK_3 + \log \frac{[His^{2-}]}{[HHis^-]} \Rightarrow 9.30 = 9.28 + \log \frac{x}{0.052\,1_9 - x}$$

$$\Rightarrow x = 0.026\,7_0$$

Total mol KOH required $= 0.052\,1_9 + 0.026\,7_0 = 0.078\,9$ mol

$\qquad\qquad = 78.9$ mL of 1.00 M KOH

9-23. (a) $\quad pH = pK_3 \text{ (citric acid)} + \log \dfrac{[C^{3-}]\,\gamma_{C^{3-}}}{[HC^{2-}]\,\gamma_{HC^{2-}}}$

$$pH = 6.396 + \log \frac{(1.00)(0.405)}{(2.00)(0.665)} = 5.88$$

(b) If the ionic strength is raised to 0.10 M,

$$pH = 6.396 + \log \frac{(1.00)(0.115)}{(2.00)(0.37)} = 5.59$$

9-24. (a) HA $\qquad\qquad$ (b) A^-

$\qquad$ (c) $pH = pK_a + \dfrac{[A^-]}{[HA]}$

$$7.00 = 7.00 + \log \frac{[A^-]}{[HA]} \Rightarrow [A^-]/[HA] = 1.0$$

$$6.00 = 7.00 + \log \frac{[A^-]}{[HA]} \Rightarrow [A^-]/[HA] = 0.10$$

9-25. (a) 4.00 $\quad$ (b) 8.00 $\quad$ (c) H_2A $\quad$ (d) HA^- $\quad$ (e) A^{2-}

9-26. (a) 9.00 $\quad$ (b) 9.00 $\quad$ (c) BH^+

$\qquad$ (d) $12.00 = 9.00 + \log \dfrac{[B]}{[BH^+]} \Rightarrow [B]/[BH^+] = 1.0 \times 10^3$

9-27.

9-28.   Fraction in form HA $= \alpha_{HA} = \dfrac{[H^+]}{[H^+] + K_a} = \dfrac{10^{-5}}{10^{-5} + 10^{-4}} = 0.091.$

Fraction in form A$^- = \alpha_{A^-} = \dfrac{K_a}{[H^+] + K_a} = 0.909.$

$\dfrac{[A^-]}{[HA]} = \dfrac{\alpha_{A^-}}{\alpha_{HA}} = 10$, which makes sense.

9-29. $\alpha_{H_2A} = \dfrac{[H^+]^2}{[H^+]^2 + [H^+]K_1 + K_1K_2}$, where $[H^+] = 10^{-7.00}$, $K_1 = 10^{-8.00}$, and

$K_2 = 10^{-10.00} \Rightarrow \alpha_{H_2A} = 0.91$

9-30.

	pH 8.00	pH 10.00
α_{H_2A}	0.877	0.0496
α_{HA^-}	0.123	0.694
$\alpha_{A^{2-}}$	4.54×10^{-4}	0.257

9-31.

pH :	1.00	1.92	6.00	6.27	10.00
α_{H_2A}	0.893	0.500	5.41×10^{-5}	2.23×10^{-5}	1.55×10^{-12}
α_{HA^-}	0.107	0.500	0.651	0.500	1.86×10^{-4}
$\alpha_{A^{2-}}$	5.76×10^{-7}	2.23×10^{-5}	0.349	0.500	0.9998

9-32.   (a)   The derivation follows the outline of Equations 9-19 through 9-21.   The
results are

$$\alpha_{H_3A} = \frac{[H_3A]}{F} = \frac{[H^+]^3}{[H^+]^3 + [H^+]^2K_1 + [H^+]K_1K_2 + K_1K_2K_3}$$

$$\alpha_{H_2A^-} = \frac{[H_2A^-]}{F} = \frac{[H^+]^2 K_1}{[H^+]^3 + [H^+]^2K_1 + [H^+]K_1K_2 + K_1K_2K_3}$$

$$\alpha_{HA^{2-}} = \frac{[HA^{2-}]}{F} = \frac{[H^+] K_1K_2}{[H^+]^3 + [H^+]^2K_1 + [H^+]K_1K_2 + K_1K_2K_3}$$

$$\alpha_{A^{3-}} = \frac{[A^{3-}]}{F} = \frac{K_1K_2K_3}{[H^+]^3 + [H^+]^2K_1 + [H^+]K_1K_2 + K_1K_2K_3}$$

(b)   For phosphoric acid, $pK_1 = 2.148$, $pK_2 = 7.198$, and $pK_3 = 12.375$. At
pH $= 7.00$, the previous expressions give $\alpha_{H_3A} = 8.6 \times 10^{-6}$, $\alpha_{H_2A^-} =$
0.61, $\alpha_{HA^{2-}} = 0.39$, and $\alpha_{A^{3-}} = 1.6 \times 10^{-6}$.

9-33. $pH = pK_{NH_4^+} + \log \dfrac{[NH_3]}{[NH_4^+]} \Rightarrow 9.00 = 9.245 + \log \dfrac{[NH_3]}{[NH_4^+]} \Rightarrow \dfrac{[NH_3]}{[NH_4^+]} = 0.56_9$

Fraction unprotonated $= \dfrac{[NH_3]}{[NH_3] + [NH_4^+]} = \dfrac{0.56_9}{0.56_9 + 1} = 0.36$

9-34. The quantity of morphine in the solution is negligible compared to the quantity of cacodylic acid. The pH is determined by the reaction of cacodylic acid (HA) with NaOH:

$$\text{HA} \quad + \quad \text{OH}^- \quad \rightarrow \quad \text{A}^- \quad + \quad \text{H}_2\text{O}$$

Initial mmol:	1.000	0.800	—	—
Final mmol:	0.200	—	0.800	—

$$pH = pK_a + \log \frac{[\text{A}^-]}{[\text{HA}]} = 6.19 + \log \frac{0.800}{0.200} = 6.79$$

For morphine (B), $K_a = K_w/K_b = 1.0 \times 10^{-14}/1.6 \times 10^{-6} = 6.2_5 \times 10^{-9}$

$$\Rightarrow pK_a = 8.20$$

At pH 6.79, we can write $pH = pK_{\text{BH}^+} + \log \frac{[\text{B}]}{[\text{BH}^+]} \Rightarrow$

$$6.79 = 8.20 + \log \frac{[\text{B}]}{[\text{BH}^+]} \Rightarrow \frac{[\text{B}]}{[\text{BH}^+]} = 0.039 \Rightarrow [\text{B}] = 0.039\,[\text{BH}^+]$$

$$\text{Fraction in form BH}^+ = \frac{[\text{BH}^+]}{[\text{B}] + [\text{BH}^+]} = \frac{[\text{BH}^+]}{0.039\,[\text{BH}^+] + [\text{BH}^+]} = 96\%$$

9-35.

	A	B	C	D	E	F
1	Fractional composition for diprotic acid					
2						
3	K1 =	pH	[H+]	α(H2A)	α(HA⁻)	α(A²⁻)
4	9.55E-04	1	0.1	9.91E-01	0.00946	3.13E-06
5	K2 =	2	0.01	0.912562	0.087149	0.000289
6	3.31E-05	3	0.001	0.503369	0.480713	0.015918
7	pK1 =	4	0.0001	0.072928	0.696455	0.230618
8	3.02	5	0.00001	0.002423	0.231386	0.766191
9	pK2 =	6	0.000001	3.07E-05	0.029313	0.970656
10	4.48	7	1E-07	3.15E-07	0.003011	0.996989
11						
12	A4 = 10^-A8	D4 = $C4^2/($C4^2+$C4*$A$4+$A$4*$A$6)				
13	A6 = 10^-A10	E4 = $C4*$A$4/($C4^2+$C4*$A$4+$A$4*$A$6)				
14	C4 = 10^-B4	F4 = $A$4*$A$6/($C4^2+$C4*$A$4+$A$4*$A$6)				

9-36.

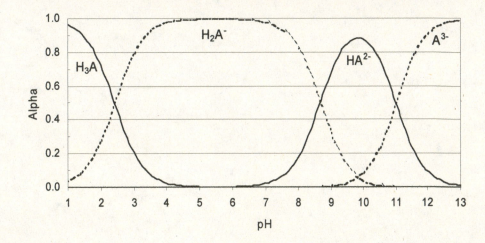

	A	B	C	D	E	F	G
1	Fractional composition for triprotic acid						
2							
3	K1 =	pH	[H+]	α(H3A)	α(H2A⁻)	α(HA²⁻)	α(A³⁻)
4	3.89E-03	1	1.00E-01	9.63E-01	3.74E-02	8.01E-10	7.82E-20
5	K2 =	2	1.00E-02	7.20E-01	2.80E-01	5.99E-08	5.85E-17
6	2.14E-09	3	1.00E-03	2.04E-01	7.96E-01	1.70E-06	1.66E-14
7	K3 =	4	1.00E-04	2.51E-02	9.75E-01	2.08E-05	2.04E-12
8	9.77E-12	5	1.00E-05	2.56E-03	9.97E-01	2.13E-04	2.08E-10
9	pK1 =	6	1.00E-06	2.56E-04	9.98E-01	2.13E-03	2.08E-08
10	2.41	7	1.00E-07	2.52E-05	9.79E-01	2.09E-02	2.05E-06
11	pK2 =	8	1.00E-08	2.12E-06	8.24E-01	1.76E-01	1.72E-04
12	8.67	9	1.00E-09	8.14E-08	3.17E-01	6.77E-01	6.61E-03
13	pK3 =	10	1.00E-10	1.05E-09	4.09E-02	8.74E-01	8.54E-02
14	11.01	11	1.00E-11	6.07E-12	2.36E-03	5.05E-01	4.93E-01
15		12	1.00E-12	1.12E-14	4.34E-05	9.28E-02	9.07E-01
16		13	1.00E-13	1.22E-17	4.74E-07	1.01E-02	9.90E-01
17	A4 = 10^-A10						
18	C4 = 10^-B4						
19	D4 = $C4^3/($C4^3+$C4^2*$A$4+$C4*$A$4*$A$6+$A$4*$A$6*$A$8)						
20	E4 = $C4^2*$A$4/($C4^3+$C4^2*$A$4+$C4*$A$4*$A$6+$A$4*$A$6*$A$8)						
21	F4 = $C4*$A$4*$A$6/($C4^3+$C4^2*$A$4+$C4*$A$4*$A$6+$A$4*$A$6*$A$8)						
22	G4 = $A$4*$A$6*$A$8/($C4^3+$C4^2*$A$4+$C4*$A$4*$A$6+$A$4*$A$6*$A$8)						

9-37.   (a) Fractional composition in tetraprotic system

	A	B	C	D	E	F	G	H	I
1	Fractional composition in tetraprotic system								
2									
3	Ka1 =	pH	[H+]	Denom.	Alph(H4A)	Alph(H3A)	Alph(H2A)	Alph(HA)	Alph(A)
4	1.58E-04	1	1E-01	1.0E-04	1.0E+00	1.6E-03	6.3E-09	2.5E-14	9.9E-25
5	Ka2 =	2	1E-02	1.0E-08	9.8E-01	1.6E-02	6.2E-07	2.5E-11	9.8E-21
6	3.98E-07	3	1E-03	1.2E-12	8.6E-01	1.4E-01	5.4E-05	2.2E-08	8.6E-17
7	Ka3 =	4	1E-04	2.6E-16	3.9E-01	6.1E-01	2.4E-03	9.7E-06	3.9E-13
8	3.98E-07	5	1E-05	1.7E-19	5.7E-02	9.1E-01	3.6E-02	1.4E-03	5.7E-10
9	Ka4 =	6	1E-06	2.5E-22	4.1E-03	6.4E-01	2.5E-01	1.0E-01	4.0E-07
10	3.98E-12	7	1E-07	3.3E-24	3.0E-05	4.8E-02	1.9E-01	7.6E-01	3.0E-05
11		8	1E-08	2.6E-25	3.9E-08	6.2E-04	2.4E-02	9.7E-01	3.9E-04
12		9	1E-09	2.5E-26	4.0E-11	6.3E-06	2.5E-03	9.9E-01	4.0E-03
13		10	1E-10	2.6E-27	3.8E-14	6.1E-08	2.4E-04	9.6E-01	3.8E-02
14		11	1E-11	3.5E-28	2.9E-17	4.5E-10	1.8E-05	7.2E-01	2.8E-01
15		12	1E-12	1.2E-28	8.0E-21	1.3E-12	5.0E-07	2.0E-01	8.0E-01
16		13	1E-13	1.0E-28	9.8E-25	1.5E-15	6.2E-09	2.5E-02	9.8E-01
17									
18	C4 = 10^-B4								
19	D4 = C4^4+$A$4*C4^3+$A$4*$A$6*C4^2+$A$4*$A$6*$A$8*C4								
20		+$A$4*$A$6*$A$8*$A$10							
21	E4 = C4^4/D4								
22	F4 = $A$4*C4^3/D4				H4 = $A$4*$A$6*$A$8*C4/D4				
23	G4 = $A$4*$A$6*C4^2/D4				I4 = $A$4*$A$6*$A$8*$A$10/D4				

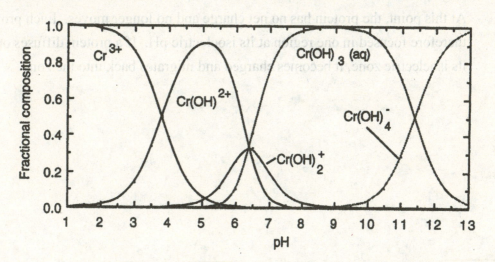

(b) $K = 10^{-6.84} = [\text{Cr(OH)}_3(aq)]$, so $[\text{Cr(OH)}_3(aq)] = 10^{-6.84}\ M = 1.4_5 \times 10^{-7}\ M$

(c) $K_{a3} = 10^{-6.40} = \dfrac{[\text{Cr(OH)}_3(aq)][\text{H}^+]}{[\text{Cr(OH)}_2^+]} = \dfrac{[10^{-6.84}][10^{-4.00}]}{[\text{Cr(OH)}_2^+]}$

$$\Rightarrow [\text{Cr(OH)}_2^+] = \dfrac{[10^{-6.84}][10^{-4.00}]}{10^{-6.40}} = 10^{-4.44}\ M$$

$$K_{a2} = 10^{-6.40} = \frac{[Cr(OH)_2^+][H^+]}{[Cr(OH)^{2+}]} = \frac{[10^{-4.44}][10^{-4.00}]}{[Cr(OH)^{2+}]}$$

$$\Rightarrow [Cr(OH)^{2+}] = \frac{[10^{-4.44}][10^{-4.00}]}{10^{-6.40}} = 10^{-2.04} \text{ M}$$

9-38. The isoelectric pH is the pH at which the protein has no net charge, even though it has many positive and negative sites. The isoionic pH is the pH of a solution containing only protein, H^+, and OH^-.

9-39. The <u>average</u> charge is zero. There is no pH at which <u>all</u> molecules have zero charge.

9-40. Isoionic $[H^+] = \sqrt{\dfrac{K_1 K_2(0.010) + K_1 K_w}{K_1 + (0.010)}} \Rightarrow$ pH = 5.72

Isoelectric pH $= \dfrac{pK_1 + pK_2}{2} = 5.59$

9-41. A mixture of proteins is exposed to a strong electric field in a medium with a pH gradient. Positively charged molecules move toward the negative pole and negatively charged molecules move toward the positive pole. Each protein migrates until it reaches the point where the pH is the same as its isoelectric pH. At this point, the protein has no net charge and no longer moves. Each protein is therefore focused in one region at its isoelectric pH. If a protein diffuses out of its isoelectric zone, it becomes charged and migrates back into the zone.

CHAPTER 10
ACID-BASE TITRATIONS

10-1. The equivalence point occurs when the quantity of titrant is exactly the stoichiometric amount needed for complete reaction with analyte. The end point occurs when there is an abrupt change in a physical property, such as pH or indicator color. Ideally, the end point is chosen to occur at the equivalence point.

10-2.

V_a	0	1	5	9	9.9	10	10.1	12
pH	13.00	12.95	12.68	11.96	10.96	7.00	3.04	1.75

Representative calculations:

0 mL: $\quad$ $pH = -\log \dfrac{K_w}{[OH^-]} = -\log \dfrac{10^{-14}}{0.100} = 13.00$

1 mL: $\quad$ $[OH^-] = \dfrac{9}{10}(0.100)\dfrac{100}{101} = 0.0891\ M \Rightarrow pH = 12.95$

10 mL: $\quad$ $[OH^-] = [H^+] = 10^{-7}\ M$

10.1 mL: $\quad$ $[H^+] = \left(\dfrac{0.1}{110.1}\right)(1.00) = 9.08 \times 10^{-4}\ M \Rightarrow pH = 3.04$

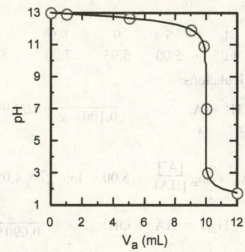

10-3. Consider the titration curve near the equivalence point. If we titrate strong acid with strong base, the concentration of H^+ is close to 1% of its initial value when we are 99% of the way to the equivalence point (i.e., when $V_b = 0.99 V_e$). (This statement would be exactly true if there were no dilution occurring. We will neglect dilution.) If the initial acid concentration were, say, 0.1 M, then $[H^+] =$ 1% of 0.1 M = 0.001 M at $V_b = 0.99\ V_e$. The pH is $-\log(0.001) = 3$. At 99.9% completion, $[H^+] = 0.1\%$ of 0.1 M = 0.0001 M and the pH is 4. When the titration is 0.1% past the equivalence point, $[OH^-] = 0.0001$ M and the pH is $-\log(K_w/0.0001) = 10$. The pH jumps from 4 to 10 in the interval from $V_b =$

$0.999V_e$ to $1.001V_e$. Even though the concentration of H^+ hardly changes, its logarithm changes rapidly around the equivalence point because $[H^+]$ decreases by orders of magnitude with tiny additions of OH^- when there is hardly any H^+ present.

10-4. The sketch should look like Figure 10-2. Before base is added, the pH is determined by the acid dissociation reaction of HA. Between the initial point and the equivalence point, each mole of OH^- converts an equivalent quantity of HA into A^-. The resulting buffer containing HA and A^- determines the pH. At the equivalence point, all HA has been converted to A^-. The pH is controlled by the base hydrolysis reaction of A^- with H_2O. After the equivalence point, excess OH^- is being added to the solution. To a good approximation, the pH is determined just by the concentration of excess OH^-.

10-5. If the analyte is too weak or too dilute, there is very little change in pH at the equivalence point.

10-6.

V_b	0	1	5	9	9.9	10	10.1	12
pH	3.00	4.05	5.00	5.95	7.00	8.98	10.96	12.25

Representative calculations:

<u>0 mL:</u> $HA = H^+ + A^-$ $\dfrac{x^2}{0.100 - x} = 10^{-5.00} \Rightarrow x = 9.95 \times 10^{-4} \text{ M}$
          $0.100 - x \quad x \quad x$

$\Rightarrow \text{pH} = 3.00$

<u>1 mL:</u> $\text{pH} = pK_a + \log\dfrac{[A^-]}{[HA]} = 5.00 + \log\dfrac{1}{9} = 4.05$

<u>10 mL:</u> $A^- + H_2O = HA + OH^-$ $\dfrac{x^2}{0.0909 - x} = \dfrac{K_w}{K_a}$

$\left(\dfrac{100}{110}\right)(0.100) - x \qquad x \qquad x$

$\Rightarrow x = 9.53 \times 10^{-6}$

$\Rightarrow [H^+] = \dfrac{K_w}{x} \Rightarrow \text{pH} = 8.98$

<u>10.1 mL:</u> $[OH^-] = \left(\dfrac{0.1}{110.1}\right)(1.00) = 9.08 \times 10^{-4} \text{ M} \Rightarrow \text{pH} = 10.96$

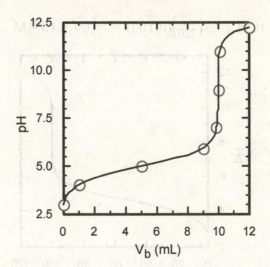

10-7. $\quad \text{pH} = \text{p}K_a + \log \dfrac{[A^-]}{[HA]}$ $\qquad \text{p}K_a - 1 = \text{p}K_a + \log \dfrac{[A^-]}{[HA]} \Rightarrow \dfrac{[A^-]}{[HA]} = \dfrac{1}{10}$

If the ratio $\dfrac{[A^-]}{[HA]}$ is to be $\dfrac{1}{10}$, then $\dfrac{1}{11}$ of the initial HA must remain as HA.

At this point, $[A^-]/[HA] = (1/11)/(10/11) = 1/10$. So $\text{pH} = \text{p}K_a - 1$ when $V_b = V_e/11$.

In a similar manner, $\text{pH} = \text{p}K_a + 1$ when $V_b = 10V_e/11$.

For anilinium ion, $\text{p}K_a = 4.601$. For the titration of 100 mL of 0.100 M

anilinium ion with 0.100 M OH⁻, the reaction is

$$\text{C}_6\text{H}_5\text{-}\overset{+}{\text{N}}\text{H}_3 + \text{OH}^- \;\rightarrow\; \text{C}_6\text{H}_5\text{-}\text{NH}_2 + \text{H}_2\text{O} \quad \text{and } V_e = 100 \text{ mL}.$$

$\underline{0 \text{ mL}}:\quad \text{C}_6\text{H}_5\text{-}\overset{+}{\text{N}}\text{H}_3 \;\rightleftharpoons\; \text{C}_6\text{H}_5\text{-}\text{NH}_2 + \text{H}^+$

$\qquad\qquad\quad 0.100 - x \qquad\qquad\quad x \qquad\quad x$

$$\frac{x^2}{0.100 - x} = K_a = 10^{-4.60} \Rightarrow x = 1.57 \times 10^{-3} \Rightarrow \text{pH} = 2.80$$

$\underline{V_e/11 = 9.09 \text{ mL}}:\ \text{pH} = \text{p}K_a - 1 = 3.60$

$\underline{V_e/2 = 50.0 \text{ mL}}:\ \text{pH} = \text{p}K_a = 4.60$

$\underline{10V_e/11 = 90.91 \text{ mL}}:\ \text{pH} = \text{p}K_a + 1 = 5.60$

$\underline{V_e = 100.0 \text{ mL}}:\ \text{BH}^+ \text{ has been converted to B.}$

$$\text{B} + \text{H}_2\text{O} \rightleftharpoons \text{BH}^+ + \text{OH}^- \qquad K_b = \frac{K_w}{K_a} = \frac{x^2}{\left(\frac{100}{200}\right)(0.100) - x}$$

$\qquad\quad F - x \qquad\qquad\quad x \qquad\quad x$

$$\Rightarrow x = 4.46 \times 10^{-6} \text{ M} \qquad \text{pH} = -\log \frac{K_w}{x} = 8.65$$

$\underline{1.2V_e = 120.0 \text{ mL}}:$ There are 20.0 mL of excess NaOH.

$$[OH^-] = \left(\frac{20}{220}\right)(0.100) = 9.09 \times 10^{-3}\ M \quad \Rightarrow \quad pH = 11.96$$

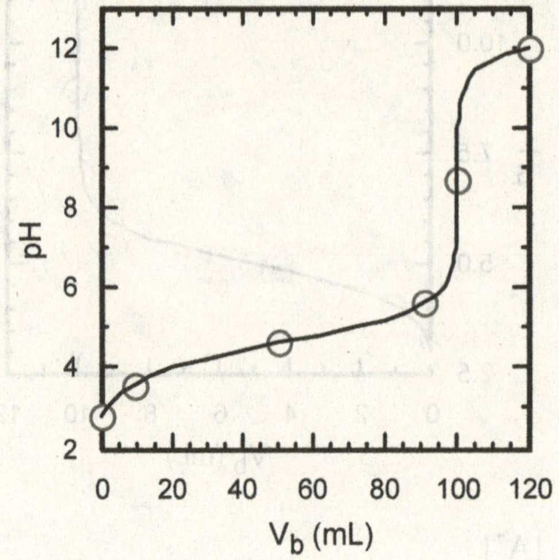

10-8. The titration reaction is $HA + OH^- \rightarrow A^- + H_2O$. A volume of V mL of HA will require $2V$ mL of KOH to reach the equivalence point, because $[HA] = 0.100$ M and $[KOH] = 0.050\,0$ M. The formal concentration of A^- at the equivalence point will be $\left(\frac{V}{V+2V}\right)(0.100) = 0.033\,3$ M. The pH is found by writing

$$A^- + H_2O \rightleftharpoons HA + OH^-$$
$$0.033\,3 - x \qquad\qquad x \qquad\quad x$$

$$\frac{x^2}{0.033\,3 - x} = K_b = \frac{K_w}{K_a} = \frac{1.0 \times 10^{-14}}{1.48 \times 10^{-4}}$$
$$\Rightarrow x = 1.50 \times 10^{-6}\ M \Rightarrow pH = 8.18$$

10-9.

$$K_a = 10^{-6.27}$$
$$1/K_w = 10^{14}$$

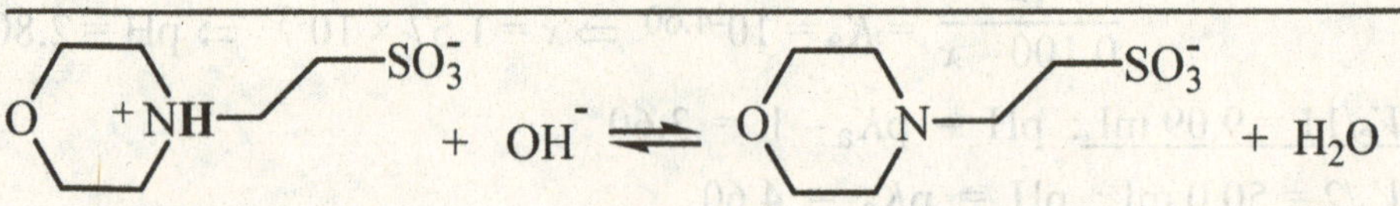

$$K = \frac{K_a}{K_w} = 5.4 \times 10^7$$

10-10.

	HA	+	OH$^-$	$\rightarrow$	A$^-$	+	H$_2$O
Initial mmol:	5.857		x		—		
Final mmol:	$5.857 - x$		—		x		

$$pH = 9.24 = pK_a + \log\frac{[A^-]}{[HA]} = 9.39 + \log\frac{x}{5.857 - x} \Rightarrow x = 2.4_{28}\ mmol$$

$$[OH^-] = \frac{2.4_{28}\ mmol}{22.63\ mL} = 0.107\ M$$

10-11. $(CH_3)_3NH^+ + OH^- \rightarrow (CH_3)_3N + H_2O$

Initial mmol:	1.00	0.40	—	
Final mmol:	0.60	—	0.40	

First, find the ionic strength:

$[(CH_3)_3NH^+] = 0.60\ mmol/14.0\ mL = 0.042\ 86\ M$

$[Br^-] = 1.00\ mmol/14.0\ mL = 0.071\ 43\ M$

$[Na^+] = 0.40\ mmol/14.0\ mL = 0.028\ 57\ M$

$\mu = \frac{1}{2}\Sigma c_i z_i^2 = 0.071\ M$

$$pH = pK_a + \log\frac{[B]\ \gamma_B}{[BH^+]\ \gamma_{BH^+}}$$

$$pH = 9.799 + \log\frac{(0.028\ 6)(1.00)}{(0.042\ 9)(0.80)} = 9.72$$

In the previous calculation, we used the size of $(CH_3)_3NH^+$ (400 pm) and the activity coefficient interpolated from Table 7-1.

10-12. The sketch should look like Figure 10-9. Before base is added, the pH is determined by the base hydrolysis reaction of B with H_2O. Between the initial point and the equivalence point, each mole of H^+ converts an equivalent quantity of B into BH^+. The resulting buffer containing B and BH^+ determines the pH. At the equivalence point, all B has been converted to BH^+. The pH is controlled by the acid dissociation reaction of BH^+. After the equivalence point, excess H^+ is being added to the solution. To a good approximation, the pH is determined just by the concentration of excess H^+.

10-13. At the equivalence point, the weak base, B, is converted completely to the conjugate acid, BH^+, which is necessarily acidic.

10-14.

V_a	0	1	5	9	9.9	10	10.1	12
pH	11.00	9.95	9.00	8.05	7.00	5.02	3.04	1.75

Representative calculations:

$\underline{0\ mL}$: $\begin{array}{c} B + H_2O \rightleftharpoons BH + OH^- \\ 0.100-x \qquad\quad x \qquad x \end{array}$ $\dfrac{x^2}{0.100-x} = 10^{-5.00}\ M \Rightarrow x = 9.95 \times 10^{-4}\ M$

$$[H^+] = \frac{K_w}{x} \Rightarrow pH = 11.00$$

$\underline{1 \text{ mL}}$: $\text{pH} = \text{p}K_{\text{BH}^+} + \log \dfrac{[\text{B}]}{[\text{BH}^+]} = 9.00 + \log \dfrac{9}{1} = 9.95$

$\underline{10 \text{ mL}}$: $\text{BH}^+ \ \rightleftharpoons \ \text{B} + \text{H}^+ \qquad \dfrac{x^2}{0.090\,9 - x} = 10^{-9.00}\ \text{M} \Rightarrow x = 9.53 \times 10^{-6}\ \text{M}$

$\left(\dfrac{100}{110}\right)(0.100) - x \qquad x \qquad x \qquad\qquad\qquad [\text{H}^+] = x \Rightarrow \text{pH} = 5.02$

$\underline{10.1 \text{ mL}}$: $[\text{H}^+] = \left(\dfrac{0.1}{110.1}\right)(1.00) = 9.08 \times 10^{-4}\ \text{M} \Rightarrow \text{pH} = 3.04$

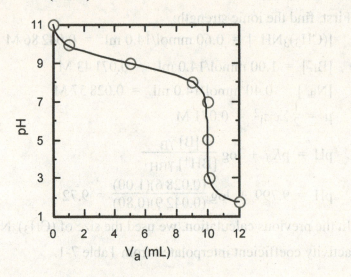

10-15. The maximum buffer capacity is reached when
$V = \tfrac{1}{2} V_e$, at which time $\dfrac{[\text{B}]}{[\text{BH}^+]} = 1$ and $\text{pH} = \text{p}K_a$ (for BH^+).

10-16. $\langle\bigcirc\rangle\text{--CH}_2\text{NH}_2 + \text{H}^+ \rightleftharpoons \langle\bigcirc\rangle\text{--CH}_2\overset{+}{\text{N}}\text{H}_3$ $\begin{array}{l}\text{Reverse of}\\ K_a \text{ reaction}\end{array}$

$K = 1/K_a$ (for $\text{C}_6\text{H}_5\text{CH}_2\overset{+}{\text{N}}\text{H}_3$) $= 2.2 \times 10^9$

10-17. Titration reaction: $\text{B} + \text{H}^+ \rightarrow \text{BH}^+$. To find the equivalence point, we write
$(50.0)(0.031\,9) = (V_e)(0.050\,0) \Rightarrow V_e = 31.9$ mL.

$\underline{0 \text{ mL}}$: $\text{B} + \text{H}_2\text{O} \rightleftharpoons \text{BH}^+ + \text{OH}^- \qquad \dfrac{x^2}{0.031\,9 - x} = K_b = \dfrac{K_w}{K_a} = 2.22 \times 10^{-5}$

$\qquad\qquad 0.031\,9 - x \qquad\quad x \qquad x \qquad\qquad \Rightarrow x = 8.31 \times 10^{-4}\ \text{M} \Rightarrow \text{pH} = 10.92$

$\underline{12.0 \text{ mL}}$:

	B	+	H$^+$	$\rightarrow$	BH$^+$
Initial :	31.9		12.0		—
Final:	19.9		—		12.0

$\text{pH} = \text{p}K_a + \log \dfrac{[\text{B}]}{[\text{BH}^+]} = 9.35 + \log \dfrac{19.9}{12.0} = 9.57$

$\underline{1/2 V_e}$: $\text{pH} = \text{p}K_a = 9.35$

30 mL: $pH = pK_a + \log \frac{1.9}{30.0} = 8.15$

V_e: B has been converted to BH^+ at a concentration of $\left(\frac{50.0}{81.9}\right)(0.031\,9)$

$= 0.019\,5\ M$

$BH^+ \rightleftharpoons B + H^+$ $\frac{x^2}{0.019\,5 - x} = K_a \Rightarrow x = 2.96 \times 10^{-6}\ M$

$0.019\,5 - x \quad x \quad x$ $\Rightarrow pH = 5.53$

35.0 mL: $[H^+] = \left(\frac{3.1}{85.0}\right)(0.050\,0) = 1.82 \times 10^{-3}\ M \Rightarrow pH = 2.74$

10-18.   Titration reaction: $CN^- + H^+ \rightarrow HCN$

At the equivalence point, moles of CN^- = moles of H^+

$(0.100\ M)\,(50.00\ mL) = (0.438\ M)\,(V_e) \Rightarrow V_e = 11.42\ mL$

(a) CN^- + H^+ $\rightarrow$ HCN

Initial:	11.42	4.20	—
Final:	7.22	—	4.20

$pH = pK_a + \log \frac{7.22}{4.20} = 9.45$

(b)   11.82 mL is 0.40 mL past the equivalence point.

$[H^+] = \left(\frac{0.40}{61.82}\right)(0.438\ M) = 2.83 \times 10^{-3}\ M \Rightarrow pH = 2.55$

(c)   At the equivalence point, we have made HCN at a formal concentration of

$\left(\frac{50.00}{61.42}\right)(0.100) = 0.081\,4\ M.$

$HCN \rightleftharpoons H^+ + CN^-$ $\frac{x^2}{0.081\,4 - x} = K_a \Rightarrow x = 7.1 \times 10^{-6}$

$0.0814 - x \qquad x \qquad x$ $\Rightarrow pH = 5.15$

10-19.

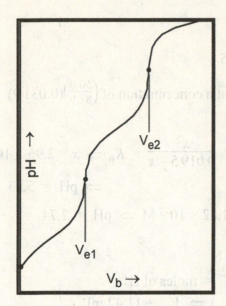

The pH of the initial solution before base is added is determined by the first acid dissociation reaction of H_2A. As base is added, it converts H_2A into an equivalent amount of HA^-. The buffer consisting of H_2A and HA^- governs the pH. At the first equivalence point, we have a solution of "pure" HA^-, the intermediate form of a diprotic acid. The pH is determined by the competitive acid and base reactions of HA^-. Between the two equivalence points there is a mixture of HA^- and A^{2-}, which is another buffer. At the second equivalence point, we have converted all HA^- into A^{2-}, whose base hydrolysis reaction determines the pH. After the second equivalence point, the excess OH^- added from the buret is mainly responsible for determining the pH, with negligible contribution from A^{2-}.

10-20. The protein has an average charge of 0 at the isoelectric point. Since the isoionic point occurs at a lower pH, the average charge of the protein must be positive at the isoionic point.

10-21. The equivalence point could be attained by mixing pure HA plus NaCl. Neglecting effects of ionic strength, the pH is equivalent to that of a solution of pure HA. This is the isoionic pH.

10-22. (a)

$$HA \rightleftharpoons A^- + H^+ \qquad\qquad K_a$$
$$\underline{B + H^+ \rightleftharpoons BH^+ \qquad\qquad K = K_b/K_w}$$
$$B + HA \rightleftharpoons BH^+ + A^- \qquad K = K_aK_b/K_w$$
$$= 10^{-2.86}\,10^{-3.36}\,/\,10^{-14.00} = 10^{7.78}$$

(b) In the upper curve, $\frac{3}{2}V_e$ is halfway between the first and second equivalence points. The pH is simply pK_2, since there is a 1:1 mixture of HA^- and A^{2-}. In the lower curve, pK_2 ($= pK_{BH^+}$) occurs when there is a 1:1 mixture of B and BH^+. To achieve this condition, all of B is first transformed into BH^+ by reaction with HA until V_e is reached. Then, at $2V_e$ one more equivalent of B has been added, giving a 1:1 mole ratio B:BH^+, so pH = pK_{BH^+}.

10-23.

V_a	0	1	5	9	10	11	15	19	20	22
pH	11.49	10.95	10.00	9.05	8.00	6.95	6.00	5.05	3.54	1.79

Representative calculations:

<u>0 mL</u>: $B + H_2O \underset{}{\overset{K_{b1}}{\rightleftharpoons}} BH^+ + HO^-$ $\dfrac{x^2}{0.100 - x} = 10^{-4.00} \Rightarrow x = 3.11 \times 10^{-3}$ M

 0.100 $- x$ x x pH $= -\log \dfrac{K_W}{x} = 11.49$

<u>1 mL</u>: pH $= pK_{BH^+} + \log \dfrac{[B]}{[BH^+]} = 10.00 + \log \dfrac{9}{1} = 10.95$

<u>10 mL</u>: Predominant form is BH^+ with formal concentration $\dfrac{100}{110}(0.100) = 0.090\,9$ M

$[H^+] \approx \sqrt{\dfrac{10^{-6.00}\,10^{-10.00}\,(0.090\,9) + 10^{-6.00}\,10^{-14.00}}{10^{-6.00} + 0.090\,9}}$

$ = 1.00 \times 10^{-8} \Rightarrow$ pH $= 8.00$

<u>11 mL</u>: pH $= pK_{BH_2^{2+}} + \log \dfrac{[BH^+]}{[BH_2^{2+}]} = 6.00 + \log \dfrac{9}{1} = 6.95$

<u>20 mL</u>: $BH_2^{2+} \rightleftharpoons BH^+ + H^+$ $\dfrac{x^2}{0.083\,3 - x} = 10^{-6.00} \Rightarrow x = 2.88 \times 10^{-4}$

 $\dfrac{100}{120}(0.100) - x$ x x $\Rightarrow$ pH $= 3.54$

<u>22 mL</u>: $[H^+] = \left(\dfrac{2}{122}\right)(1.00) = 1.64 \times 10^{-2}$ M $\Rightarrow$ pH $= 1.79$

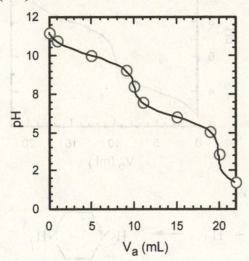

10-24.

V_b	0	1	5	9	10	11	15	19	20	22
pH	2.51	3.05	4.00	4.95	6.00	7.05	8.00	8.95	10.46	12.21

Representative calculations:

<u>0 mL</u>: $H_2A \rightleftharpoons HA^- + H^+$ $\dfrac{x^2}{0.100 - x} = 10^{-4.00} \Rightarrow x = 3.11 \times 10^{-3}$ M

 0.100 $- x$ x x $\Rightarrow$ pH $= 2.51$

$\underline{1\ mL}$: $pH = pK_1 + \log \dfrac{[HA^-]}{[H_2A]} = 4.00 + \log \dfrac{1}{9} = 3.05$

$\underline{10\ mL}$: Predominant form is HA^- with formal concentration $\left(\dfrac{100}{110}\right)(0.100)$

$\qquad\qquad = 0.090\ 9\ M.$

$$[H^+] \approx \sqrt{\dfrac{10^{-4.00}\ 10^{-8.00}\ (0.090\ 9) + 10^{-4.00}\ 10^{-14.00}}{10^{-4.00} + 0.090\ 9}}$$

$\qquad\qquad = 9.99 \times 10^{-7} \Rightarrow pH = 6.00$

$\underline{11\ mL}$: $pH = pK_2 + \log \dfrac{[A^{2-}]}{[HA^-]} = 8.00 + \log \dfrac{1}{9} = 7.05$

$\underline{20\ mL}$: $A^{2-} + H_2O \rightleftharpoons HA^- + OH^-$ $\dfrac{x^2}{0.083\ 3 - x} = \dfrac{K_w}{K_2} \Rightarrow x = 2.88 \times 10^{-4}\ M$

$\qquad \left(\dfrac{100}{120}\right)(0.100) - x \qquad\qquad x \qquad x$

$\qquad\qquad\qquad\qquad\qquad\qquad\qquad\qquad\qquad\qquad pH = -\log \dfrac{K_w}{x} = 10.46$

$\underline{22\ mL}$: $[OH^-] = \left(\dfrac{2}{122}\right)(1.00) = 1.64 \times 10^{-2}\ M \Rightarrow pH = 12.21$

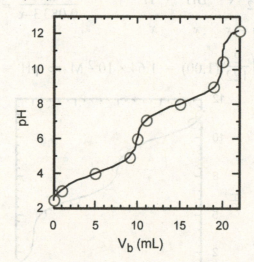

10-25.   Titration reactions:

$\qquad$ HN⬡NH + H⁺ → H₂N⁺⬡NH $V_e = 40.0\ mL$

$\qquad$ H₂N⁺⬡NH + H⁺ → H₂N⁺⬡N⁺H₂ $V_e = 80.0\ mL$

$\underline{0\ mL}$: $B + H_2O \rightleftharpoons BH^+ + OH^-$ $\dfrac{x^2}{0.100 - x} = K_{b1} = \dfrac{K_w}{K_{a2}} = \dfrac{1.0 \times 10^{-14}}{1.86 \times 10^{-10}}$

$\qquad\qquad 0.100 - x \qquad\qquad x \qquad x$

$\qquad\qquad\qquad\qquad\qquad\qquad\qquad\qquad \Rightarrow x = 2.29 \times 10^{-3}\ M \Rightarrow pH = 11.36$

<u>10.0 mL</u>: pH = pK_2 + log $\frac{[B]}{[BH^+]}$ = 9.731 + log $\frac{3}{1}$ = 10.21

<u>20.0 mL</u>: pH = pK_2 = 9.73

<u>30.0 mL</u>: pH = pK_2 + log $\frac{1}{3}$ = 9.25

<u>40.0 mL</u>: B has been converted to BH^+ at a formal concentration of

$$F = \left(\frac{40.0}{80.0}\right)(0.100) = 0.050\,0\ M$$

$$[H^+] = \sqrt{\frac{K_1 K_2 F + K_1 K_w}{K_1 + F}}$$

$$= \sqrt{\frac{(4.65 \times 10^{-6})(1.86 \times 10^{-10})(0.050\,0) + (4.65 \times 10^{-6})(1.0 \times 10^{-14})}{4.65 \times 10^{-6} + 0.050\,0}}$$

$$\Rightarrow pH = 7.53$$

<u>50.0 mL</u>: pH = pK_1 + log $\frac{[BH^+]}{[BH_2^{2+}]}$ = 5.333 + log $\frac{3}{1}$ = 5.81

<u>60.0 mL</u>: pH = pK_1 = 5.33

<u>70.0 mL</u>: pH = pK_1 + log $\frac{1}{3}$ = 4.86

<u>80.0 mL</u>: B has been converted to BH_2^{2+} at a formal concentration of

$$\left(\frac{40.0}{120.0}\right)(0.100) = 0.033\,3\ M$$

$$\begin{array}{cccc} BH_2^{2+} & \rightleftharpoons & BH^+ & + & H^+ \\ 0.0333 - x & & x & & x \end{array} \qquad \frac{x^2}{0.033\,3 - x} = K_1 = 4.65 \times 10^{-6} \Rightarrow$$

$$x = 3.91 \times 10^{-4}\ M \Rightarrow pH = 3.41$$

<u>90.0 mL</u>: $[H^+]$ = $\left(\frac{10.0}{130.0}\right)(0.100) \Rightarrow pH = 2.11$

<u>100.0 mL</u>: $[H^+]$ = $\left(\frac{20.0}{140.0}\right)(0.100) \Rightarrow pH = 1.85$

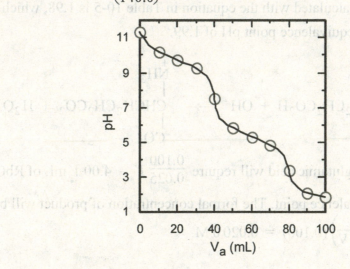

10-26.

Initial mmol:	0.500	0.164	—
Final mmol:	0.336	—	0.164

$$pH = pK_1 + \log\frac{[B]}{[BH^+]} = 4.70 + \log\frac{0.336}{0.164} = 5.01$$

10-27.   (a)   Titration reactions:

$$H_2NCH_2CO_2^- + H^+ \rightarrow H_3^+NCH_2CO_2^- \qquad V_e = 50.0 \text{ mL}$$

$$H_3^+NCH_2CO_2^- + H^+ \rightarrow H_3^+NCH_2CO_2H \qquad V_e = 100.0 \text{ mL}$$

At the second equivalence point, the formal concentration of

$H_3^+NCH_2CO_2H$ is $\left(\dfrac{50.0}{150.0}\right)(0.100) = 0.033\,3 \text{ M}$

$$H_3^+NCH_2CO_2H \rightleftharpoons H_3^+NCH_2CO_2^- + H^+ \qquad \frac{x^2}{0.0333 - x} = K_1 = 0.004\,47 \Rightarrow$$

$$0.0333 - x \qquad\qquad x \qquad\qquad x \qquad\qquad x = 1.02 \times 10^{-2} \text{ M} \Rightarrow pH = 1.99$$

(b)   At $V_a = 90.0$ mL, the approximation gives $pH = pK_1 + \log\dfrac{[HG]}{[H_2G^+]} = 2.35 +$

$\log\dfrac{1}{4} = 1.75$, which is <u>lower</u> than the correct value at 100.0 mL. The pH

calculated with the equation in Table 10-5 is 2.16, which is higher than the

equivalence point pH of 1.99.

At $V_a = 101.0$ mL, the approximation gives $[H^+] = \left(\dfrac{1.0}{151.0}\right)(0.100) = 6.62$

$\times 10^{-4}$ M $\Rightarrow$ pH $= 3.18$, which is <u>higher</u> than the correct value at 100.0 mL.

The pH calculated with the equation in Table 10-5 is 1.98, which is lower

than the equivalence point pH of 1.99.

10-28.   (a)

(b) V mL of glutamic acid will require $\dfrac{0.100}{0.025} V = 4.00\,V$ mL of RbOH to reach

the equivalence point. The formal concentration of product will be

$\left(\dfrac{V}{V + 4.00\,V}\right)(0.100) = 0.020\,0 \text{ M}.$

$$[H^+] = \sqrt{\frac{K_2 K_3 F + K_2 K_w}{K_2 + F}}$$

$$= \sqrt{\frac{(5.0 \times 10^{-5})(1.1 \times 10^{-10})(0.020\,0) + (5.0 \times 10^{-5})(1.0 \times 10^{-14})}{(5.0 \times 10^{-5}) + 0.020\,0}}$$

$$= 7.42 \times 10^{-8}\ M \Rightarrow pH = 7.13$$

10-29.

One volume of tyrosine (0.010 0 M) requires 2.5 volumes of $HClO_4$ (0.004 00 M), so the formal concentration of tyrosine at the equivalence point is $\left(\frac{1}{1+2.5}\right)(0.010\,0\ M) = 0.002\,86\ M$. The pH is calculated from the acid dissociation of H_3T^+.

$$\begin{array}{cccc} H_3T^+ & \rightleftharpoons & H_2T & + & H^+ \\ 0.002\,86 - x & & x & & x \end{array} \qquad \frac{x^2}{0.002\,86 - x} = K_1 = 3.9 \times 10^{-3} \Rightarrow$$

$$x = 0.001\,92\ M \Rightarrow pH = 2.72$$

10-30. (a) $C^{2-} + H^+ \rightarrow HC^-$. $V_e = 20.0$ mL. At the equivalence point, the formal concentration of HC^- is $\left(\frac{40.0}{60.0}\right)(0.030\,0) = 0.020\,0$ M.

$$[H^+] = \sqrt{\frac{K_2 K_3 F + K_2 K_w}{K_2 + F}}$$

$$= \sqrt{\frac{(4.4 \times 10^{-9})(1.82 \times 10^{-11})(0.020\,0) + (4.4 \times 10^{-9})(1.0 \times 10^{-14})}{(4.4 \times 10^{-9}) + 0.020\,0}}$$

$$= 2.86 \times 10^{-10}\ M \Rightarrow pH = 9.54$$

(b) $\begin{array}{cccc} H_3C^+ & \rightleftharpoons & H_2C & + & H^+ \\ 0.0500 - x & & x & & x \end{array} \qquad \frac{x^2}{0.0500 - x} = K_1 = 0.02$

$$\Rightarrow x = 0.023\ M \Rightarrow pH = 1.64$$

$$pH = pK_3 + \log \frac{[C^{2-}]}{[HC^-]}$$

$$1.64 = 10.74 + \log \frac{[C^{2-}]}{[HC^-]} \Rightarrow \frac{[C^{2-}]}{[HC^-]} = 7.9 \times 10^{-10}$$

10-31. The two values of pK_a for oxalic acid are 1.250 and 4.266. At a pH of 4.40, the $C_2O_4^{2-}$ has not yet been half-neutralized.

	$C_2O_4^{2-}$ +	H^+	→	$HC_2O_4^-$
Initial mmol:	x	16.0		—
Final mmol:	$x - 16.0$	—		16.0

$$pH = 4.40 = pK_2 + \log \frac{[C_2O_4^{2-}]}{[HC_2O_4^-]} = 4.266 + \log \frac{x - 16.0}{16.0}$$

$$\Rightarrow x = 37.8 \text{ mmol of } K_2C_2O_4 = 6.28 \text{ g}$$

10-32. Neutral alanine is designated HA.

	HA +	OH^-	→	A^- +	H_2O
Initial mmol:	1.260 5	0.516		—	
Final mmol:	0.744 5	—		0.516	

$$pH = pK_2 + \log \frac{[A^-]\,\gamma_{A^-}}{[HA]\,\gamma_{HA}} \qquad 9.57 = pK_2 + \log \frac{(0.516)(0.77)}{(0.744\,5)(1)} \Rightarrow pK_2 = 9.84$$

10-33. A Gran plot allows us to find the equivalence point by extrapolating from points measured prior to the equivalence point.

10-34. It is evident from the following table of data that the end point is near 23.4 mL, where the derivative dpH/dV_b is greatest. A graph of $V_b 10^{-pH}$ versus V_b follows.

The points from 21.01 to 23.30 mL were fit by the method of least squares to give the equation shown in the graph. The intercept is found by setting $y = 0$ in the equation, giving $x = V_e = 23.39$ mL.

V_b (mL)	$V_b 10^{-pH}$	V_b (mL)	$V_b 10^{-pH}$	V_b (mL)	$V_b 10^{-pH}$
21.01	15.22×10^{-6}	22.10	8.40×10^{-6}	22.97	2.76×10^{-6}
21.10	14.94	22.27	7.37	23.01	2.41
21.13	14.62	22.37	6.60	23.11	1.79
21.20	14.33	22.48	5.91	23.17	1.46
21.30	13.75	22.57	5.29	23.21	1.16
21.41	12.90	22.70	4.53	23.30	0.75
21.51	12.10	22.76	4.14	23.32	0.42
21.61	11.61	22.80	3.78	23.40	0.12
21.77	10.42	22.85	3.46	23.46	0.01
21.93	9.35	22.91	3.16	23.55	0.003

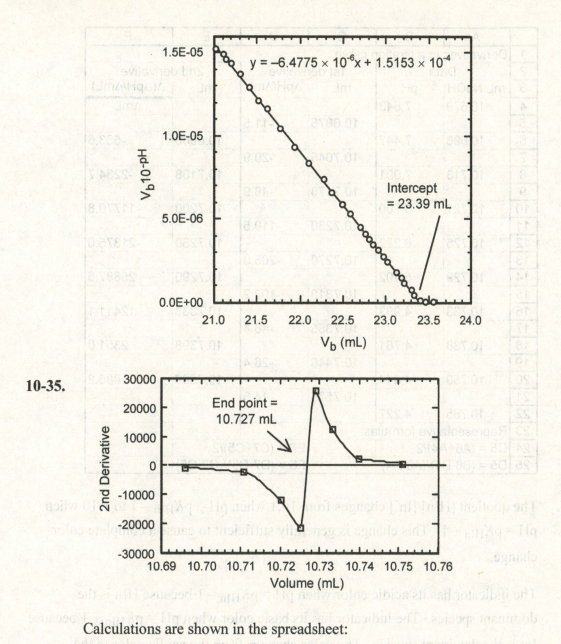

10-35.

Calculations are shown in the spreadsheet:

	A	B	C	D	E	F
1	Derivatives in a titration curve					
2	Data		1st derivative		2nd derivative	
3	mL NaOH	pH	mL	$\Delta pH/\Delta mL$	mL	$\Delta(\Delta pH/\Delta mL)$
4	10.679	7.643				ΔmL
5			10.6875	-11.5		
6	10.696	7.447			10.6960	-553.6
7			10.7045	-20.9		
8	10.713	7.091			10.7108	-2234.7
9			10.7170	-48.9		
10	10.721	6.700			10.7200	-11770.8
11			10.7230	-119.5		
12	10.725	6.222			10.7250	-21375.0
13			10.7270	-205.0		
14	10.729	5.402			10.7290	25687.5
15			10.7310	-102.2		
16	10.733	4.993			10.7333	12411.1
17			10.7355	-46.4		
18	10.738	4.761			10.7398	2351.0
19			10.7440	-26.4		
20	10.750	4.444			10.7508	885.2
21			10.7575	-14.5		
22	10.765	4.227				
23	Representative formulas:					
24	C5 = (A6+A4)/2			E6 = (C7+C5)/2		
25	D5 = (B6-B4)/(A6-A4)			F6 = (D7-D5)/(C7-C5)		

10-36. The quotient [HIn]/[In⁻] changes from 10:1 when $pH = pK_{HIn} - 1$ to 1:10 when $pH = pK_{HIn} + 1$. This change is generally sufficient to cause a complete color change.

10-37. The indicator has its acidic color when $pH = pK_{HIn} - 1$ because HIn is the dominant species. The indicator has its basic color when $pH = pK_{HIn} + 1$ because In⁻ is the dominant species. The color changes from the acidic color to the intermediate color to the basic color as the pH rises through the range $pK_{HIn} - 1$ to $pK_{HIn} + 1$. If the indicator is chosen correctly for the titration, this indicator pH transition range coincides with the steep part of the titration curve. The color change occurs near the equivalence point, which is the center of the steep portion of the titration curve.

10-38. The Henderson-Hasselbalch equation for the indicator, HIn, is
$$pH = pK_{HIn} + \log \frac{[In^-]}{[HIn]}.$$ If we know pK_{HIn} and we measure $\frac{[In^-]}{[HIn]}$ spectroscopically, then we can calculate the pH.

10-39.   Strong acids, such as H_2SO_4, HCl, HNO_3, and $HClO_4$ have $pK_a < 0$.

10-40.   yellow, green, blue

10-41.   (a)   red   (b)   orange   (c)   yellow

10-42.   (a)   red   (b)   orange   (c)   yellow   (d)   red

10-43.   No. When a weak acid is titrated with a strong base, the solution contains A^- at the equivalence point. A solution of A^- must have a pH above 7.

10-44.   (a)   The titration reaction is $F^- + H^+ \rightarrow HF$.

If V mL of NaF are used, $V_e = \frac{1}{2}V$, since the concentration of $HClO_4$ is twice as great as the concentration of NaF. The formal concentration of HF at the equivalence point is $\left(\dfrac{V}{V+\frac{1}{2}V}\right)(0.030\,0) = 0.020\,0$ M.

The pH is determined by the acid dissociation of HF.

$$\begin{array}{ccccc} HF & \rightleftharpoons & H^+ & + & F^- \\ 0.0200 - x & & x & & x \end{array} \qquad \frac{x^2}{0.020\,0 - x} = K_a \Rightarrow x = 3.36 \times 10^{-3}$$

$$\Rightarrow pH = 2.47$$

(b)   The pH is so low that there would not be much (if any) break (inflection) in the titration curve at the equivalence point. A sharp change in indicator color will not be seen.

10-45.   (a)   violet (red + blue)   (b)   blue   (c)   yellow

10-46.   (a) $\begin{array}{ccccc} NH_4^+ & \rightleftharpoons & NH_3 & + & H^+ \\ 0.010 - x & & x & & x \end{array} \qquad \dfrac{x^2}{0.010 - x} = K_a \Rightarrow x = 2.38 \times 10^{-6}$ M

$$\Rightarrow pH = 5.62$$

(b)   One possible indicator is methyl red, using the yellow end point.

For a more complete analysis of this problem, we could compute the titration curve for a mixture of HCl and NH_4^+. For simplicity, we consider the mixture to be a "diprotic" acid with $K_1 = 100$ (i.e., a "strong" acid) and $K_2 = 5.7 \times 10^{-10}$ for the ammonium ion. We use the spreadsheet equation in Table 10-5 for titrating H_2A with strong base, taking $C_a = 0.01$ M and $C_b = 0.1$ M. An indicator with a color change in the range ~4.5–7.0 would find the HCl end point without titrating a significant amount of NH_4^+.

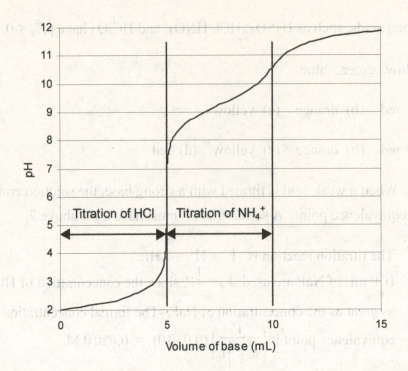

10-47.   Grams of cleaner titrated $= \left(\frac{4.373}{10.231 + 39.466}\right)(10.231\ g) = 0.900\ 3\ g$

mol HCl used = mol NH$_3$ present $= (0.014\ 22\ L)(0.106\ 3\ M) = 1.512$ mmol

1.512 mmol NH$_3$ $= 25.74$ mg NH$_3$

wt% NH$_3$ $= \dfrac{2.574 \times 10^{-2}\ g}{0.900\ 3\ g} \times 100 = 2.859\%$

10-48.   *Alkalinity* is the capacity of natural water to react with H$^+$ to reach pH 4.5, which

is the second equivalence point in the titration of carbonate (CO$_3^{2-}$) with H$^+$.

Alkalinity measures [OH$^-$] + [CO$_3^{2-}$] + [HCO$_3^-$] plus any other bases that are

present. Bromocresol green is blue above pH 5.4 and yellow below pH 3.8. It

will be at its green end-point color between these two pH values, which

approximates pH 4.5.

10-49.   Tris(hydroxymethyl) aminomethane (H$_2$NC(CH$_2$OH)$_3$), mercuric oxide (HgO),

sodium carbonate (Na$_2$CO$_3$), and borax (NaB$_4$O$_7 \cdot$ 10H$_2$O) can be used to

standardize HCl. Potassium acid phthalate (HO$_2$C-C$_6$H$_4$-CO$_2^-$K$^+$), HCl azeotrope,

potassium hydrogen iodate (KH(IO$_3$)$_2$), sulfosalicylic acid double salt

(C$_7$H$_5$SO$_6$K$\cdot$C$_7$H$_4$SO$_6$K$_2$), and sulfamic acid ($^+$H$_3$NSO$_3^-$) can be used to

standardize NaOH.

10-50. The greater the equivalent mass, the more primary standard is required. There is less relative error in weighing a large mass of reagent than a small mass.

10-51. Potassium acid phthalate is dried at 105° and weighed accurately into a flask. It is titrated with NaOH, using a pH electrode or phenolphthalein to observe the end point.

10-52. Grams of tris titrated $= \dfrac{4.963}{(1.023 + 99.367)} (1.023) = 0.050\,57 = 0.417\,5$ mmol

Concentration of $HNO_3 = \dfrac{0.417\,5 \text{ mmol}}{5.262 \text{ g solution}} = 0.079\,34$ mol/kg solution

10-53. True mass $= m = \dfrac{(1.023)\left(1 - \dfrac{0.001\,2}{8.0}\right)}{\left(1 - \dfrac{0.001\,2}{1.33}\right)} = 1.023_8$ g

Failure to account for buoyancy introduces a systematic error of

$100 \times (1.023_8 - 1.023) / 1.023 = 0.08\%$ in the calculated molarity of HCl.

The true mass is higher than the measured mass of Tris, so the calculated HCl molarity is too low.

10-54. The mmoles of HgO in $0.194\,7$ g $= 0.898\,9$, which will make 1.798 mmol of OH^- by reaction with Br^- plus H_2O. HCl molarity $= 1.798$ mmol/17.98 mL $= 0.100\,0$ M.

10-55. 30 mL of 0.05 M $OH^- = 1.5$ mmol $= 0.31$ g of potassium acid phthalate.

10-56. (a) From a graph of weight percent vs pressure,
HCl $= 20.254\%$ when $P = 746$ Torr.

(b) We need $0.100\,00$ mole of HCl $= 3.646\,1$ g

$\dfrac{3.646\,1 \text{ g HCl}}{0.202\,54 \text{ g HCl/g solution}} = 18.001_9$ g of solution.

The mass required (weighed in air) is given by Equation 2-1.

$m' = \dfrac{(18.001_9)\left(1 - \dfrac{0.001\,2}{1.096}\right)}{\left(1 - \dfrac{0.001\,2}{8.0}\right)} = 17.985$ g

10-57. (a) For a rectangular distribution of uncertainty in atomic mass, divide the uncertainty listed in the periodic table by $\sqrt{3}$ to find the standard uncertainty:

C: $12.010\,7 \pm 0.000\,8/\sqrt{3} = 12.010\,7 \pm 0.000\,4_6$

H: $1.007\,94 \pm 0.000\,07/\sqrt{3} = 1.007\,94 \pm 0.000\,04_0$

O: $15.999\,4 \pm 0.000\,3/\sqrt{3} = 15.999\,4 \pm 0.000\,1_7$

K: $39.098\,3 \pm 0.000\,1/\sqrt{3} = 39.098\,3 \pm 0.000\,0_{58}$

8C:	$8(12.010\,7 \pm 0.000\,4_6)$	=	$96.085\,6 \pm 0.003\,7$
5H:	$5(1.007\,94 \pm 0.000\,04_0)$	=	$5.039\,7 \pm 0.000\,20$
4O:	$4(15.999\,4 \pm 0.000\,1_7)$	=	$63.997\,6 \pm 0.000\,69$
1K:	$1(39.098\,3 \pm 0.000\,0_{58})$	=	$39.098\,3 \pm 0.000\,0_{58}$

$C_8H_5O_4K$: $204.221\,2 \pm ?$

Uncertainty $= \sqrt{0.003\,7^2 + 0.000\,20^2 + 0.000\,69^2 + 0.000\,058^2} = 0.003\,8$

Answer: 204.221 ± 0.004 g/mol

(b) For a rectangular distribution, divide the stated uncertainty by $\sqrt{3}$ to find the standard uncertainty. Purity $= 1.000\,00 \pm 0.000\,05/\sqrt{3} = 1.000\,00 \pm 0.000\,03$

10-58. 5.00 mL of 0.033 6 M HCl $= 0.168\,0$ mmol. 6.34 mL of 0.010 0 M NaOH $= 0.063\,4$ mmol. HCl consumed by $NH_3 = 0.168\,0 - 0.063\,4 = 0.104\,6$ mmol $= 1.465$ mg of nitrogen. 256 μL of protein solution contains 9.702 mg protein. 1.465 mg of N/9.702 mg protein $= 15.1$ wt%.

10-59. When an acid that is stronger than H_3O^+ is added to H_2O, it reacts to give H_3O^+ and is "leveled" to the strength of H_3O^+. Similarly, bases stronger than OH^- are leveled to the strength of OH^-.

10-60. Methanol and ethanol have nearly the same acidity as water. Both of the following equilibria are driven to the right because of the high concentration of H_2O:

$CH_3O^- + H_2O \rightarrow CH_3OH + OH^-$

$CH_3CH_2O^- + H_2O \rightarrow CH_3CH_2OH + OH^-$

10-61. (a) In acetic acid, strong acids are not leveled to the strength of $CH_3CO_2H_2^+$. Therefore, very weak bases can be titrated in acetic acid.

(b) If tetrabutylammonium hydroxide were added to an acetic acid solution, most of the hydroxide would react with acetic acid instead of analyte. However, OH^- will not react with pyridine, so this solvent would be suitable.

10-62. Sodium amide and phenyl lithium are stronger bases than OH^-. Each reacts with

H_2O to give OH^-:

$$NH_2^- + H_2O \rightarrow NH_3 + OH^-$$

$$C_6H_5^- + H_2O \rightarrow C_6H_6 + OH^-$$

10-63.   The reaction of pyridine with acid is

Methanol is less polar than water. If methanol is added to the aqueous solution, the neutral pyridine molecule will tend to be favored over the protonated pyridinium cation. It will take a higher concentration of acid (a lower pH) to protonate pyridine in the mixed solvent. pK_a for pyridinium ion is lowered when methanol is added to the solution.

10-64.   Titration reaction: $K^+HP^- + Na^+OH^- \rightarrow K^+Na^+P^{2-} + H_2O$

Begin with C_aV_a moles of K^+HP^- and add C_bV_b moles of NaOH

Fraction of titration $= \phi = \dfrac{C_bV_b}{C_aV_a}$

Charge balance: $[H^+] + [Na^+] + [K^+] = [HP^-] + 2[P^{2-}] + [OH^-]$

Substitutions: $[K^+] = \dfrac{C_aV_a}{V_a+V_b}$ $[Na^+] = \dfrac{C_bV_b}{V_a+V_b}$

$[HP^-] = \alpha_{HP^-}\dfrac{C_aV_a}{V_a+V_b}$ $[P^{2-}] = \alpha_{P^{2-}}\dfrac{C_aV_a}{V_a+V_b}$

Putting these expressions into the charge balance gives

$$[H^+] + \frac{C_bV_b}{V_a+V_b} + \frac{C_aV_a}{V_a+V_b} = \alpha_{HP^-}\frac{C_aV_a}{V_a+V_b} + 2\alpha_{P^{2-}}\frac{C_aV_a}{V_a+V_b} + [OH^-]$$

Multiply by $V_a + V_b$ and collect terms:

$$[H^+]V_a + [H^+]V_b + C_bV_b + C_aV_a = \alpha_{HP^-}C_aV_a + 2\alpha_{P^{2-}}C_aV_a + [OH^-]V_a + [OH^-]V_b$$

$$V_a([H^+] + C_a - \alpha_{HP^-}C_a - 2\alpha_{P^{2-}}C_a - [OH^-]) = V_b([OH^-] - [H^+] - C_b)$$

$$\frac{V_b}{V_a} = \frac{\alpha_{HP^-}C_a + 2\alpha_{P^{2-}}C_a - C_a - [H^+] + [OH^-]}{C_b + [H^+] - [OH^-]}$$

Multiply both sides by $\dfrac{1/C_a}{1/C_b}$:

$$\phi = \frac{C_bV_b}{C_aV_a} = \frac{\alpha_{HP^-} + 2\alpha_{P^{2-}} - 1 - \dfrac{[H^+] - [OH^-]}{C_a}}{1 + \dfrac{[H^+] - [OH^-]}{C_b}}$$

10-65.

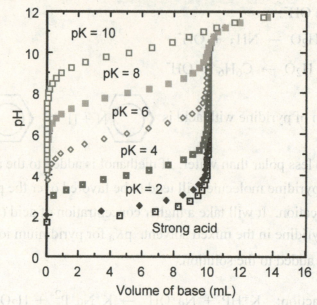

10-66.

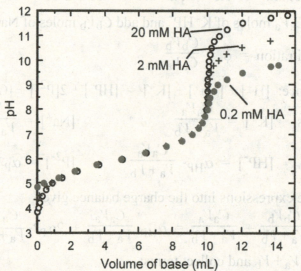

10-67.

	A	B	C	D	E	F	G
1	Effect of pKb in the titration of weak base with strong acid						
2							
3	Ca =	pH	[H+]	[OH-]	Alpha(BH+)	Phi	Va (mL)
4	0.1	2.00	1.00E-02	1.00E-12	9.90E-01	1.66E+00	16.557
5	Cb =	2.90	1.26E-03	7.94E-12	9.26E-01	1.00E+00	10.020
6	0.02	3.50	3.16E-04	3.16E-11	7.60E-01	7.78E-01	7.780
7	Vb =	4.00	1.00E-04	1.00E-10	5.00E-01	5.06E-01	5.055
8	50	4.50	3.16E-05	3.16E-10	2.40E-01	2.42E-01	2.419
9	K(BH+) =	6.00	1.00E-06	1.00E-08	9.90E-03	9.95E-03	0.100
10	1E-04	8.15	7.08E-09	1.41E-06	7.08E-05	5.17E-07	0.000
11	Kw =						
12	1E-14			E4 = C4/(C4+$A$10)			
13		C4 = 10^-B4		F4 = (E4+(C4-D4)/$A$6)/(1-(C4-D4)/$A$4)			
14		D4 = $A$12/C4		G4 = F4*$A$6*$A$8/$A$4			

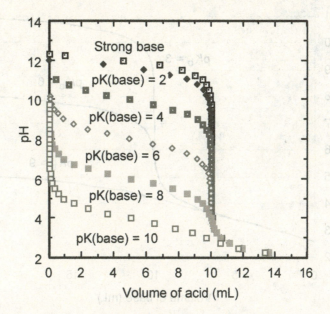

10-68. (a)

	A	B	C	D	E	F	G	H
1	Titrating weak acid with weak base							
2								
3	Cb =	pH	[H+]	[OH-]	Alpha(A-)	Alpha(BH+)	Phi	Vb (mL)
4	0.1	2.86	1.4E-03	7.2E-12	6.76E-02	1.00E+00	-1.4E-03	-0.01
5	Ca =	3.00	1.0E-03	1.0E-11	9.09E-02	1.00E+00	4.1E-02	0.41
6	0.02	4.00	1.0E-04	1.0E-10	5.00E-01	1.00E+00	4.9E-01	4.95
7	Va =	5.00	1.0E-05	1.0E-09	9.09E-01	9.99E-01	9.1E-01	9.09
8	50	6.00	1.0E-06	1.0E-08	9.90E-01	9.90E-01	1.0E+00	10.00
9	Ka =	7.00	1.0E-07	1.0E-07	9.99E-01	9.09E-01	1.1E+00	10.99
10	1E-04	8.00	1.0E-08	1.0E-06	1.00E+00	5.00E-01	2.0E+00	20.00
11	Kw =							
12	1E-14		A16 = $A$12/$A$14			D4 = $A$12/C4		
13	Kb =		C4 = 10^-B4			E4 = $A$10/(C4+$A$10)		
14	1E-06		F4 = C4/(C4+$A$16)					
15	K(BH+) =		G4 = (E4-(C4-D4)/$A$6)/(F4+(C4-D4)/$A$4)					
16	1E-08		H4 = G4*$A$6*$A$8/$A$4					

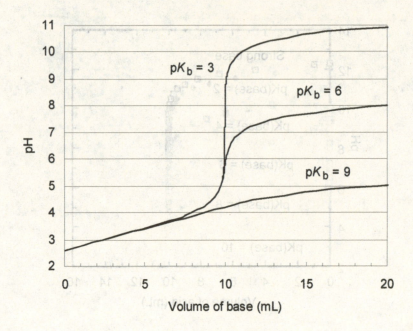

(b)   HA   +   B   ⇌   A⁻ + BH⁺

$K_a = 1.75 \times 10^{-5}$ $K_b = 1.59 \times 10^{-10}$

$V_a = 212$ mL $K_{BH^+} = 6.28 \times 10^{-5}$

$C_a = 0.200$ M $V_b = 325$ mL

                                   $C_b = 0.0500$ M

To find the equilibrium constant we write

HA   ⇌   A⁻ + H⁺ K_a

H⁺ + B   ⇌   BH⁺ $1/K_{BH^+}$
_____

HA + B ⇌ A⁻ + BH⁺ $K = K_a/K_{BH^+} = 0.279$

A pH of 4.16 gives $V_b = 325.0$ mL in the following spreadsheet:

	A	B	C	D	E	F	G	H
1	Mixing acetic acid and sodium benzoate							
2								
3	Cb =	pH	[H+]	[OH-]	Alpha(A-)	Alpha(BH+)	Phi	Vb (mL)
4	0.05	4.00	1.0E-04	1.0E-10	1.49E-01	6.14E-01	2.4E-01	204.28
5	Ca =	4.2	6.3E-05	1.6E-10	2.17E-01	5.01E-01	4.3E-01	365.98
6	0.2	4.1	7.9E-05	1.3E-10	1.81E-01	5.58E-01	3.2E-01	272.78
7	Va =	4.15	7.1E-05	1.4E-10	1.98E-01	5.30E-01	3.7E-01	315.79
8	212	4.1598	6.9E-05	1.4E-10	2.02E-01	5.24E-01	3.8E-01	325.03
9	Ka =							
10	1.750E-05		A16 = $A$12/$A$14					
11	Kw =		C4 = 10^-B4					
12	1.E-14		D4 = $A$12/C4					
13	Kb =		E4 = $A$10/(C4+$A$10)					
14	1.592E-10		F4 = C4/(C4+$A$16)					
15	K(BH+) =		G4 = (E4-(C4-D4)/$A$6)/(F4+(C4-D4)/$A$4)					
16	6.281E-05		H4 = G4*$A$6*$A$8/$A$4					

10-69.

		A	B	C	D	E	F	G	H
	1	Titrating diprotic acid with strong base							
	2								
	3	Cb =	pH	[H+]	[OH-]	Alpha(HA-)	Alpha(A2-)	Phi	Vb (mL)
	4	0.1	2.865	1.4E-03	7.3E-12	6.83E-02	5.00E-07	5.0E-05	0.000
	5	Ca =	4.00	1.0E-04	1.0E-10	5.00E-01	5.00E-05	4.9E-01	4.946
	6	0.02	6.00	1.0E-06	1.0E-08	9.80E-01	9.80E-03	1.0E+00	9.999
	7	Va =	8.00	1.0E-08	1.0E-06	5.00E-01	5.00E-01	1.5E+00	15.000
	8	50	10.0	1.0E-10	1.0E-04	9.90E-03	9.90E-01	2.0E+00	19.971
	9	Kw =	12.0	1.0E-12	1.0E-02	1.00E-04	1.00E+00	2.8E+00	27.777
	10	1E-14							
	10	K1 =		C4 = 10^-B4			D4 = $A$12/C4		
	12	1E-4		E4 = C4*$A$12/(C4^2+C4*$A$12+$A$12*$A$14)					
	13	K2 =		F4 = $A$12*$A$14/(C4^2+C4*$A$12+$A$12*$A$14)					
	14	1.E-08		G4 = (E4+2*F4-(C4-D4)/$A$6)/(1+(C4-D4)/$A$4)					
	15			H4 = G4*$A$6*$A$8/$A$4					

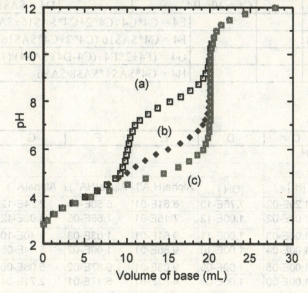

10-70.

	A	B	C	D	E	F	G	H
1	Titrating nicotine with strong acid							
2								
3	Cb =	pH	[H+]	[OH-]	Alpha(BH2)	Alpha(BH)	Phi	Va (mL)
4	0.1	1.75	1.8E-02	5.6E-13	9.62E-01	3.83E-02	2.6E+00	26.023
5	Ca =	2.00	1.0E-02	1.0E-12	9.34E-01	6.61E-02	2.3E+00	22.599
6	0.1	3.00	1.0E-03	1.0E-11	5.86E-01	4.14E-01	1.6E+00	16.117
7	Vb =	4.00	1.0E-04	1.0E-10	1.24E-01	8.76E-01	1.1E+00	11.258
8	10	5.00	1.0E-05	1.0E-09	1.39E-02	9.85E-01	1.0E+00	10.127
9	Kw =	6.00	1.0E-06	1.0E-08	1.39E-03	9.85E-01	9.9E-01	9.875
10	1.E-14	7.00	1.0E-07	1.0E-07	1.24E-04	8.76E-01	8.8E-01	8.764
11	KB1 =	8.00	1.0E-08	1.0E-06	5.86E-06	4.14E-01	4.1E-01	4.145
12	7.079E-7	9.00	1.0E-09	1.0E-05	9.34E-08	6.61E-02	6.6E-02	0.660
13	KB2 =	10.00	1.0E-10	1.0E-04	9.93E-10	7.03E-03	6.0E-03	0.060
14	1.41E-11	10.42	3.8E-11	2.6E-04	1.44E-10	2.68E-03	5.4E-05	0.001
15	KA1 =							
16	7.077E-4		C4 = 10^-B4			D4 = $A$12/C4		
17	KA2 =		E4 = C4*C4/(C4^2+C4*$A$16+$A$16*$A$18)					
18	1.413E-8		F4 = C4*$A$16/(C4^2+C4*$A$16+$A$16*$A$18)					
19			G4 = (F4+2*E4+(C4-D4)/$A$4)/(1-(C4-D4)/$A$6)					
20			H4 = G4*$A$4*$A$8/$A$6					

10-71.

	A	B	C	D	E	F	G	H	I
1	Titrating H_3A with NaOH								
2									
3	C_b =	pH	$[H^+]$	$[OH^-]$	Alpha(H_2A^-)	Alpha(HA^{2-})	Alpha(A^{3-})	Phi	V_b (mL)
4	0.1	1.89	1.29E-02	7.76E-13	6.61E-01	5.50E-05	2.24E-12	1.50E-02	0.150
5	C_a =	2.00	1.00E-02	1.00E-12	7.15E-01	7.66E-05	4.02E-12	1.96E-01	1.958
6	0.02	3.00	1.00E-03	1.00E-11	9.61E-01	1.03E-03	5.40E-10	9.04E-01	9.037
7	V_a =	4.00	1.00E-04	1.00E-10	9.86E-01	1.06E-02	5.54E-08	1.00E+00	10.006
8	50	5.00	1.00E-05	1.00E-09	9.03E-01	9.67E-02	5.08E-06	1.10E+00	10.958
9	K_w =	6.00	1.00E-06	1.00E-08	4.83E-01	5.17E-01	2.71E-04	1.52E+00	15.176
10	1E-14	7.00	1.00E-07	1.00E-07	8.50E-02	9.10E-01	4.78E-03	1.92E+00	19.198
11	K_1 =	8.00	1.00E-08	1.00E-06	8.79E-03	9.42E-01	4.94E-02	2.04E+00	20.407
12	2.51E-02	9.00	1.00E-09	1.00E-05	6.12E-04	6.55E-01	3.44E-01	2.34E+00	23.441
13	K_2 =	10.00	1.00E-10	1.00E-04	1.49E-05	1.60E-01	8.40E-01	2.85E+00	28.478
14	1.07E-06	11.00	1.00E-11	1.00E-03	1.75E-07	1.87E-02	9.81E-01	3.06E+00	30.619
15	K_3 =	12.00	1.00E-12	1.00E-02	1.77E-09	1.90E-03	9.98E-01	3.89E+00	38.868
16	5.25E-10		C4 = 10^-B4						
17	pK_1 =		D4 = $A$10/C4						
18	1.60		E4 = $C4^2*$A$12/($C4^3+$C4^2*$A$12+$C4*$A$12*$A$14+$A$12*$A$14*$A$16)						
19	pK_2 =		F4 = $C4*$A$12*$A$14/($C4^3+$C4^2*$A$12+$C4*$A$12*$A$14+$A$12*$A$14*$A$16						
20	5.97		G4 = $A$12*$A$14*$A$16/($C4^3+$C4^2*$A$12+$C4*$A$12*$A$14+$A$12*$A$14*$A$16)						
21	pK_3 =		H4 = (E4+2*F4+3*G4-(C4-D4)/$A$6)/(1+(C4-D4)/$A$4)						
22	9.28		I4 = H4*$A$6*$A$8/$A$4						

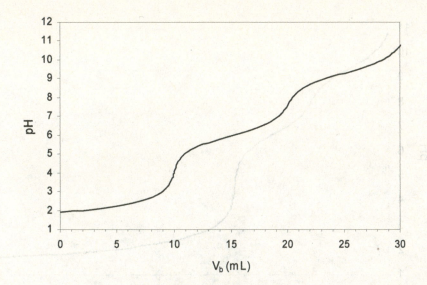

10-72. $$\phi = \frac{C_a V_a}{C_b V_b} = \frac{\alpha_{BH^+} + 2\alpha_{BH_2^{2+}} + 3\alpha_{BH_3^{3+}} + 4\alpha_{BH_4^{4+}} + \dfrac{[H^+] - [OH^-]}{C_b}}{1 - \dfrac{[H^+] - [OH^-]}{C_a}}$$

	A	B	C	D	E	F	G	H	I	J
1	Titrating Tetrabasic B with H⁺									
2										
3	C_b =	pH	$[H^+]$	$[OH^-]$	$\alpha(BH^+)$	$\alpha(BH_2^{2+})$	$\alpha(BH_3^{3+})$	$\alpha(BH_4^{4+})$	Phi	V_b (mL)
4	0.02	10.9	1.3E-11	7.9E-04	3.1E-02	1.9E-06	4.6E-15	4.6E-25	-0.009	-0.090
5	C_a =	10.0	1.0E-10	1.0E-04	2.0E-01	1.0E-04	1.9E-12	1.5E-21	0.196	1.957
6	0.1	9.0	1.0E-09	1.0E-05	7.1E-01	3.6E-03	6.8E-10	5.4E-18	0.719	7.193
7	V_b =	8.0	1.0E-08	1.0E-06	9.2E-01	4.6E-02	8.8E-08	7.0E-15	1.009	10.094
8	50	7.0	1.0E-07	1.0E-07	6.6E-01	3.3E-01	6.3E-06	5.0E-12	1.330	13.303
9	K_w =	6.0	1.0E-06	1.0E-08	1.7E-01	8.3E-01	1.6E-04	1.3E-09	1.834	18.338
10	1E-14	5.0	1.0E-05	1.0E-09	2.0E-02	9.8E-01	1.9E-03	1.5E-07	1.983	19.830
11	K_{a1} =	4.0	1.0E-04	1.0E-10	2.0E-03	9.8E-01	1.9E-02	1.5E-05	2.024	20.238
12	1.26E-01	3.0	1.0E-03	1.0E-11	1.7E-04	8.4E-01	1.6E-01	1.3E-03	2.235	22.345
13	K_{a2} =	2.0	1.0E-02	1.0E-12	6.5E-06	3.3E-01	6.2E-01	5.0E-02	3.580	35.804
14	5.25E-03	1.7	2.0E-02	5.0E-13	1.9E-06	1.9E-01	7.0E-01	1.1E-01	4.902	49.022
15	K_{a3} =									
16	2.00E-07		C12 = 10^-A20							
17	K_{a4} =		C4 = 10^-B4		D4 = $A$10/C4					
18	3.98E-10		Denominator = ($C4^4+$C4^3*$A$12+$C4^2*$A$12*$A$14							
19	pK_1 =		+$C4*$A$12*$A$14*$A$16+$A$12*$A$14*$A$16*$A$18)							
20	0.90		E4 = $C4*$A$12*$A$14*$A$16/Denominator							
21	pK_2 =		F4 = $C4^2*$A$12*$A$14/Denominator							
22	2.28		G4 = $C4^3*$A$12//Denominator							
23	pK_3 =		H4 = $C4^4/Denominator							
24	6.70		I4 = (E4+2*F4+3*G4+4*H4+(C4-D4)/$A$4)/(1-(C4-D4)/$A$6)							
25	pK_4 =		J4 = I4*$A$4*$A$8/$A$6							
26	9.40									

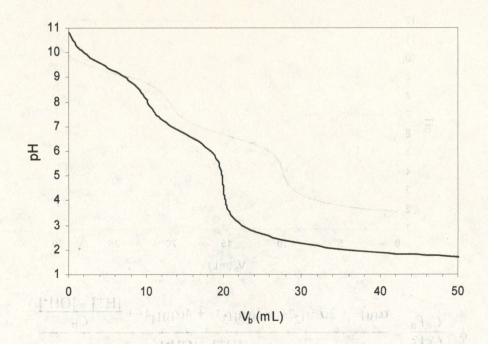

$$V_b \text{ (mL)}$$

10-73. $A_{604} = \varepsilon_{In^-}[In^-](1.00) \Rightarrow [In^-] = \dfrac{0.118}{4.97 \times 10^4} = 2.37 \times 10^{-6} \text{ M}$

Since the indicator was diluted with KOH solution, the formal concentration of indicator is 0.700×10^{-5} M.

$[HIn] = 7.00 \times 10^{-6} - 2.37 \times 10^{-6} = 4.63 \times 10^{-6}$ M

$pH = pK_{In} + \log \dfrac{[In^-]}{[HIn]} = 7.95 + \log \dfrac{2.37}{4.63} = 7.66$

Call benzene-1,2,3-tricarboxylic acid H_3A, with $pK_1 = 2.88$, $pK_2 = 4.75$, and $pK_3 = 7.13$. Since the pH is 7.66, the main species is A^{3-} and the second main species is HA^{2-}. Enough KOH to react with H_3A and H_2A^- must have been added, and there is enough KOH to react with part of the HA^{2-}.

	HA^{2-}	+	OH^-	$\longrightarrow$	A^{3-}	+	H_2O
Initial mmol:	1.00		x		—		
Final mmol:	$1.00 - x$		—		x		

$pH = pK_3 + \log \dfrac{[A^{3-}]}{[HA^{2-}]}$

$7.66 = 7.13 + \log \dfrac{x}{1.00 - x} \Rightarrow x = 0.77_2 \text{ mmol of OH}^-$

The total KOH added is 2.77_2 mmol. The molarity is $\dfrac{2.77_2 \text{ mmol}}{20.0 \text{ mL}} = 0.139$ M.

10-74. The pH of the solution is 7.50, and the total concentration of indicator is 5.00×10^{-5} M. At pH 7.50, there is a negligible amount of H_2In, since $pK_1 = 1.00$. We can write

$$[HIn^-] + [In^{2-}] = 5.0 \times 10^{-5}$$

$$pH = pK_2 + \log \frac{[In^{2-}]}{[HIn^-]}$$

$$7.50 = 7.95 + \log \frac{[In^{2-}]}{5.00 \times 10^{-5} - [In^{2-}]} \Rightarrow [In^{2-}] = 1.31 \times 10^{-5} \text{ M}$$

$$[HIn] = 3.69 \times 10^{-5} \text{ M}$$

$$A_{435} = \varepsilon_{435}[HIn^-] + \varepsilon_{435}[In^{2-}]$$

$$= (1.80 \times 10^4)(3.69 \times 10^{-5}) + (1.15 \times 10^4)(1.31 \times 10^{-5}) = 0.815$$

CHAPTER 11
EDTA TITRATIONS

11-1. The chelate effect is the observation that multidentate ligands form more stable metal complexes than do similar, monodentate ligands.

11-2. $\alpha_{Y^{4-}}$ gives the fraction of all free EDTA in the form Y^{4-}.

    (a) At pH 3.50:

$$\alpha_{Y^{4-}} = \frac{10^{-0.0}10^{-1.5}10^{-2.00}10^{-2.69}10^{-6.13}10^{-10.37}}{(10^{-3.50})^6 + (10^{-3.50})^5 10^{-0.0} + ... + 10^{-0.0}10^{-1.5}...10^{-10.37}} = 2.7 \times 10^{-10}$$

    (b) At pH 10.50:

$$\alpha_{Y^{4-}} = \frac{10^{-0.0}10^{-1.5}10^{-2.00}10^{-2.69}10^{-6.13}10^{-10.37}}{(10^{-10.50})^6 + (10^{-10.50})^5 10^{-0.0} + ... + 10^{-0.0}10^{-1.5}...10^{-10.37}} = 0.57$$

11-3. (a) $K_f' = \alpha_{Y^{4-}} K_f = 0.041 \times 10^{8.79} = 2.5 \times 10^7$

    (b)

$$Mg^{2+} + EDTA \rightleftharpoons MgY^{2-}$$

              x x $0.050 - x$

$$\frac{0.050 - x}{x^2} = 2.5 \times 10^7 \Rightarrow [Mg^{2+}] = 4.5 \times 10^{-5} \, M$$

11-4. $[Ca^{2+}] = 10^{-9.00} \, M$, so essentially all calcium in solution is CaY^{2-}.

$$[CaY^{2-}] = \frac{1.95 \, g}{(200.12 \, g/mol) \, (0.500 \, L)} = 0.019 \, 49 \, M$$

$$K_f' = (0.041)(10^{10.65}) = \frac{[CaY^{2-}]}{[EDTA] \, [Ca^{2+}]} = \frac{(1.949 \times 10^{-2})}{[EDTA] \, (10^{-9.00})}$$

$$\Rightarrow [EDTA] = 0.010 \, 6 \, M$$

Total EDTA needed $=$ mol CaY^{2-} $+$ mol free EDTA

$$= (0.019 \, 49 \, M) \, (0.500 \, L) + (0.010 \, 6 \, M) \, (0.500 \, L) = 0.015 \, 0_4 \, mol$$

$$= 5.60 \, g \, Na_2EDTA \cdot 2 \, H_2O$$

11-5.

Neutral H_5DTPA has 2 carboxylic acid protons and 3 ammonium protons. We are not given the pK_a values, but, by analogy with EDTA, we expect carboxyl pK_a values to be below ~3 and ammonium pK_a values to be above ~6. At pH 14, we expect all acidic protons of DTPA to be dissociated, so the predominant species will be $DTPA^{5-}$. At pH 3-4, nitrogen should be protonated, but carboxyl groups should be deprotonated. The predominant species is probably H_3DTPA^{2-}.

For HSO_4^-, $pK_a = 2.0$. At pH 14 and at pH 3, sulfate is in the form SO_4^{2-}.

At pH 14, $DTPA^{5-}$ is apparently a strong enough ligand to chelate Ba^{2+} and dissolve $BaSO_4(s)$. At pH 3-4, H_3DTPA^{2-} is not a strong enough ligand to dissolve $BaSO_4(s)$. An equivalent statement is that H^+ at a concentration of 10^{-3}-10^{-4} M competes with Ba^{2+} for binding sites on DTPA, but H^+ at a concentration of 10^{-14} M does not compete with Ba^{2+} for binding sites on DTPA.

Now that you have seen my reasoning, I'll provide some more information. The pK_a values for DTPA, beginning with the fully protonated H_8DTPA^{3+}, are

H_8DTPA^{3+}	$pK_1 = -0.1$	CO_2H	H_4DTPA^-	$pK_5 = 2.7$	CO_2H
H_7DTPA^{2+}	$pK_2 = 0.7$	CO_2H	H_3DTPA^{2-}	$pK_6 = 4.3$	NH^+
H_6DTPA^+	$pK_3 = 1.6$	CO_2H	H_2DTPA^{3-}	$pK_7 = 8.6$	NH^+
H_5DTPA	$pK_4 = 2.0$	CO_2H	$HDTPA^{4-}$	$pK_8 = 10.5$	NH^+

As pH is lowered from 14, the three nitrogen atoms are 50% protonated at pH 10.5, 8.6, and 4.3. The third nitrogen atom is not quite fully protonated at pH 3-4. The predominant species is H_3DTPA^{2-}, as I guessed correctly. The species H_4DTPA^- and H_2DTPA^{3-} are also present to some extent in the pH range 3-4.

11-6. (a) mmol EDTA = mmol M^{n+}

$(V_e)(0.0500\ M) = (100.0\ mL)(0.0500\ M) \Rightarrow V_e = 100.0\ mL$

(b) $[M^{n+}] = \left(\dfrac{1}{2}\right) \cdot (0.0500\ M) \cdot \left(\dfrac{100}{150}\right) = 0.01667\ M$

    fraction   original   dilution
    remaining   concentration   factor

(c) 0.041 (Table 11-1)

(d) $K_f' = (0.041)(10^{12.00}) = 4.1 \times 10^{10}$

(e) $[MY^{n-4}] = (0.0500\ M)\left(\dfrac{100}{200}\right) = 0.0250\ M$

$\dfrac{[MY^{n-4}]}{[M^{n+}][EDTA]} = \dfrac{0.0250 - x}{x^2} = 4.1 \times 10^{10} \Rightarrow x = [M^{n+}] = 7.8 \times 10^{-7}\ M$

(f) $[EDTA] = (0.0500\ M)\left(\dfrac{10.0}{210.0}\right) = 2.38 \times 10^{-3}\ M$

$[MY^{n-4}] = (0.0500\ M)\left(\dfrac{100.0}{210.0}\right) = 2.38 \times 10^{-2}\ M$

$\dfrac{[MY^{n-4}]}{[M^{n+}][EDTA]} = \dfrac{(2.38 \times 10^{-2})}{[M^{n+}](2.38 \times 10^{-3})} = 4.1 \times 10^{10} \Rightarrow [M^{n+}] = 2.4 \times 10^{-10}\ M$

11-7. Co^{2+} + EDTA $\rightleftharpoons$ CoY^{2-} $\alpha_{Y^{4-}} K_f$ = $(1.8 \times 10^{-5})(10^{16.45})$ = 5.1×10^{11}

$$V_e = (25.00)\left(\frac{0.020\,26\ M}{0.038\,55\ M}\right) = 13.14\ mL$$

(a)   <u>12.00 mL</u>: $[Co^{2+}]$ = $\left(\frac{13.14 - 12.00}{13.14}\right)(0.020\,26\ M)\left(\frac{25.00}{37.00}\right)$

$$= 1.19 \times 10^{-3}\ M \Rightarrow pCo^{2+} = 2.93$$

(b)   <u>V_e</u>: Formal concentration of CoY^{2-} is $\left(\frac{25.00}{38.14}\right)(0.020\,26\ M)$ = $1.33 \times 10^{-2}\ M$

$$
\begin{array}{ccccc}
Co^{2+} & + & EDTA & \rightleftharpoons & CoY^{2-} \\
x & & x & & 1.33 \times 10^{-2} - x
\end{array}
$$

$$\frac{1.33 \times 10^{-2} - x}{x^2} = \alpha_{Y^{4-}} K_f \Rightarrow x = 1.6 \times 10^{-7}\ M \Rightarrow pCo^{2+} = 6.79$$

(c)   <u>14.00 mL</u>: Formal concentration of CoY^{2-} is $\left(\frac{25.00}{39.00}\right)(0.020\,26\ M)$

$$= 1.30 \times 10^{-2}\ M$$

Formal concentration of EDTA is $\left(\frac{14.0 - 13.14}{39.00}\right)(0.038\,55\ M)$ = $8.50 \times 10^{-4}\ M$

$$[Co^{2+}] = \frac{[CoY^{2-}]}{[EDTA]\,K_f'} = \frac{1.30 \times 10^{-2}}{8.50 \times 10^{-4}\,(5.1 \times 10^{11})} = 3.0 \times 10^{-11}\ M$$

$$\Rightarrow pCo^{2+} = 10.52$$

11-8.   Titration reaction: Mn^{2+} + EDTA $\rightleftharpoons$ MnY^{2-}

$$K_f' = \alpha_{Y^{4-}} K_f = \ = (4.2 \times 10^{-3})(10^{13.89}) = 3.3 \times 10^{11}$$

The equivalence point is 50.0 mL.   Sample calculations:

<u>20.0 mL</u>: The fraction of Mn^{2+} that has reacted is 2/5 and the fraction remaining is 3/5.

$$[Mn^{2+}] = \left(\frac{30.0}{50.0}\right)(0.020\,0\ M)\left(\frac{25.0}{45.0}\right) = 6.67 \times 10^{-3}\ M \Rightarrow pMn^{2+} = 2.18$$

<u>50.0 mL</u>: The formal concentration of MnY^{2-} is

$$[MnY^{2-}] = \left(\frac{25.0}{75.0}\right)(0.020\,0\ M) = 0.006\,67\ M$$

$$
\begin{array}{ccccc}
Mn^{2+} & + & EDTA & \rightleftharpoons & MnY^{2-} \\
x & & x & & 0.006\,67 - x
\end{array}
$$

$$\frac{0.006\,67 - x}{x^2} = \alpha_{Y^{4-}} K_f \Rightarrow x = 1.4 \times 10^{-7} \Rightarrow pMn^{2+} = 6.85$$

<u>60.0 mL</u>: There are 10.0 mL of excess EDTA.

$$[EDTA] = \left(\frac{10.0}{85.0}\right)(0.010\,0\ M) = 1.176 \times 10^{-3}\ M$$

$$[MnY^{2-}] = \left(\frac{25.0}{85.0}\right)(0.020\,0\ M) = 5.88 \times 10^{-3}\ M$$

$$[Mn^{2+}] = \frac{[MnY^{2-}]}{[EDTA]K_f'} = 1.5 \times 10^{-11} \Rightarrow pMn^{2+} = 10.82$$

Volume (mL)	pMn^{2+}	Volume	pMn^{2+}	Volume	pMn^{2+}
0	1.70	49.0	3.87	50.1	8.82
20.0	2.18	49.9	4.87	55.0	10.51
40.0	2.81	50.0	6.85	60.0	10.82

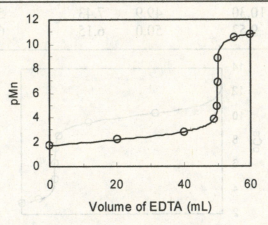

11-9. Titration reaction: $Ca^{2+} + EDTA \rightleftharpoons CaY^{2-}$

$$K_f' = \alpha_{Y^{4-}} K_f = (0.30)(10^{10.65}) = 1.3_4 \times 10^{10}$$

The equivalence point is 50.0 mL. Sample calculations:

<u>20.0 mL</u>: The fraction of EDTA consumed is 2/5.

$$[EDTA] = \left(\frac{30.0}{50.0}\right)(0.020\,0\ M)\left(\frac{25.0}{45.0}\right) = 0.006\,67\ M$$

$$[CaY^{2-}] = \left(\frac{20.0}{50.0}\right)(0.020\,0\ M)\left(\frac{25.0}{45.0}\right) = 0.004\,44\ M$$

$$[Ca^{2+}] = \frac{[CaY^{2-}]}{[EDTA]K_f'} = 4.9_7 \times 10^{-11} \Rightarrow pCa^{2+} = 10.30$$

<u>50.0 mL</u>: The formal concentration of CaY^{2-} is

$$[CaY^{2-}] = \left(\frac{25.0}{75.0}\right)(0.020\,0\text{ M}) = 0.006\,67\text{ M}$$

$$\begin{array}{ccccc}
Ca^{2+} & + & EDTA & \rightleftharpoons & CaY^{2-} \\
x & & x & & 0.006\,67 - x
\end{array}$$

$$\frac{0.006\,67 - x}{x^2} = \alpha_{Y^{4-}} \cdot K_f \Rightarrow x = 7.0_5 \times 10^{-7}\text{ M} \Rightarrow pCa^{2+} = 6.15$$

<u>50.1 mL</u>: There is an excess of 0.1 mL of Ca^{2+}.

$$[Ca^{2+}] = \left(\frac{0.1}{75.1}\right)(0.010\,0\text{ M}) = 1.33 \times 10^{-5}\text{ M} \Rightarrow pCa^{2+} = 4.88$$

Volume (mL)	pCa^{2+}	Volume	pCa^{2+}	Volume	pCa^{2+}
0	(∞)	49.0	8.44	50.1	4.88
20.0	10.30	49.9	7.43	55.0	3.20
40.0	9.52	50.0	6.15	60.0	2.93

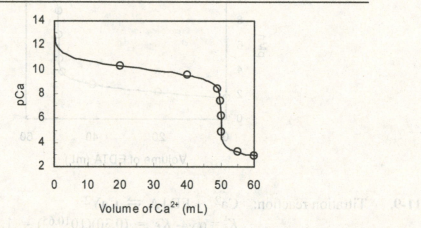

11-10. There is more VO^{2+} than EDTA in this solution.

$$[VO^{2+}] = \left(\frac{0.10}{29.9}\right)(0.010\,0\text{ M}) = 3.34 \times 10^{-5}\text{ M}$$

$$[VOY^{2-}] = \left(\frac{9.90}{29.90}\right)(0.010\,0\text{ M}) = 3.31 \times 10^{-3}\text{ M}$$

K_f for $VOY^{2-} = 10^{18.7}$; pK_6 for $H_6Y^{2+} = 10.37$; pH = 4.00

$$[Y^{4-}] = \frac{[VOY^{2-}]}{[VO^{2+}]\,K_f} = 1.98 \times 10^{-17}\text{ M}$$

$$[HY^{3-}] = \frac{[H^+]\,[Y^{4-}]}{K_6} = 4.6 \times 10^{-11}\text{ M}$$

11-11.

	A	B	C	D	E	F	G
1	Titration of V_M mL of C_M M Cu^{2+} with C(ligand) M EDTA						
2							
3	C_M =	pM	M	Phi	V(EDTA)		
4	0.001	3.0	1.00E-03	0.000	0.000		
5	V_M =	4.0	1.00E-04	0.891	0.891		
6	10	5.0	1.00E-05	0.989	0.989		
7	C(ligand) =	6.0	1.00E-06	0.999	0.999		
8	0.01	7.0	1.00E-07	1.000	1.000		
9	K_f' =	8.0	1.00E-08	1.000	1.000		
10	1.75E+12	9.0	1.00E-09	1.001	1.001		
11	$\alpha(Y^{4-})$=	10.0	1.00E-10	1.006	1.006		
12	2.90E-07	11.0	1.00E-11	1.057	1.057		
13	K_f =	12.0	1.00E-12	1.572	1.572		
14	6.0256E+18	12.3	5.01E-13	2.142	2.142		
15							
16	A10 = A12*A14						
17	C4 = 10^-B4						
18	D4 = (1+$A$10*C4-(C4+C4*C4*$A$10)/$A$4)/(C4*$A$10+(C4+C4*C4*$A$10)/$A$8)						
19	E4 = D4*$A$4*$A$6/$A$8						

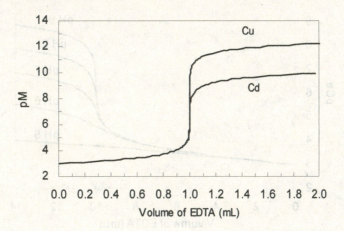

11-12.   The spreadsheet below gives representative calculations for the pH 7.

	A	B	C	D	E	F
1	Titration of 10 mL of 1 mM Ca^{2+} with 1 mM EDTA vs pH					
2	pH 7					
3	C$_M$ =	pM	M	Phi	V(ligand)	
4	0.001	3.000	1.00E-03	0.000	0.000	
5	V$_M$ =	3.250	5.62E-04	0.280	2.801	
6	10	3.500	3.16E-04	0.520	5.196	
7	C(ligand) =	3.750	1.78E-04	0.698	6.982	
8	0.001	4.000	1.00E-04	0.819	8.186	
9	K$_f'$ =	4.500	3.16E-05	0.940	9.404	
10	1.70E+07	5.000	1.00E-05	0.986	9.859	
11	α(Y^{4-}) =	5.500	3.16E-06	1.012	10.121	
12	3.80E-04	6.000	1.00E-06	1.057	10.567	
13	K$_f$ =	6.500	3.16E-07	1.185	11.855	
14	4.4668E+10	7.000	1.00E-07	1.589	15.887	
15	A10 = A12*A14					
16	C4 = 10^-B4					
17	D4 = (1+A$10*C4-(C4+C4*C4*A$10)/A$4)/(C4*A$10+(C4+C4*C4*A$10)/A$8)					
18	E4 = D4*A$4*A$6/A$8					

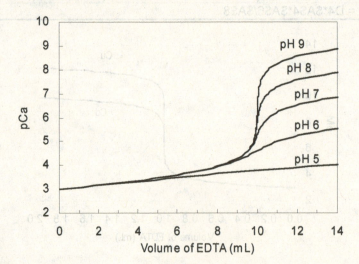

11-13.

	A	B	C	D	E	F	G
1	Titration of EDTA with metal						
2							
3	C_M =	pM	M	Phi	V_M		
4	0.08	14.640	2.29E-15	0.004	0.100		
5	V(ligand) =	12.844	1.43E-13	0.200	5.004		
6	50	12.418	3.82E-13	0.400	10.007		
7	C(ligand) =	12.066	8.59E-13	0.600	15.004		
8	0.04	11.640	2.29E-12	0.800	20.003		
9	K_f' =	10.860	1.38E-11	0.960	24.005		
10	1.75E+12	6.910	1.23E-07	1.000	25.000		
11	$\alpha(Y^{4-})$	2.978	1.05E-03	1.040	25.999		
12	2.90E-07	2.301	5.00E-03	1.200	30.000		
13	K_f =						
14	6.0256E+18						
15							
16	A10 = A12*A14						
17	C4 = 10^-B4						
18	D4 = (C4*$A$10+(C4+C4*C4*$A$10)/$A$8)/(1+$A$10*C4-(C4+C4*C4*$A$10)/$A$4)						
19	E4 = D4*$A$8*$A$6/$A$4						

11-14. An auxiliary complexing agent forms a weak complex with analyte ion, thereby keeping it in solution without interfering with the EDTA titration. For example, NH_3 keeps Zn^{2+} in solution at high pH, but is easily displaced by EDTA.

11-15. (a) $\beta_2 = K_1K_2 = \beta_1K_2 \Rightarrow K_2 = \beta_2/\beta_1 = 10^{3.63} / 10^{2.23} = 10^{1.40} = 25$

(b) $\alpha_{Cu^{2+}} = \dfrac{1}{1+\beta_1[L] + \beta_2[L]^2} = \dfrac{1}{1 + 10^{2.23}(0.100)+10^{3.63}(0.100)^2} = 0.017$

11-16. $Cu^{2+} + Y^{4-} \rightleftharpoons CuY^{2-}$ $K_f = 10^{18.78} = 6.0_3 \times 10^{18}$

$\alpha_{Y^{4-}} = 0.81$ at pH 11.00 (Table 11-1)

For Cu^{2+} and NH_3, Appendix I gives $\log\beta_1 = 3.99$, $\log\beta_2 = 7.33$, $\log\beta_3 = 10.06$, and $\log\beta_4 = 12.03$. Therefore, $\beta_1 = 9.8 \times 10^3$, $\beta_2 = 2.1 \times 10^7$, $\beta_3 = 1.15 \times 10^{10}$, and $\beta_4 = 1.07 \times 10^{12}$.

$\alpha_{Cu^{2+}} = \dfrac{1}{1 + \beta_1(1.00) + \beta_2(1.00)^2 + \beta_3(1.00)^3 + \beta_4(1.00)^4} = 9.2_3 \times 10^{-13}$

$K_f' = \alpha_{Y^{4-}} K_f = 4.8_8 \times 10^{18}$

$K_f'' = \alpha_{Y^{4-}} \alpha_{Cu^{2+}} K_f = 4.5_1 \times 10^6$

Equivalence point = 50.00 mL

(a) At 0 mL, the total concentration of copper is $C_{Cu^{2+}}$ = 0.001 00 M and

$[Cu^{2+}] = \alpha_{Cu^{2+}} C_{Cu^{2+}} = 9.2_3 \times 10^{-16}$ M $\Rightarrow$ pCu^{2+} = 15.03

(b)   At 1.00 mL, $C_{Cu^{2+}} = \left(\dfrac{49.00}{50.00}\right)(0.001\ 00\ M)\left(\dfrac{50.00}{51.00}\right) = 9.61 \times 10^{-4}\ M$

         fraction original dilution
       remaining concentration factor

$[Cu^{2+}] = \alpha_{Cu^{2+}} C_{Cu^{2+}} = 8.8_7 \times 10^{-16}\ M \Rightarrow pCu^{2+} = 15.05$

(c)   At 45.00 mL, $C_{Cu^{2+}} = \left(\dfrac{5.00}{50.00}\right)(0.001\ 00)\left(\dfrac{50.00}{95.00}\right) = 5.26 \times 10^{-5}\ M$

$[Cu^{2+}] = \alpha_{Cu^{2+}} C_{Cu^{2+}} = 5.0_4 \times 10^{-17}\ M \Rightarrow pCu^{2+} = 16.30$

(d)   At the equivalence point, we can write

$$C_{Cu^{2+}} + EDTA \rightleftharpoons CuY^{2-}$$

    x x $\left(\dfrac{50.00}{100.00}\right)(0.001\ 00) - x$

$$\frac{0.000\ 500 - x}{x^2} = K_f'' = 4.5_1 \times 10^6 \Rightarrow x = C_{Cu^{2+}} = 1.04 \times 10^{-5}\ M$$

$[Cu^{2+}] = \alpha_{Cu^{2+}} C_{Cu^{2+}} = 9.6_2 \times 10^{-18}\ M \Rightarrow pCu^{2+} = 17.02$

(e)   Past the equivalence point at 55.00 mL, we can say

$[EDTA] = \left(\dfrac{5.00}{105.00}\right)(0.001\ 00\ M) = 4.76 \times 10^{-5}\ M$

$[CuY^{2-}] = \left(\dfrac{50.00}{105.00}\right)(0.001\ 00\ M) = 4.76 \times 10^{-4}\ M$

$K_f' = \dfrac{[CuY^{2-}]}{[Cu^{2+}][EDTA]} = \dfrac{(4.76 \times 10^{-4})}{[Cu^{2+}](4.76 \times 10^{-5})}$

$\Rightarrow [Cu^{2+}] = 2.05 \times 10^{-18}\ M \Rightarrow pCu^{2+} = 17.69$

11-17.   (a) $\alpha_{ML} = \dfrac{[ML]}{C_M} = \dfrac{\beta_1[M][L]}{[M]\{1+\beta_1[L]+\beta_2[L]^2\}} = \dfrac{\beta_1[L]}{1+\beta_1[L]+\beta_2[L]^2}$

     $\alpha_{ML_2} = \dfrac{[ML_2]}{C_M} = \dfrac{\beta_2[M][L]^2}{[M]\{1+\beta_1[L]+\beta_2[L]^2\}} = \dfrac{\beta_2[L]^2}{1+\beta_1[L]+\beta_2[L]^2}$

(b)   For $[L] = 0.100\ M$, $\beta_1 = 1.7 \times 10^2$, and $\beta_2 = 4.3 \times 10^3$, we get
$\alpha_{ML} = 0.28$ and $\alpha_{ML_2} = 0.70$.

11-18.   Let T = transferrin

(a) $Fe^{3+} + T \overset{K_1}{\rightleftharpoons} FeT$ $K_1 = \dfrac{[FeT]}{[Fe^{3+}][T]}$

   $Fe^{3+} + FeT \overset{K_2}{\rightleftharpoons} Fe_2T$ $K_2 = \dfrac{[Fe_2T]}{[Fe^{3+}][FeT]}$

(b) $K_1 = \dfrac{[Fe_aT] + [Fe_bT]}{[Fe^{3+}][T]} = \dfrac{[Fe_aT]}{[Fe^{3+}][T]} + \dfrac{[Fe_bT]}{[Fe^{3+}][T]} = k_{1a} + k_{1b}$

$$\frac{1}{K_2} = \frac{[Fe^{3+}]([Fe_aT] + [Fe_bT])}{[Fe_2T]} = \frac{[Fe^{3+}][Fe_aT]}{[Fe_2T]} + \frac{[Fe^{3+}][Fe_bT]}{[Fe_2T]} = \frac{1}{k_{2b}} + \frac{1}{k_{2a}}$$

(c) $k_{1a} k_{2b} = \frac{[\cancel{Fe_aT}]}{[Fe^{3+}][T]} \frac{[Fe_2T]}{[Fe^{3+}] [\cancel{Fe_aT}]} = \frac{[\cancel{Fe_bT}]}{[Fe^{3+}][T]} \frac{[Fe_2T]}{[Fe^{3+}] [\cancel{Fe_bT}]} = k_{1b} k_{2a}$

(d)   Substituting from Eq. (A) into Eq. (C) gives

$$19._{44} = \frac{[FeT]^2}{(1 - [FeT] - [Fe_2T]) [Fe_2T]} \tag{D}$$

Substituting from Eq. (B) into Eq. (D) gives

$$19._{44} = \frac{(0.8 - 2[Fe_2T])^2}{\{1-(0.8-2[Fe_2T]) - [Fe_2T]\} [Fe_2T]} \quad \underset{\substack{\text{quadratic}\\\text{equation}}}{\overset{\text{solve}}{\Rightarrow}} \quad [Fe_2T] = 0.077_3$$

Using this value for $[Fe_2T]$ in Eqns. (A) and (B) gives $[FeT] = 0.645$ and

$[T] = 0.277_3$. Now we also know that $\frac{k_{1a}}{k_{1b}} = \frac{[Fe_aT]}{[Fe_bT]} = 6.0$, which tells us

that $[Fe_aT] = \left(\frac{6.0}{7.0}\right) [FeT] = 0.553_2$ and $[Fe_bT] = \left(\frac{1.0}{7.0}\right) [FeT] = 0.092_2$.

Final result: $[T] = 0.27_7$, $[Fe_aT] = 0.55_3$, $[Fe_bT] = 0.09_2$, $[Fe_2T] = 0.07_7$.

11-19.   In place of Equation 11-8, we write

$$M_{free} + EDTA \rightleftharpoons M(EDTA) \qquad\qquad K_f'' = \frac{[M(EDTA)]}{[M]_{free} [EDTA]}$$

where $[M]_{free}$ is the concentration of all metal not bound to EDTA. [EDTA] is
the concentration of all EDTA not bound to metal. The mass balances are

Metal: $[M]_{free} + [M(EDTA)] = \dfrac{C_M V_M}{V_M + V_{EDTA}}$

EDTA: $[EDTA] + [M(EDTA)] = \dfrac{C_{EDTA} V_{EDTA}}{V_M + V_{EDTA}}$

These equations have the same form as the first three equations in Section 11-4,
with K_f replaced by K_f'', [M] replaced by $[M]_{free}$, and [L] replaced by [EDTA].
The derivation therefore leads to Equation 11-11, with K_f replaced by K_f'', [M]
replaced by $[M]_{free}$, and C_L replaced by C_{EDTA}.

11-20. (a)

	A	B	C	D	E	F
1	Titration of 50 mL of 0.001 M Zn^{2+} with 0.001 M EDTA/pH 10 with NH_3					
2						
3	$C_M =$		pM	M	$[M]_{tot}$	ϕ	V_{EDTA}
4	0.001		8.115	7.67E-09	4.29E-04	0.400	19.9814
5	$V_M =$		12.014	9.68E-13	5.41E-08	1.000	50.0000
6	50		15.278	5.27E-16	2.95E-11	1.200	59.9965
7	$C_{EDTA} =$						
8	0.001						
9	$K_f'' =$		A10 = A12*A16*10^A14				
10	1.70E+11		A12 = 1/(1+A20*A18+B20*A18^2+C20*A18^3+D20*A18^4)				
11	$\alpha(Zn^{2+}) =$						
12	1.79E-05		C4 = 10^-B4				
13	$\log K_f =$		D4 = C4/\$A\$12				
14	16.5		E4 = (1+\$A\$10*D4-(D4+D4^2*\$A\$10)/\$A\$4)/				
15	$\alpha(Y^{4-}) =$		(D4*\$A\$10+(D4+D4^2*\$A\$10)/\$A\$8)				
16	0.30		F4 = E4*\$A\$4*\$A\$6/\$A\$8				
17	$[NH_3] =$						
18	0.1						
19	$\beta_1 =$		$\beta_2 =$	$\beta_3 =$	$\beta_4 =$		
20	1.51E+02		2.69E+04	5.50E+06	5.01E+08		

(b)

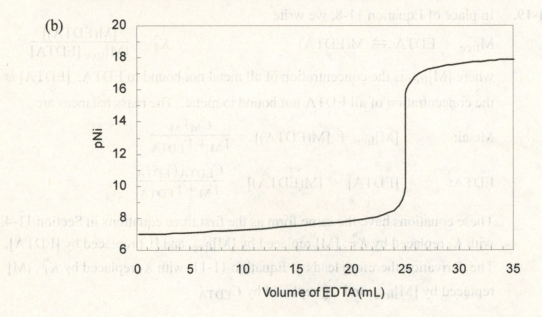

	A	B	C	D	E	F
1	Titration of 50 mL of 0.05 M Ni^{2+} with 0.1 M EDTA / pH 11 / 0.1 M Oxalate					
2						
3	C_M =	pM	M	$[M]_{tot}$	ϕ	V_{EDTA}
4	0.005	6.97	1.07E-07	4.94E-03	0.008	0.210
5	V_M =	7.00	1.00E-07	4.61E-03	0.054	1.342
6	50	7.20	6.31E-08	2.91E-03	0.324	8.106
7	C_{EDTA} =	7.50	3.16E-08	1.46E-03	0.618	15.461
8	0.01	8.00	1.00E-08	4.61E-04	0.868	21.696
9	K_f'' =	8.40	3.98E-09	1.83E-04	0.946	23.649
10	4.42E+13	8.80	1.58E-09	7.30E-05	0.978	24.456
11	$\alpha(Ni^{2+})$ =	9.50	3.16E-10	1.46E-05	0.996	24.891
12	2.17E-05	10.50	3.16E-11	1.46E-06	1.000	24.989
13	log K_f =	12.80	1.58E-13	7.30E-09	1.000	25.000
14	18.4	14.00	1.00E-14	4.61E-10	1.000	25.001
15	$\alpha(Y^{4-})$ =	15.00	1.00E-15	4.61E-11	1.000	25.012
16	0.81	16.00	1.00E-16	4.61E-12	1.005	25.123
17	$[Oxalate^{2-}]$ =	17.00	1.00E-17	4.61E-13	1.049	26.229
18	0.1	17.40	3.98E-18	1.83E-13	1.123	28.086
19	β_1 =	17.60	2.51E-18	1.16E-13	1.196	29.892
20	1.45E+05	17.80	1.58E-18	7.30E-14	1.310	32.753
21	β_2 =	17.90	1.26E-18	5.80E-14	1.390	34.760
22	3.16E+06	18.00	1.00E-18	4.61E-14	1.491	37.287
23						
24	A10 = A16*A12*10^A14					
25	A12 = 1/(1+A20*A18+A22*A18^2)					
26	C4 = 10^-B4					
27	D4 = C4/$A$12					
28	E4 = (1+$A$10*D4-(D4+D4*D4*$A$10)/$A$4)/					
29	(D4*$A$10+(D4+D4*D4*$A$10)/$A$8)					
30	F4 = E4*$A$4*$A$6/$A$8					

11-21.
$$[L] + [ML] + 2[ML_2] = \frac{C_L V_L}{V_M + V_L}$$

$$[L] + \alpha_{ML}\frac{C_M V_M}{V_M + V_L} + 2\alpha_{ML_2}\frac{C_M V_M}{V_M + V_L} = \frac{C_L V_L}{V_M + V_L}$$

Multiply both sides by $V_M + V_L$:

$$[L]V_M + [L]V_L + \alpha_{ML}C_M V_M + 2\alpha_{ML_2}C_M V_M = C_L V_L$$

Collect terms $\quad V_L([L] - C_L) = V_M(-[L] - \alpha_{ML}C_M - 2\alpha_{ML_2}C_M)$

$$\frac{V_L}{V_M} = \frac{[L] + \alpha_{ML}C_M + 2\alpha_{ML_2}C_M}{C_L - [L]}$$

Divide the denominator by C_L and divide the numerator by C_M to obtain ϕ, the fraction of the way to the equivalence point:

$$\phi = \frac{C_L V_L}{C_M V_M} = \frac{\dfrac{[L]}{C_M} + \alpha_{ML} + 2\alpha_{ML_2}}{1 - \dfrac{[L]}{C_L}}$$

11-22.

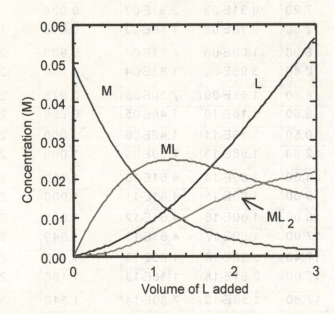

	A	B	C	D	E	F	G	H	I	J	K
1	Copper-acetate complexes ML and ML$_2$										
2											
3	$C_M =$	pL	[L]	α_M	α_{ML}	α_{ML_2}	ϕ	V(ligand)	[M]	[ML]	[ML$_2$]
4	0.05	4.0	0.0001	0.983	0.017	0.000	0.019	0.019	0.0491	0.0008	0.0000
5	$V_M =$	3.0	0.0010	0.852	0.145	0.004	0.172	0.172	0.0419	0.0071	0.0002
6	10	2.8	0.0016	0.781	0.210	0.008	0.260	0.260	0.0381	0.0103	0.0004
7	C(ligand)	2.6	0.0025	0.688	0.294	0.019	0.383	0.383	0.0331	0.0141	0.0009
8	0.5	2.4	0.0040	0.573	0.388	0.039	0.550	0.550	0.0272	0.0184	0.0019
9	$\beta_1 =$	2.2	0.0063	0.446	0.478	0.076	0.766	0.766	0.0207	0.0222	0.0035
10	170	2.0	0.0100	0.319	0.543	0.137	1.039	1.039	0.0145	0.0246	0.0062
11	$\beta_2 =$	1.9	0.0126	0.262	0.560	0.178	1.199	1.199	0.0117	0.0250	0.0080
12	4300	1.8	0.0158	0.209	0.564	0.226	1.377	1.377	0.0092	0.0248	0.0099
13		1.7	0.0200	0.164	0.556	0.280	1.579	1.579	0.0071	0.0240	0.0121
14		1.6	0.0251	0.125	0.535	0.340	1.808	1.808	0.0053	0.0226	0.0144
15		1.5	0.0316	0.094	0.504	0.403	2.073	2.073	0.0039	0.0209	0.0167
16		1.4	0.0398	0.069	0.464	0.467	2.385	2.385	0.0028	0.0187	0.0189
17		1.3	0.0501	0.049	0.419	0.532	2.761	2.761	0.0019	0.0164	0.0208
18		1.3	0.0562	0.041	0.396	0.563	2.981	2.981	0.0016	0.0152	0.0217
19											
20	C4 = 10^-B4						H4 = G4*$A$4*$A$6/$A$8				
21	D4 = 1/(1+$A$10*C4+$A$12*C4*C4)						I4 = D4*$A$4*$A$6/($A$6+H4)				
22	E4 = $A$10*C4/(1+$A$10*C4+$A$12*C4*C4)						J4 = E4*$A$4*$A$6/($A$6+H4)				
23	F4 = $A$12*C4*C4/(1+$A$10*C4+$A$12*C4*C4)						K4 = F4*$A$4*$A$6/($A$6+H4)				
24	G4 = (C4/$A$4+E4+2*F4)/(1-C4/$A$8)										

11-23. Only a small amount of indicator is employed. Most of the Mg^{2+} is not bound to the indicator. Free Mg^{2+} reacts with EDTA before MgIn reacts. Therefore, the concentration of MgIn is constant until all of Mg^{2+} has been consumed. Only when MgIn begins to react does the color change.

11-24. 1. With metal ion indicators
2. With a mercury electrode
3. With an ion-selective electrode
4. With a glass electrode

11-25. HIn^{2-}, wine red, blue

11-26. Buffer (a) (pH 6-7) will give a yellow $\rightarrow$ blue color change that will be easier to observe than the violet $\rightarrow$ blue change expected with the other buffers.

11-27. A back titration is necessary if the analyte precipitates in the absence of EDTA, if it reacts too slowly with EDTA, or if it blocks the indicator.

11-28. In a displacement titration, analyte displaces a metal ion from a complex. The displaced metal ion is then titrated. An example is the liberation of Ni^{2+} from $Ni(CN)_4^{2-}$ by the analyte Ag^+. The liberated Ni^{2+} is then titrated by EDTA to find out how much Ag^+ was present.

11-29. The Mg^{2+} in a solution of Mg^{2+} and Fe^{3+} can be titrated by EDTA if the Fe^{3+} is masked with CN^- to form $Fe(CN)_6^{3-}$, which does not react with EDTA.

11-30. Hardness refers to the total concentration of alkaline earth cations in water, which normally means $[Ca^{2+}] + [Mg^{2+}]$. Hardness gets its name from the reaction of these cations with soap to form insoluble curds. Temporary hardness, due to $Ca(HCO_3)_2$, is lost by precipitation of $CaCO_3(s)$ upon heating. Permanent hardness derived from other salts, such as $CaSO_4$, is not affected by heat.

11-31. $(50.0 \text{ mL})(0.010\,0 \text{ mmol/mL}) = 0.500 \text{ mmol } Ca^{2+}$, which requires 0.500 mmol EDTA = 10.0 mL EDTA.
0.500 mmol Al^{3+} requires the same amount of EDTA, 10.0 mL.

11-32. mmol EDTA = mmol Ni^{2+} + mmol Zn^{2+}
$$1.250 = x + 0.250 \Rightarrow 1.000 \text{ mmol } Ni^{2+} \text{ in } 50.0 \text{ mL} = 0.020\,0 \text{ M}$$

11-33. The formula mass of $MgSO_4$ is 120.37. The 50.0 mL aliquot contains

$$\left(\frac{50.0 \text{ mL}}{500 \text{ mL}}\right)\left(\frac{0.450 \text{ g}}{120.37 \text{ g/mol}}\right) = 0.373\,8 \text{ mmol of } Mg^{2+}$$

37.6 mL of EDTA reacts with this much Mg^{2+}, so the EDTA solution contains
0.373 8 mmol / 37.6 mL = 9.943×10^{-3} mmol/mL. The formula mass of $CaCO_3$
is 100.09. 1.00 mL of EDTA will react with 9.943×10^{-3} mmol of $CaCO_3$ =
0.995 mg.

11-34. 30.10 mL Ni^{2+} reacted with 39.35 mL 0.013 07 M EDTA, so Ni^{2+} molarity is

$$[Ni^{2+}] = \frac{(39.35 \text{ mL})(0.013\,07 \text{ mol/L})}{30.10 \text{ mL}} = 0.017\,09 \text{ M}.$$

25.00 mL Ni^{2+} contains 0.427 2 mmol Ni^{2+}. 10.15 mL EDTA = 0.132 7 mmol
EDTA. The Ni^{2+} which must have reacted with CN^- was 0.427 2 − 0.132 7 =
0.294 5 mmol. Cyanide reacting with Ni^{2+} must have been (4)(0.294 5 mmol) =
1.178 mmol. Original $[CN^-]$ = 1.178 mmol/12.73 mL = 0.092 54 M.

11-35. For 1.00 mL of unknown:

$$\begin{aligned}
25.00 \text{ mL of EDTA} &= 0.968\,0 \text{ mmol} \\
- \quad 23.54 \text{ mL of } Zn^{2+} &= 0.500\,7 \text{ mmol} \\
\hline
Co^{2+} + Ni^{2+} &= 0.467\,3 \text{ mmol}
\end{aligned}$$

For 2.000 mL of unknown:

$$\begin{aligned}
25.00 \text{ mL of EDTA} &= 0.968\,0 \text{ mmol} \\
- \quad 25.63 \text{ mL of } Zn^{2+} &= 0.545\,2 \text{ mmol} \\
\hline
Ni^{2+} \text{ in 2.000 mL} &= 0.422\,8 \text{ mmol}
\end{aligned}$$

Co^{2+} in 2.000 mL of unknown = 2 (0.467 3) − 0.422 8 = 0.511 8 mmol. The Co^{2+}
will react with 0.511 8 mmol of EDTA, leaving 0.968 0 − 0.511 8 = 0.456 2 mmol

EDTA.   mL Zn needed = $\dfrac{0.456\,2 \text{ mmol}}{0.021\,27 \text{ mmol/mL}}$ = 21.45 mL

11-36.

$$\begin{aligned}
\text{Total EDTA} &= (25.0 \text{ mL}) (0.045\,2 \text{ M}) = 1.130 \text{ mmol} \\
- \quad Mg^{2+} \text{ required} &= (12.4 \text{ mL}) (0.012\,3 \text{ M}) = 0.153 \text{ mmol} \\
\hline
Ni^{2+} + Zn^{2+} &= 0.977 \text{ mmol}
\end{aligned}$$

Zn^{2+} = EDTA displaced by 2,3-dimercapto-1-propanol

$$= (29.2 \text{ mL}) (0.012\,3 \text{ M}) = 0.359 \text{ mmol}$$

$$\Rightarrow \text{Ni}^{2+} = 0.977 - 0.359 = 0.618 \text{ mmol}; \quad [\text{Ni}^{2+}] = \frac{0.618 \text{ mmol}}{50.0 \text{ mL}} = 0.012\,4 \text{ M}$$

$$[\text{Zn}^{2+}] = \frac{0.359 \text{ mmol}}{50.0 \text{ mL}} = 0.007\,18 \text{ M}$$

11-37. The precipitation reaction is $\text{Cu}^{2+} + \text{S}^{2-} \rightarrow \text{CuS} \,(s)$.

Total Cu^{2+} used $= (25.00 \text{ mL}) (0.043\,32 \text{ M}) = 1.083\,0 \text{ mmol}$

$\overline{}$ Excess Cu^{2+} $= (12.11 \text{ mL}) (0.039\,27 \text{ M}) = 0.475\,6 \text{ mmol}$

 mmol of S^{2-} $= 0.607\,4 \text{ mmol}$

$$[\text{S}^{2-}] = 0.607\,4 \text{ mmol}/25.00 \text{ mL} = 0.024\,30 \text{ M}$$

11-38. mmol Bi in reaction $= (25.00 \text{ mL}) (0.086\,40 \text{ M}) = 2.160 \text{ mmol}$

EDTA required $= (14.24 \text{ mL}) (0.043\,7 \text{ M}) = 0.622 \text{ mmol}$

mmol Bi that reacted with Cs $= 2.160 - 0.622 = 1.538 \text{ mmol}$

Since 2 mol Bi react with 3 mol Cs to give $\text{Cs}_3\text{Bi}_2\text{I}_9$,

mmol Cs^+ in unknown $= \frac{3}{2} (1.538) = 2.307 \text{ mmol}$

$$[\text{Cs}^+] = \frac{2.307 \text{ mmol}}{25.00 \text{ mL}} = 0.092\,28 \text{ M}.$$

11-39. Total standard $\text{Ba}^{2+} + \text{Zn}^{2+}$ added to the sulfate was $(5.000 \text{ mL})(0.014\,63 \text{ M}$ $\text{BaCl}_2) + (1.000 \text{ mL})(0.010\,00 \text{ M ZnCl}_2) = 0.083\,15 \text{ mmol}$. Total EDTA required was $(2.39 \text{ mL})(0.009\,63 \text{ M}) = 0.023\,0_2 \text{ mmol}$. Therefore, the original solid must have contained $0.083\,15 - 0.023\,0_2 = 0.060\,1_3 \text{ mmol}$ sulfur (which made $0.060\,1_3$ mmol sulfate that precipitated $0.060\,1_3$ mmol Ba^{2+}). The mass of sulfur was $(0.060\,1_3 \text{ mmol})(32.066 \text{ mg/mmol}) = 1.92_8$ mg. wt% S $= 100 \times$ $(1.92_8 \text{ mg S}/5.89 \text{ mg sphalerite}) = 32.7$ wt%. Theoretical wt% S in pure ZnS $=$ $100 \times (32.066 \text{ g S} / 97.46 \text{ g ZnS}) = 32.90$ wt%.

CHAPTER 12
ADVANCED TOPICS IN EQUILIBRIUM

12-1. As pH is lowered, $[H^+]$ increases. H^+ reacts with basic anions to increase the solubility of their salts. Dissolution of minerals such as galena and cerussite increases the concentration of Pb^{2+} in the environment.

Galena: $PbS(s) + H^+ \rightleftharpoons Pb^{2+} + HS^-$

Cerussite: $PbCO_3(s) + H^+ \rightleftharpoons Pb^{2+} + HCO_3^-$

12-2. (a) Hydroxybenzene = HA with $pK_{HA} = 9.997$

Mixture contains 0.010 0 mol HA and 0.005 0 mol KOH in 1.00 L.

Chemical reactions:

$HA \rightleftharpoons A^- + H^+$ $K_{HA} = \dfrac{[H^+][A^-]}{[HA]} = 10^{-9.997}$

$H_2O \rightleftharpoons H^+ + OH^-$ $K_w = [H^+][OH^-] = 10^{-14.00}$

Charge balance:

$[H^+] + [K^+] = [OH^-] + [A^-]$

Mass balances:

$[K^+] = 0.005\ 0\ M$

$[HA] + [A^-] = 0.010\ 0\ M = F_A$

We have 5 equations and 5 chemical species.

Fractional composition equations:

$[HA] = \alpha_{HA} F_A = \dfrac{[H^+]F_A}{[H^+] + K_{HA}}$

$[A^-] = \alpha_{A^-} F_A = \dfrac{K_{HA}F_A}{[H^+] + K_{HA}}$

Substitute concentration expressions into the charge balance:

$[H^+] + [0.005\ 0] = K_w/[H^+] + \alpha_{A^-} F_A$ (A)

We could solve Equation A for $[H^+]$ by using the solution to a quadratic equation. Instead, we will use Solver in the following spreadsheet, with an initial guess of pH = 10 in cell H9. Select Solver and choose Options. Set Precision to 1e-16 and click OK. In the Solver window, Set Target Cell <u>E12</u> Equal To Value of <u>0</u> By Changing Cells <u>H9</u>. Click Solve and Solver finds pH = 9.98 in cell H9, giving a net charge near 0 in cell E12.

	A	B	C	D	E	F	G	H	I
1	Mixture of 0.010 M HA and 0.005 M NaOH								
2									
3	$F_A =$	0.010		$[K^+] =$	0.005				
4	$pK_{HA} =$	9.997		$pK_w =$	14.000				
5	$K_{HA} =$	1.01E-10		$K_w =$	1.00E-14				
6									
7	Species in charge balance:						Other concentrations:		
8	$[H^+] =$	1.05E-10		$[A^-] =$	4.90E-03		$[HA] =$	5.10E-03	
9	$[K^+] =$	5.00E-03		$[OH^-] =$	9.56E-05		pH =	9.980	
10							↑ initial value is a guess		
11									
12	Positive charge minus negative charge				-3.00E-18		= B8+B9-E8-E9		
13	Formulas:								
14	B5 = 10^-B4			B8 = 10^-H9					
15	E5 = 10^-E4			B9 = E3					
16	E8 = B5*B3/(B8+B5)			E9 = E5/B8					
17	H8 = B8*B3/(B8+B5)								

(b) In Chapter 8, we would have said that there is enough KOH to neutralize half of the HA. Therefore, $[HA] = [A^-]$.

$pH = pK_a + \log([A^-]/[HA]) = pK_a + \log(1) = pK_a = 10.00$. The systematic treatment of equilibrium in the spreadsheet gave pH = 9.98.

(c) If we dilute HA to 0.000 10 M and KOH to 0.000 50 in cells B3 and E3, then Solver finds a pH of 9.45. It makes sense that as the solution becomes more dilute, the pH must move toward 7.

12-3. We use effective equilibrium constants, K', defined as follows:

$$K_{HA} = \frac{[H^+]\gamma_{H^+}[A^-]\gamma_{A^-}}{[HA]\gamma_{HA}} \Rightarrow K'_{HA} = \frac{[H^+][A^-]}{[HA]} = K_{HA}\frac{\gamma_{HA}}{\gamma_{H^+}\gamma_{A^-}}$$

$$K_w = [H^+]\gamma_{H^+}[OH^-]\gamma_{OH^-} = 10^{-13.995}$$

$$K'_w = \frac{K_w}{\gamma_{H^+}\gamma_{OH^-}} = [H^+][OH^-] \Rightarrow [OH^-] = K'_w/[H^+]$$

$$pH = -\log([H^+]\gamma_{H^+})$$

The following spreadsheet shows the beginning of the first iteration with an initial ionic strength of 0 in cell C17. Execute Solver to find the pH in cell H13 that produces a net charge of 0 in cell E16. Solver finds pH = 9.98 and ionic strength = 0.005 0 in cell C18. Write 0.0050 in cell C17 and execute Solver again to find pH = 9.95 and ionic strength = 0.0050 in cell C18. Since the ionic strength did not change after the 2nd iteration, we are finished. The pH is 9.95.

	A	B	C	D	E	F	G	H	I
1	Mixture of 0.010 M HA and 0.005 M NaOH with activity coefficients								
2									
3	F_A =	0.010		[K$^+$] =	0.005				
4	pK_{HA} =	9.997		pK_w =	13.995				
5	K_{HA}' =	1.01E-10		K_w' =	1.01E-14				
6									
7	Activity coefficients:								
8	H$^+$ =	1.00		A$^-$	1.00				
9	OH$^-$ =	1.00		HA	1.00				
10									
11	Species in charge balance:						Other concentrations:		
12	[H$^+$] =	1.00E-10		[A$^-$] =	5.02E-03		[HA] =	4.98E-03	
13	[K$^+$] =	0.005		[OH$^-$] =	1.01E-04		pH =	10.000	
14							↑ initial value is a guess		
15									
16	Positive charge minus negative charge =				-1.18E-04		= B12+B13-E12-E13		
17	Ionic strength =		0.0000	← initial value is 0					
18	New ionic strength =		0.0051	← substitute this value into cell C17 for next iteration					
19									
20	Formulas:								
21	B5 = (10^-B4)*E9/(B8*E8)				E9 = 1		B13 = E3		
22	E5 = (10^-E4)/(B8*B9)								
23	B8 = B9 = E8 = 10^(-0.51*1^2*(SQRT($C$17)/(1+SQRT($C$17))-0.3*$C$17))								
24	B12 = (10^-H13)/B8								
25	E12 = B5*B3/(B12+B5)				E13 = E5/B12				
26	C18 = 0.5*(B12+B13+E12+E13)				H12 = B12*B3/(B12+B5)				

12-4. Abbreviating the protonated form of glycine as H_2G^+, we write

$$H_2G^+ \rightleftharpoons HG + H^+ \qquad K_1 = \frac{[HG]\gamma_{HG}[H^+]\gamma_{H^+}}{[H_2G^+]\gamma_{H_2G^+}} \qquad pK_1 = 2.350$$

$$HG \rightleftharpoons G^- + H^+ \qquad K_2 = \frac{[G^-]\gamma_{G^-}[H^+]\gamma_{H^+}}{[HG]\gamma_{HG}} \qquad pK_2 = 9.778$$

At $\mu = 0.1$ M, the activity coefficient of a monovalent ion is 0.78. The activity coefficient of a neutral molecule is 1. Putting these coefficients into the expressions for K_1 and K_2 gives

$$K_1' = \frac{[HG][H^+]}{[H_2G^+]} \text{ (at } \mu = 0.1 \text{ M)} = K_1\frac{\gamma_{H_2G^+}}{\gamma_{HG}\gamma_{H^+}} = 10^{-2.350}\frac{0.78}{(1)(0.78)} = 10^{-2.350}$$

$$K_2' = \frac{[G^-][H^+]}{[HG]} \text{ (at } \mu = 0.1 \text{ M)} = K_2\frac{\gamma_{HG}}{\gamma_{G^-}\gamma_{H^+}} = 10^{-9.778}\frac{1}{(0.78)(0.78)} = 10^{-9.562}$$

The predicted values are $pK_1' = 2.350$ and $pK_2' = 9.562$. Values from fitting the data in the spreadsheet are 2.312 and 9.625. The change from pK_1 to pK_1' is expected to be zero and it is observed to be -0.038. The change from pK_2 to pK_2' is expected to be -0.216 and it is observed to be -0.153.

12-5. Ethylenediamine = B from diprotic H_2B^{2+} $pK_1 = 6.848$ $pK_2 = 9.928$

Mixture contains 0.100 mol B and 0.035 mol HBr in 1.00 L.

Chemical reactions:

$H_2B^{2+} \rightleftharpoons HB^+ + H^+$ $K_1 = 10^{-6.848}$

$HB^+ \rightleftharpoons B + H^+$ $K_2 = 10^{-9.928}$

$H_2O \rightleftharpoons H^+ + OH^-$ $K_w = 10^{-14.00}$

Charge balance:

$[H^+] + 2[H_2B^{2+}] + [HB^+] = [OH^-] + [Br^-]$

Mass balances:

$[Br^-] = 0.035\ M; \quad [H_2B^+] + [HB] + [B] = 0.100\ M = F_B$

We have 6 equations and 6 chemical species, so there is enough information.

Fractional composition equations:

$$[H_2B^{2+}] = \alpha_{H_2B^{2+}} F_B = \frac{[H^+]^2 F_B}{[H^+]^2 + [H^+]K_1 + K_1K_2}$$

$$[HB^+] = \alpha_{HB^+} F_B = \frac{K_1[H^+]F_B}{[H^+]^2 + [H^+]K_1 + K_1K_2}$$

$$[B] = \alpha_B F_B = \frac{K_1K_2F_B}{[H^+]^2 + [H^+]K_1 + K_1K_2}$$

Substitute into charge balance:

$[H^+] + 2\alpha_{H_2B^{2+}} F_B + \alpha_{HB^+} F_B = K_w/[H^+] + [0.035\ M]$ (A)

We solve Equation A for $[H^+]$ by using Solver in the following spreadsheet, with an initial guess of pH = 10 in cell H11. Select Solver and choose Options. Set Precision to 1E-16 and click OK. In the Solver window, Set Target Cell E14 Equal To Value of 0 By Changing Cells H11. Click Solve and Solver finds pH = 10.194 in cell H11, giving a net charge of $\sim 10^{-17}$ in cell E14.

	A	B	C	D	E	F	G	H	I
1	Mixture of 0.100 M B and 0.035 M HBr								
2									
3	F_B =	0.100		[Br⁻] =	0.035				
4	pK_1 =	6.848		pK_w =	14.000				
5	pK_2 =	9.928		K_w =	1.00E-14				
6	K_1 =	1.42E-07							
7	K_2 =	1.18E-10							
8									
9	Concentrations:								
10	[H⁺] =	6.39E-11		[H₂B²⁺] =	1.58E-05		[B] =	6.49E-02	
11	[Br⁻] =	3.50E-02		[HB⁺] =	3.51E-02		pH =	10.194	
12	[OH⁻] =	1.56E-04					↑ initial value is a guess		
13									
14	Positive charge minus negative charge				-1.70E-17				
15	Formulas:								
16	B6 = 10^-B4		B7 = 10^-B5		E5 = 10^-E4				
17	B10 = 10^-H11		B11 = E3		B12 = E5/B10				
18	E10 = B10^2*$B$3/(B10^2+B10*$B$6+$B$6*$B$7)								
19	E11 = B10*$B$6*$B$3/(B10^2+B10*$B$6+$B$6*$B$7)								
20	H10 = $B$6*$B$7*$B$3/(B10^2+B10*$B$6+$B$6*$B$7)								

In your earlier life, you would have solved this problem by noting that 0.035 mol HBr converts 0.035 mol B into 0.035 mol HB⁺, leaving (0.100 − 0.035) mol B.

$$pH = pK_2 + \log \frac{[B]}{[HB^+]} = 9.928 + \log \frac{0.065}{0.035} = 10.197 \text{ (close to spreadsheet answer)}$$

12-6. Benzene-1,2,3-tricarboxylic acid = H_3A with pK_1 = 2.86, pK_2 = 4.30, pK_3 = 6.28; Imidazole = HB from diprotic H_2B^+ with pK_1 = 6.993, pK_2 = 14.5

Mixture contains 0.040 mol H_3A, 0.030 mol HB, and 0.035 mol NaOH in 1.00 L.

Charge balance:

$$[H^+] + [H_2B^+] + [Na^+] = [OH^-] + [H_2A^-] + 2[HA^{2-}] + 3[A^{3-}] + [B^-]$$

Substitute fractional composition equations into charge balance:
$$[H^+] + \alpha_{H_2B^+} F_B + [0.035]$$
$$= K_w/[H^+] + \alpha_{H_2A^-} F_A + 2\alpha_{H_2A^{2-}} F_A + 3\alpha_{HA^{3-}} F_A + \alpha_{B^-} F_B$$

We solve for [H⁺] with the following spreadsheet, with an initial guess of pH = 7 in cell H14. Select Solver and choose Options. Set Precision to 1e-16 and click OK. In Solver, Set Target Cell E16 Equal To Value of 0 By Changing Cells H14. Click Solve and Solver finds pH = 4.52 in cell H14, giving a net charge of ~10⁻¹⁶ in cell E16.

	A	B	C	D	E	F	G	H	I
1	Mixture of 0.040 M H_3A, 0.030 M HB, and 0.035 M NaOH								
2									
3	$F_A =$	0.040		$F_B =$	0.030		$[Na^+] =$	0.035	
4	$pK_1 =$	2.86		$pK_{H2B} =$	6.993		$pK_w =$	14.000	
5	$pK_2 =$	4.30		$pK_{HB} =$	14.5		$K_w =$	1.00E-14	
6	$pK_3 =$	6.28		$K_{H2B} =$	1.02E-07				
7	$K_1 =$	1.4E-03		$K_{HB} =$	3.16E-15				
8	$K_2 =$	5.0E-05							
9	$K_3 =$	5.2E-07							
10									
11	Concentrations								
12	$[H^+] =$	3.05E-05		$[A^{3-}] =$	4.21E-04		$[OH^-] =$	3.28E-10	
13	$[H_3A] =$	3.27E-04		$[H_2B^+] =$	2.99E-02		$[Na^+] =$	0.035	
14	$[H_2A^-] =$	1.48E-02		$[HB] =$	9.98E-05		$pH =$	4.516	
15	$[HA^{2-}] =$	2.44E-02		$[B^-] =$	1.04E-14		↑ initial value is a guess		
16	Positive charge minus negative charge				-2.42E-17				
17									
18	Formulas:								
19	B7 = 10^-B4			B8 = 10^-B5			B9 = 10^-B6		
20	E6 = 10^-E4			E7 = 10^-E5					
21	B12 = 10^-H14			H12 = H5/B12			H13 = H3		
22	B13 = $B$12^3*$B$3/($B$12^3+$B$12^2*$B$7+$B$12*$B$7*$B$8+$B$7*$B$8*$B$9)								
23	B14 = $B$12^2*$B$7*$B$3/($B$12^3+$B$12^2*$B$7+$B$12*$B$7*$B$8+$B$7*$B$8*$B$9)								
24	B15 = $B$12*$B$7*$B$8*$B$3/($B$12^3+$B$12^2*$B$7+$B$12*$B$7*$B$8+$B$7*$B$8*B								
25	E12 = B$7*B$8*B$9*B$3/(B12^3+B12^2*B$7+B$12*B$7*B$8+B$7*B$8*B$9)								
26	E13 = $B$12^2*$E$3/($B$12^2+$B$12*$E$6+$E$6*$E$7)								
27	E14 = $B$12*$E$6*$E$3/($B$12^2+$B$12*$E$6+$E$6*$E$7)								
28	E15 = $E$6*$E$7*$E$3/($B$12^2+$B$12*$E$6+$E$6*$E$7)								

12-7. Arginine = HA from H_3A^{2+} with $pK_1 = 1.823$, $pK_2 = 8.991$, $pK_3 = 12.1$; glutamic acid = H_2G from H_3G^+ with $pK_1 = 2.160$, $pK_2 = 4.30$, $pK_3 = 9.96$

Mixture contains 0.020 mol arginine, 0.030 mol glutamic acid, and 0.005 mol KOH in 1.00 L. $F_A = 0.020$ M and $F_G = 0.030$ M.

Charge balance:

$$[H^+] + 2[H_3A^{2+}] + [H_2A^+] + [H_3G^+] + [K^+] = [OH^-] + [A^-] + [HG^-] + 2[G^{2-}]$$

Substitute fractional composition equations into charge balance:

$$[H^+] + 2\alpha_{H_3A^{2+}} F_A + \alpha_{H_2A^+} F_A + \alpha_{H_3G^+} F_G + [0.005]$$

$$= K_w/[H^+] + \alpha_{A^-} F_A + \alpha_{HG^-} F_G + 2\alpha_{G^{2-}} F_G$$

	A	B	C	D	E	F	G	H	I	J
1	Mixture of 0.020 M arginine, 0.030 M glutamic acid, and 0.005 M KOH									
2										
3	F_A =	0.020		F_B =	0.030		$[K^+]$ =	0.005		
4	pK_1 =	1.823		pK_{H3G} =	2.160		pK_w =	14.00		
5	pK_2 =	8.991		pK_{H2G} =	4.30		K_w =	1.00E-14		
6	pK_3 =	12.1		pK_{HG} =	9.96					
7	K_1 =	1.5E-02		K_{H3G} =	6.9E-03					
8	K_2 =	1.0E-09		K_{H2G} =	5.0E-05					
9	K_3 =	7.9E-13		K_{HG} =	1.1E-10					
10										
11	Concentrations									
12	$[H_3A^{2+}]$ =	1.32E-05		$[H_3G^+]$ =	7.14E-06		$[H^+]$ =	9.94E-06		
13	$[H_2A^+]$ =	2.00E-02		$[H_2G]$ =	4.96E-03		$[OH^-]$ =	1.01E-09		
14	$[HA]$ =	2.05E-06		$[HG^-]$ =	2.50E-02		$[K^+]$ =	0.005		
15	$[A^-]$ =	1.64E-13		$[G^{2-}]$ =	2.76E-07		pH =	5.003		
16								↑ initial value is a guess		
17	Positive charge minus negative charge =	-9.94E-17								
18					= 2*B12+B13+E12+H12+H14-B15-E14-2*E15-H13					
19	Formulas:									
20	B7 = 10^-B4			B8 = 10^-B5			B9 = 10^-B6			
21	E7 = 10^-E4			E8 = 10^-E5			E9 = 10^-E6			
22	H12 = 10^-H15			H13 = H5/H12			H14 = H3			
23	B12 = $H$12^3*$B$3/($H$12^3+$H$12^2*$B$7+$H$12*$B$7*$B$8+$B$7*$B$8*$B$9)									
24	B13 = $H$12^2*$B$7*$B$3/($H$12^3+$H$12^2*$B$7+$H$12*$B$7*$B$8+$B$7*$B$8*$B$9)									
25	B14 = $H$12*$B$7*$B$8*$B$3/($H$12^3+$H$12^2*$B$7+$H$12*$B$7*$B$8+$B$7*$B$8*$B$9)									
26	B15 = $B$7*$B$8*$B$9*$B$3/($H$12^3+$H$12^2*$B$7+$H$12*$B$7*$B$8+$B$7*$B$8*$B$9)									
27	E12 = $H$12^3*$E$3/($H$12^3+$H$12^2*$E$7+$H$12^$E$7*$E$8+$E$7*$E$8*$E$9)									
28	E13 = $H$12^2*$E$7*$E$3/($H$12^3+$H$12^2*$E$7+$H$12^$E$7*$E$8+$E$7*$E$8*$E$9)									
29	E14 = $H$12*$E$7*$E$8*$E$3/($H$12^3+$H$12^2*$E$7+$H$12*$E$7*$E$8+$E$7*$E$8*$E$9)									
30	E15 = $E$7*$E$8*$E$9*$E$3/($H$12^3+$H$12^2*$E$7+$H$12*$E$7*$E$8+$E$7*$E$8*$E$9)									

We solve for $[H^+]$ with the spreadsheet, with an initial guess of pH = 3 in cell H15. Select Solver and choose Options. Set Precision to 1e-16 and click OK. In Solver, Set Target Cell E17 Equal To Value of 0 By Changing Cells H55. Click Solve and Solver finds pH = 5.00 in cell H15, giving a net charge of ~10^{-16} in cell E17.

12-8. $H_3A^{2+} \rightleftharpoons H_2A^+ + H^+$ $K_1 = \dfrac{[H_2A^+]\gamma_{H_2A^+}[H^+]\gamma_{H^+}}{[H_3A^{2+}]\gamma_{H_3A^{2+}}} = 10^{-1.823}$

$H_2A^+ \rightleftharpoons HA + H^+$ $K_2 = \dfrac{[HA]\gamma_{HA}[H^+]\gamma_{H^+}}{[H_2A^+]\gamma_{H_2A^+}} = 10^{-8.991}$

$HA \rightleftharpoons A^- + H^+$ $K_3 = \dfrac{[A^-]\gamma_{A^-}[H^+]\gamma_{H^+}}{[HA]\gamma_{HA}} = 10^{-12.1}$

$$K_1' = K_1\left(\frac{\gamma_{H_3A^{2+}}}{\gamma_{H_2A^+}\gamma_{H^+}}\right) = \frac{[H_2A^+][H^+]}{[H_3A^{2+}]} \qquad K = K_2\left(\frac{\gamma_{H_2A^+}}{\gamma_{HA}\gamma_{H^+}}\right) = \frac{[HA][H^+]}{[H_2A^+]}$$

$$K_3' = K_3\left(\frac{\gamma_{HA}}{\gamma_{A^-}\gamma_{H^+}}\right) = \frac{[A^-][H^+]}{[HA]}$$

$$H_3G^+ \rightleftharpoons H_2G + H^+ \qquad K_{H_3G} = \frac{[H_2G]\gamma_{H_2G}[H^+]\gamma_{H^+}}{[H_3G^+]\gamma_{H_3G^+}} = 10^{-2.160}$$

$$H_2G \rightleftharpoons HG^- + H^+ \qquad K_{H_2G} = \frac{[HG^-]\gamma_{HG^-}[H^+]\gamma_{H^+}}{[H_2G]\gamma_{H_2G}} = 10^{-4.30}$$

$$HG^- \rightleftharpoons G^{2-} + H^+ \qquad K_{HG} = \frac{[G^{2-}]\gamma_{G^{2-}}[H^+]\gamma_{H^+}}{[HG^-]\gamma_{HG^-}} = 10^{-9.96}$$

$$K_{H_3G}' = K_{H_3G}\left(\frac{\gamma_{H_3G^+}}{\gamma_{H_2G}\gamma_{H^+}}\right) = \frac{[H_2G][H^+]}{[H_3G^+]} \qquad K_{H_2G}' = K_{H_2G}\left(\frac{\gamma_{H_2G}}{\gamma_{HG^-}\gamma_{H^+}}\right) = \frac{[HG^-][H^+]}{[H_2G]}$$

$$K_{HG}' = K_{HG}\left(\frac{\gamma_{HG^-}}{\gamma_{G^{2-}}\gamma_{H^+}}\right) = \frac{[G^{2-}][H^+]}{[HG^-]}$$

In the next spreadsheet, $[H^+]$ in cell H18 is computed from $(10^{-pH})/\gamma_{H^+}$. Don't forget that activity coefficient! Ionic strength in cell D24 was initially set to 0, and activity coefficients are computed in cells A13:H15. From the activity coefficients, effective equilibrium constants are computed in cells H4:H10. pH = 3 is *guessed* in cell H21. From pH and the *K'* values, concentrations are computed in cells A18:H21. Solver is used to vary pH in cell H21 until the net charge in cell H23 is 0. If Solver does not find an answer, try a different initial value for pH or increase Precision in the Options window of Solver. A value of 1e-16 generally works, but for some problems I need larger numbers, such as 1E-10. The ionic strength computed in cell D25 is then entered into cell D24 and Solver is executed again. The process is complete when ionic strength no longer changes. pH computed with activities is 4.95 and μ = 0.025 M. When we found pH without activities in the previous problem, the pH was 5.00.

Iteration	ionic strength	pH
1	0	5.003
2	0.025 0	4.939
3	0.025 1	4.939

	A	B	C	D	E	F	G	H	I
1	Mixture of 0.020 M arginine, 0.030 M glutamic acid, and 0.005 M KOH								
2	Solved with Davies activity coefficients								
3	F_A =	0.020		F_B =	0.030		$[K^+]$ =	0.005	
4	pK_1 =	1.823		pK_{H3G} =	2.160		K_1' =	1.1E-02	
5	pK_2 =	8.991		pK_{H2G} =	4.30		K_2' =	1.0E-09	
6	pK_3 =	12.1		pK_{HG} =	9.96		K_3' =	1.1E-12	
7	K_1 =	1.5E-02		K_{H3G} =	6.9E-03		K_{H3G}' =	6.9E-03	
8	K_2 =	1.0E-09		K_{H2G} =	5.0E-05		K_{H2G}' =	6.8E-05	
9	K_3 =	7.9E-13		K_{HG} =	1.1E-10		K_{HG}' =	2.01E-10	
10	pK_w =	13.995		K_w =	1.01E-14		K_w' =	1.37E-14	
11									
12	Davies activity coefficients								
13	H_3A^{2+} =	0.54		H_3G^+ =	0.86				
14	H_2A^+ =	0.86		HG^- =	0.86		H^+ =	0.86	
15	A^- =	0.86		G^{2-} =	0.54		OH^- =	0.86	
16									
17	Concentrations								
18	$[H_3A^{2+}]$ =	2.41E-05		$[H_3G^+]$ =	9.58E-06		$[H^+]$ =	1.34E-05	
19	$[H_2A^+]$ =	2.00E-02		$[H_2G]$ =	4.95E-03		$[OH^-]$ =	1.02E-09	
20	$[HA]$ =	1.52E-06		$[HG^-]$ =	2.50E-02		$[K^+]$ =	0.005	
21	$[A^-]$ =	1.22E-13		$[G^{2-}]$ =	3.76E-07		pH =	4.939	
22							↑ initial value is a guess		
23	Positive charge minus negative charge =			-8.86E-18					
24	Ionic strength =			0.0251	← initial value is 0				
25	New ionic strength =			0.0251	← substitute this value into cell D24				
26						for next interation			
27	Formulas								
28	K_1' = B7*B13/(B14*H14)			K_2' = B8*B14/(H14)			K_3' = B9/(B15*H14)		
29	K_{H3B}' = E7*E13/H14			K_{H2B}' =E8/(E14*H14)			K_{HB}' = E9*E14/(E15*H14)		
30	K_w' = E10/(H14*H15)			[H+] = (10^-H21)/H14			[OH-] = H10/H18		
31	Activity coefficient = 10^(-0.51*charge^2*(SQRT($D$24)/(1+SQRT($D$24))-0.3*$D$24))								
32	Denom1 = ($H$18^3+$H$18^2*$H$4+$H$18*$H$4*$H$5+$H$4*$H$5*$H$6)								
33	$[H_3A^{2+}]$ = $H$18^3*$B$3/Denom1						$[H_2A^+]$ = $H$18^2*$H$4*$B$3/Denom1		
34	$[HA]$ = $H$18*$H$4*$H$5*$B$3/Denom1						$[A^-]$ = $H$4*$H$5*$H$6*$B$3/Denom1		
35	Denom2 = ($H$18^3+$H$18^2*$H$7+$H$18*$H$7*$H$8+$H$7*$H$8*$H$9)								
36	$[H_3G^+]$ = $H$18^3*$E$3/Denom2						$[H_2G]$ = $H$18^2*$H$7*$E$3/Denom2		
37	$[HG^-]$ = $H$18*$H$7*$H$8*$E$3/Denom2						$[G^{2-}]$ = $H$7*$H$8*$H$9*$E$3/Denom2		
38	E23 = 2*B18+B19+E18+H18+H20-B21-E20-2*E21-H19								
39	D25 = 0.5*(4*B18+B19+B21+E18+E20+4*E21+H18+H19+H20)								

12-9. (a) This is the same problem that was worked in Section 12-2, but with $[KH_2PO_4]$ = 0.008 695 m and $[Na_2HPO_4]$ = 0.030 43 m in cells B3 and B4 of the spreadsheet in Figure 12-4. Beginning with ionic strength = 0 in cell C19 and pH = 7 in cell H15, we find the value of pH required to make the net charge 0 in cell E18 by using Solver with Precision = 1e-16. The first iteration gives pH = 7.742. The ionic strength in cell C20 after the first (and all subsequent) iterations is 0.100 m. Typing this value in cell C19 and performing a second iteration gives pH = 7.420.

(b) To use the Debye-Hückel equation, we can compute activity coefficients with ion size parameters from Table 7-1 or we can just use activity coefficients from Table 7-1 for $\mu = 0.1$ M:

	A	B	C	D	E	F	G	H
8	Activity coefficients from table iin textbook:							
9	H^+ =	0.83		H_3P =	1.00	(fixed at 1)	HP^{2-} =	0.36
10	OH^- =	0.76		H_2P^- =	0.78		P^{3-} =	0.10

These activity coefficients produce a pH of 7.403 after executing Solver to reduce the net charge to 0 in cell E18.

12-10. EDTA = H_4A from hexaprotic H_6A^{2+} with $pK_1 = 0.0$, $pK_2 = 1.5$, $pK_3 = 2.00$, $pK_4 = 2.69$, $pK_5 = 6.13$, $pK_6 = 10.37$; Lysine = HL from triprotic H_3L^{2+} with $pK_1 = 1.77$, $pK_2 = 9.07$, $pK_3 = 10.82$

Mixture contains 0.040 mol H_4A, 0.030 mol HL, and 0.050 mol NaOH in 1.00 L.

Charge balance:

$[H^+] + 2[H_6A^{2+}] + [H_5A^+] + 2[H_3L^{2+}] + [H_2L^+] + [Na^+]$
$= [OH^-] + [H_3A^-] + 2[H_2A^{2-}] + 3[HA^{3-}] + 4[A^{4-}] + [L^-]$

Substitute fractional composition equations into charge balance:

$[H^+] + 2\alpha_{H_6A^{2+}} F_A + \alpha_{H_5A^+} F_A + 2\alpha_{H_3L^{2+}} F_L + \alpha_{H_2L^+} F_L + [0.050]$
$= K_w/[H^+] + \alpha_{H_3A^-} F_A + 2\alpha_{H_2A^{2-}} F_A + 3\alpha_{HA^{3-}} F_A + 4\alpha_{A^{4-}} F_A + \alpha_{L^-} F_L$

We solve this equation for pH with the following spreadsheet, with an initial guess of pH = 7 in cell H18. In Solver, Set Target Cell E19 Equal To Value of 0 By Changing Cells H18. Click OK and Solver finds pH = 4.44 in cell H18.

	A	B	C	D	E	F	G	H	I
1	Mixture of 0.040 M H_4A, 0.030 M HL, and 0.05 M NaOH								
2									
3	F_A =	0.040		F_L =	0.030		$[Na^+]$ =	0.050	
4	pK_1 =	0.0		pK_{L1} =	1.77				
5	pK_2 =	1.5		pK_{L2} =	9.07				
6	pK_3 =	2.0		pK_{L3} =	10.82		pK_w =	14.00	
7	pK_4 =	2.69		K_{L1} =	1.70E-02		K_w =	1.00E-14	
8	pK_5 =	6.13		K_{L2} =	8.51E-10				
9	pK_6 =	10.37		K_{L3} =	1.51E-11				
10	K_1 =	1.00E+00		K_3 =	1.00E-02		K_5 =	7.41E-07	
11	K_2 =	3.16E-02		K_4 =	2.04E-03		K_6 =	4.27E-11	
12									
13	Species in charge balance:								
14	$[H^+]$ =	3.62E-05		$[HA^{3-}]$ =	7.88E-04		$[OH]$ =	2.76E-10	
15	$[H_6A^{2+}]$ =	1.03E-13		$[A^{4-}]$ =	9.28E-10		$[Na^+]$ =	0.050	
16	$[H_5A^+]$ =	2.84E-09		$[H_3L^{2+}]$ =	6.39E-05		$[H_4A]$ =	2.48E-06	
17	$[H_3A^-]$ =	6.84E-04		$[H_2L^+]$ =	2.99E-02		$[HL]$ =	7.03E-07	
18	$[H_2A^{2-}]$ =	3.85E-02		$[L]$ =	2.94E-13		pH =	4.441	← initial value
19	Positive charge minus negative charge :	-9.77E-17							is a guess
20	Formulas:								
21	B10 = 10^-B4 with analogous formulas for K_2 - K_6								
22	E7 = 10-E4 with analogous formulas for K_{L2} and K_{L3}								
23	B14 = 10^-H18		H14 = H7/B14				H15 = H3		
24	Denom1 = \$B\$14^6+\$B\$14^5*\$B\$10+\$B\$14^4*\$B\$10*\$B\$11+\$B\$14^3*\$B\$10*\$B\$11*\$E\$10								
25	+\$B\$14^2*\$B\$10*\$B\$11*\$E\$10*\$E\$11+\$B\$14*\$B\$10*\$B\$11*\$E\$10*\$E\$11*\$H\$10								
26	+\$B\$10*\$B\$11*\$E\$10*\$E\$11*\$H\$10*\$H\$11								
27	Denom2 = \$B\$14^3+\$B\$14^2*\$E\$7+\$B\$14*\$E\$7*\$E\$8+\$E\$7*\$E\$8*\$E\$9								
28	B15 = \$B\$14^6*\$B\$3/Denom1								
29	B16 = \$B\$14^5*\$B\$10*\$B\$3/Denom1								
30	B17 = \$B\$14^3*\$B\$10*\$B\$11*\$E\$10*\$B\$3/Denom1								
31	B18 = \$B\$14^2*\$B\$10*\$B\$11*\$E\$10*\$E\$11*\$B\$3/Denom1								
32	E14 = \$B\$14*\$B\$10*\$B\$11*\$E\$10*\$E\$11*\$H\$10*\$B\$3/Denom1								
33	E15 = \$B\$10*\$B\$11*\$E\$10*\$E\$11*\$H\$10*\$H\$11*\$B\$3/Denom1								
34	H16 = \$B\$14^4*\$B\$10*\$B\$11*\$B\$3/Denom1								
35	E16 = \$B\$14^3*\$E\$3/Denom2								
36	E17 = \$B\$14^2*\$E\$7*\$E\$3/Denom2								
37	E18 = \$E\$7*\$E\$8*\$E\$9*\$E\$3/Denom2								
38	H17 = \$B\$14*\$E\$7*\$E\$8*\$E\$3/Denom2								
39	E19 = B14+2*B15+B16+2*E16+E17+H15-B17-2*B18-3*E14-4*E15-E18-H14								

12-11. (a) $Fe^{3+} + SCN^- \rightleftharpoons Fe(SCN)^{2+}$ $\qquad [Fe(SCN)^{2+}] = \beta_1'[Fe^{3+}][SCN^-]$

$Fe^{3+} + 2SCN^- \rightleftharpoons Fe(SCN)_2^+$ $\qquad [Fe(SCN)_2^+] = \beta_2'[Fe^{3+}][SCN^-]^2$

$Fe^{3+} + H_2O \rightleftharpoons FeOH^{2+} + H^+$ $\qquad [FeOH^{2+}] = K_a'[Fe^{3+}]/[H^+]$

$H_2O \rightleftharpoons H^+ + OH^-$ $\qquad [OH^-] = K_w'/[H^+]$

$$\beta_1' = \beta_1 \frac{\gamma_{Fe^{3+}}\gamma_{SCN^-}}{\gamma_{Fe(SCN)^{2+}}} \qquad \beta_2' = \beta_2 \frac{\gamma_{Fe^{3+}}\gamma_{SCN^-}^2}{\gamma_{Fe(SCN)_2^+}} \qquad K_a' = K_a \frac{\gamma_{Fe^{3+}}}{\gamma_{FeOH^{2+}}\gamma_{H^+}}$$

$$K_w' = \frac{K_w}{\gamma_{H^+}\gamma_{OH^-}} \qquad \beta_1 = 10^{3.03} \qquad \beta_2 = 10^{4.6} \qquad K_a = 10^{-2.195}$$

(b) Charge balance: $[OH^-] + [SCN^-] + [NO_3^-] =$

$[H^+] + 3[Fe^{3+}] + 2[Fe(SCN)^{2+}] + [Fe(SCN)_2^+] + 2[FeOH^{2+}] + [Na^+]$

(c) Mass balances:

Total iron $\equiv F_{Fe} = 5.0 \text{ mM} = [Fe^{3+}] + [Fe(SCN)^{2+}] + [Fe(SCN)_2^+] + [FeOH^{2+}]$

Total thiocyanate $\equiv F_{SCN} = 5.0 \text{ } \mu M = [Fe(SCN)^{2+}] + 2[Fe(SCN)_2^+] + [SCN^-]$

$[Na^+] = 5.0 \text{ } \mu M$

$[NO_3^-] = 3(5.0 \text{ mM}) + 15.0 \text{ mM} = 30.0 \text{ mM}$

(d) $[Fe(SCN)^{2+}] + 2[Fe(SCN)_2^+] + [SCN^-] = F_{SCN}$

$\beta_1'[Fe^{3+}][SCN^-] + 2\beta_2'[Fe^{3+}][SCN^-]^2 + [SCN^-] = F_{SCN}$

$[Fe^{3+}](\beta_1'[SCN^-] + 2\beta_2'[SCN^-]^2) = F_{SCN} - [SCN^-]$

$$[Fe^{3+}] = \frac{F_{SCN} - [SCN^-]}{\beta_1'[SCN^-] + 2\beta_2'[SCN^-]^2}$$

(e) $[Fe^{3+}] + [Fe(SCN)^{2+}] + [Fe(SCN)_2^+] + [FeOH^{2+}] = F_{Fe}$

$[Fe^{3+}] + \beta_1'[Fe^{3+}][SCN^-] + \beta_2'[Fe^{3+}][SCN^-]^2 + K_a'[Fe^{3+}]/[H^+] = F_{Fe}$

$$[H^+] = \frac{K_a'[Fe^{3+}]}{F_{Fe} - [Fe^{3+}] - \beta_1'[Fe^{3+}][SCN^-] - \beta_2'[Fe^{3+}][SCN^-]^2}$$

or $[H^+] = \dfrac{K_a'[Fe^{3+}]}{F_{Fe} - [Fe^{3+}] - [Fe(SCN)^{2+}] - [Fe(SCN)_2^+]}$ (which is easier to use)

(f), (g), (h), (i) The spreadsheet on the next page shows the following results:

	A	B	C	D	E	F	G	H
13	[SCN⁻]	[Fe³⁺]	[H⁺]	[OH⁻]	[FeSCN²⁺]	[Fe(SCN)₂⁺]	[FeOH²⁺]	Net charge
14	2.03E-06	4.20E-03	1.58E-02	9.20E-13	2.97E-06	1.06E-10	8.02E-04	-2.78E-17

Ionic strength = 0.043 4 M

pH = 1.88

$$\frac{[Fe(SCN)^{2+}]}{\{[Fe^{3+}] + [FeOH^{2+}]\}[SCN^-]} = 293 \text{ (The graph in the textbook gives 270.)}$$

(j) The next spreadsheet shows the results when the solution also contains 0.20 M KNO_3:

	A	B	C	D	E	F	G	H
13	[SCN⁻]	[Fe³⁺]	[H⁺]	[OH⁻]	[FeSCN²⁺]	[Fe(SCN)₂⁺]	[FeOH²⁺]	Net charge
14	2.81E-06	4.45E-03	1.55E-02	1.18E-12	2.19E-06	6.82E-11	5.46E-04	0.00E+00

Ionic strength = 0.243 9 M

pH = 1.94

$$\frac{[Fe(SCN)^{2+}]}{\{[Fe^{3+}] + [FeOH^{2+}]\}[SCN^-]} = 156 \text{ (The graph in the textbook gives 150.)}$$

Spreadsheet for 0 M KNO₃:

	A	B	C	D	E	F	G	H
1	Composition of Fe(III)-thiocyanate solution							
2								
3	log β₁ =	3.03	β₁ =	1.07E+03	β₁' =	1.75E+02	F_Fe =	5.0E-03
4	log β₂ =	4.6	β₂ =	3.98E+04	β₂' =	1.94E+03	F_SCN =	5.0E-06
5	log Kₐ =	-2.195	Kₐ =	6.38E-03	Kₐ' =	1.90E-03	[Na⁺] =	5.0E-06
6	log K_w =	-14.00	K_w =	1.00E-14	K_w' =	1.83E-14	[NO₃⁻] =	2.3E-01
7							[K⁺] =	2.0E-01
8	Davies activity coefficients:							
9	Fe³⁺	0.07		FeOH²⁺	0.30		OH⁻	0.74
10	Fe(SCN)²⁺	0.30		SCN⁻	0.74		H⁺	0.74
11	Fe(SCN)₂⁺	0.74						
12								
13	[SCN⁻]	[Fe³⁺]	[H⁺]	[OH⁻]	[FeSCN²⁺]	[Fe(SCN)₂⁺]	[FeOH²⁺]	Net charge
14	2.81E-06	4.45E-03	1.55E-02	1.18E-12	2.19E-06	6.82E-11	5.46E-04	0.00E+00
15								
16	Ionic strength =	0.2439	← Initial value is 0					
17	New ionic strength =	0.2439		[FeSCN²⁺]/({[Fe³⁺]+[FeOH²⁺]}[SCN⁻])=				
18	pH =	1.94		156				
19	Total Fe =	5.000E-03	= [Fe³⁺] + [FeSCN²⁺] + [Fe(SCN)₂⁺] + [FeOH²⁺]					
20	Total SCN =	5.000E-06	= [SCN⁻] + [FeSCN²⁺] + 2[Fe(SCN)₂⁺]					
21								
22	Activity coefficients = 10^(-0.51*charge^2*(SQRT($C$16)/(1+SQRT($C$16))-0.3*$C$16))							
23	β₁' = D3*B9*E10/B10			Kₐ' = D5*B9/(E9*H10)				
24	β₂' = D4*B9*E10^2/B11			K_w' = D6/(H9*H10)				
25	[Fe³⁺] = (H4-A14)/(F3*A14+2*F4*A14^2)							
26	[H⁺] = F5*B14/(H3-B14-E14-F14)			[OH⁻] = F6/C14				
27	[FeSCN²⁺] = F3*B14*A14			[Fe(SCN)₂⁺] = F4*B14*A14^2				
28	[FeOH²⁺] = F5*B14/C14			pH =-LOG10(C14*H10)				
29	Net charge = 3*B14+C14+ 2*E14+F14+2*G14+H5-A14-D14-H6+H7							
30	New ionic strength = 0.5*(H5+H6+A14+9*B14+C14+D14+4*E14+F14+4*G14+H7)							

Spreadsheet for 0.20 M KNO$_3$:

	A	B	C	D	E	F	G	H
1	Composition of Fe(III)-thiocyanate solution							
2								
3	log β$_1$ =	3.03	β$_1$ =	1.07E+03	β$_1$' =	1.75E+02	F$_{Fe}$ =	5.0E-03
4	log β$_2$ =	4.6	β$_2$ =	3.98E+04	β$_2$' =	1.94E+03	F$_{SCN}$ =	5.0E-06
5	log K$_a$ =	-2.195	K$_a$ =	6.38E-03	K$_a$' =	1.90E-03	[Na$^+$] =	5.0E-06
6	log K$_w$ =	-14.00	K$_w$ =	1.00E-14	K$_w$' =	1.83E-14	[NO$_3^-$] =	2.3E-01
7							[K$^+$] =	2.0E-01
8	Davies activity coefficients:							
9	Fe^{3+}	0.07		FeOH^{2+}		0.30	OH$^-$	0.74
10	Fe(SCN)$^{2+}$	0.30		SCN$^-$		0.74	H$^+$	0.74
11	Fe(SCN)$_2^+$	0.74						
12								
13	[SCN$^-$]	[Fe^{3+}]	[H$^+$]	[OH$^-$]	[FeSCN^{2+}]	[Fe(SCN)$_2^+$]	[FeOH^{2+}]	Net charge
14	2.81E-06	4.45E-03	1.55E-02	1.18E-12	2.19E-06	6.82E-11	5.46E-04	0.00E+00
15								
16	Ionic strength =		0.2439	← Initial value is 0				
17	New ionic strength =	0.2439			[FeSCN^{2+}]/({[Fe^{3+}]+[FeOH^{2+}]}[SCN$^-$])=			
18		pH =	1.94			156		
19		Total Fe =	5.000E-03	= [Fe^{3+}] + [FeSCN^{2+}] + [Fe(SCN)$_2^+$] + [FeOH^{2+}]				
20		Total SCN =	5.000E-06	= [SCN$^-$] + [FeSCN^{2+}] + 2[Fe(SCN)$_2^+$]				

12-12. (a) $La^{3+} + SO_4^{2-} \rightleftharpoons La(SO_4)^+$ $\qquad$ $[La(SO_4)^+] = \beta_1'[La^{3+}][SO_4^{2-}]$

$La^{3+} + 2SO_4^{2-} \rightleftharpoons La(SO_4)_2^-$ $\qquad$ $[La(SO_4)_2^-] = \beta_2'[La^{3+}][SO_4^{2-}]^2$

$La^{3+} + H_2O \rightleftharpoons LaOH^{2+} + H^+$ $\qquad$ $[LaOH^{2+}] = K_a'[La^{3+}]/[H^+]$

$H_2O \rightleftharpoons H^+ + OH^-$ $\qquad\qquad\qquad$ $[OH^-] = K_w'/[H^+]$

$$\beta_1' = \beta_1 \frac{\gamma_{La^{3+}}\gamma_{SO_4^{2-}}}{\gamma_{La(SO_4)^+}} \qquad \beta_2' = \beta_2 \frac{\gamma_{La^{3+}}\gamma_{SO_4^{2-}}^2}{\gamma_{La(SO_4)_2^-}} \qquad K_a' = K_a \frac{\gamma_{La^{3+}}}{\gamma_{LaOH^{2+}}\gamma_{H^+}}$$

$$K_w' = \frac{K_w}{\gamma_{H^+}\gamma_{OH^-}} \qquad \beta_1 = 10^{3.64} \qquad \beta_2 = 10^{5.3} \qquad K_a = 10^{-8.5}$$

Charge balance:

$[OH^-] + 2[SO_4^{2-}] + [La(SO_4)_2^-] = [H^+] + 3[La^{3+}] + [La(SO_4)^+] + 2[LaOH^{2+}]$

Mass balances:

Lanthanum $\equiv F_{La} = 2.0$ mM $= [La^{3+}] + [La(SO_4)^+] + [La(SO_4)_2^-] + [LaOH^{2+}]$

Sulfate $\equiv F_{SO_4} = 3.0$ mM $= [La(SO_4)^+] + 2[La(SO_4)_2^-] + [SO_4^{2-}]$

Express $[La^{3+}]$ in terms of $[SO_4^{2-}]$:

$[La(SO_4)^+] + 2[La(SO_4)_2^-] + [SO_4^{2-}] = F_{SO_4}$

$\beta_1'[La^{3+}][SO_4^{2-}] + 2\beta_2'[La^{3+}][SO_4^{2-}]^2 + [SO_4^{2-}] = F_{SO_4}$

$$[La^{3+}](\beta_1'[SO_4^{2-}] + 2\beta_2'[SO_4^{2-}]^2) = F_{SO_4} - [SO_4^{2-}]$$

$$[La^{3+}] = \frac{F_{SO_4} - [SO_4^{2-}]}{\beta_1'[SO_4^{2-}] + 2\beta_2'[SO_4^{2-}]^2}$$

Express $[H^+]$ in terms of $[La^{3+}]$ and $[SO_4^{2-}]$:

$$[La^{3+}] + [La(SO_4)^+] + [La(SO_4)_2^-] + [LaOH^{2+}] = F_{La}$$

$$[La^{3+}] + \beta_1'[La^{3+}][SO_4^{2-}] + \beta_2'[La^{3+}][SO_4^{2-}]^2 + K_a'[La^{3+}]/[H^+] = F_{La}$$

$$[H^+] = \frac{K_a'[La^{3+}]}{F_{La} - [La^{3+}] - \beta_1'[La^{3+}][SO_4^{2-}] - \beta_2'[La^{3+}][SO_4^{2-}]^2}$$

or $[H^+] = \dfrac{K_a'[La^{3+}]}{F_{La} - [La^{3+}] - [La(SO_4)^+] - [La(SO_4)_2^-]}$ (which is easier to use)

The spreadsheet is shown after (e).

(b) If $La_2(SO_4)_3$ were a strong electrolyte, $\mu = \frac{1}{2}\{[La^{3+}]\cdot(+3)^2 + [SO_4^{2-}]\cdot(-2)^2\}$

$= \frac{1}{2}\{(2.0\ \text{mM}\cdot 9) + (3.0\ \text{mM}\cdot 4)\} = 15.0\ \text{mM}$. The actual ionic strength in

cell C17 is 6.3 mM.

(c) $[La^{3+}]/F_{La} = 0.285$

(d) We expected the solution to have a pH near neutral. pK_a for HSO_4^- is 1.99.

Therefore, we did not expect very much HSO_4^- to be present. With a pH near

6, the fraction on sulfate that is protonated is $\sim 10^{-4}$.

(e) Evaluate the solubility product expression for $La(OH)_3$:

$$[La^{3+}][OH^-]^3\gamma_{La^{3+}}\gamma_{OH^-}^3 = (5.69 \times 10^{-4})(1.04 \times 10^{-8})^3(0.47)(0.92)^3$$

$= 2.3 \times 10^{-28} < K_{sp}$ for $La(OH)_3 = 2 \times 10^{-21}$

$La(OH)_3(s)$ does not precipitate.

	A	B	C	D	E	F	G	H
1	Composition of La(III)-sulfate solution							
2								
3	log β_1 =	3.64	β_1 =	4.37E+03	β_1' =	1.59E+03	F_{La} =	2.0E-03
4	log β_2 =	5.3	β_2 =	2.00E+05	β_2' =	5.20E+04	F_{SO4} =	3.0E-03
5	log K_a =	-8.5	K_a =	3.16E-09	K_a' =	2.26E-09		
6	log K_w =	-14.00	K_w =	1.00E-14	K_w' =	1.18E-14		
7								
8	Davies activity coefficients:							
9	La^{3+}	0.47		$LaOH^{2+}$	0.71		OH^-	0.92
10	$La(SO_4)^+$	0.92		SO_4^{2-}	0.71		H^+	0.92
11	$La(SO_4)_2^-$	0.92						
12								
13	$[SO_4^{2-}]$	$[La^{3+}]$	$[H^+]$	$[OH^-]$	$[La(SO_4)^+]$	$[La(SO_4)_2]$	$[LaOH^{2+}]$	Net charge
14	1.50E-03	5.69E-04	1.14E-06	1.04E-08	1.36E-03	6.69E-05	1.13E-06	3.12E-19
15								
16	Ionic strength =		0.00629	← Initial value is 0				
17	New ionic strength =		0.00629		$[La^{3+}]/F_{La}$ =	0.285		$[SO_4^{2-}]/F_{SO4}$ =
18		pH =	5.98		$[La(SO_4)^+]/F_{La}$ =	0.681	0.501	
19		Total La =	2.000E-03		$[La(SO_4)_2]/F_{La}$ =	0.033		
20		Total SO_4^{2-} =	3.000E-03		$[LaOH^{2+}]/F_{La}$ =	5.6E-04		
21								
22	Activity coefficients = 10^(-0.51*charge^2*(SQRT($C$16)/(1+SQRT($C$16))-0.3*$C$16))							
23	β_1' = D3*B9*E10/B10			K_a' = D5*B9/(E9*H10)				
24	β_2' = D4*B9*E10^2/B11			K_w' = D6/(H9*H10)				
25	$[La^{3+}]$ = (H4-A14)/(F3*A14+2*F4*A14^2)							
26	$[H^+]$ = F5*B14/(H3-B14-E14-F14)			$[OH^-]$ = F6/C14				
27	$[La(SO_4)^+]$ = F3*B14*A14			$[La(SO_4)_2]$ = F4*B14*A14^2				
28	$[LaOH^{2+}]$ = F5*B14/C14			pH = -LOG10(C14*H10)				
29	Net charge = 3*B14+C14+ E14+2*G14-2*A14-D14-F14							
30	New ionic strength = 0.5*(4*A14+9*B14+C14+D14+E14+F14+4*G14)							

12-13.

$$CaSO_4(s) \underset{}{\overset{K_{sp}}{\rightleftharpoons}} Ca^{2+} + SO_4^{2-} \qquad\qquad K_{sp} = 2.4 \times 10^{-5} \qquad\qquad (A)$$

$$CaSO_4(s) \underset{}{\overset{K_{ion\,pair}}{\rightleftharpoons}} CaSO_4(aq) \qquad\qquad K_{ion\,pair} = 5.0 \times 10^{-3} \qquad\qquad (B)$$

$$Ca^{2+} + H_2O \underset{}{\overset{K_{acid}}{\rightleftharpoons}} CaOH^+ + H^+ \qquad\qquad K_{acid} = 2.0 \times 10^{-13} \qquad\qquad (C)$$

$$SO_4^{2-} + H_2O \underset{}{\overset{K_{base}}{\rightleftharpoons}} HSO_4^- + OH^- \qquad\qquad K_{base} = 9.8 \times 10^{-13} \qquad\qquad (D)$$

$$H_2O \underset{}{\overset{K_w}{\rightleftharpoons}} H^+ + OH^- \qquad\qquad K_w = 1.0 \times 10^{-14} \qquad\qquad (E)$$

Charge balance: $2[Ca^{2+}] + [CaOH^+] + [H^+] = 2[SO_4^{2-}] + [HSO_4^-] + [OH^-]$ $\qquad\qquad$ (F)

Mass balance: [total calcium] = [total sulfate]

$$[Ca^{2+}] + [\cancel{CaSO_4(aq)}] + [CaOH^+] = [SO_4^{2-}] + [HSO_4^-] + [\cancel{CaSO_4(aq)}] \qquad (G)$$

Write expressions for effective equilibrium constants K'.

$$K_{sp} = [Ca^{2+}]\gamma_{Ca^{2+}}[SO_4^{2-}]\gamma_{SO_4^{2-}} \Rightarrow K'_{sp} = K_{sp}\frac{1}{\gamma_{Ca^{2+}}\gamma_{SO_4^{2-}}} = [Ca^{2+}][SO_4^{2-}] \tag{H}$$

$$K_{acid} = \frac{[CaOH^+]\gamma_{CaOH^+}[H^+]\gamma_{H^+}}{[Ca^{2+}]\gamma_{Ca^{2+}}} \Rightarrow K'_{acid} = K_{acid}\frac{\gamma_{Ca^{2+}}}{\gamma_{CaOH^+}\gamma_{H^+}} = \frac{[CaOH^+][H^+]}{[Ca^{2+}]} \tag{I}$$

$$K_{base} = \frac{[HSO_4^-]\gamma_{HSO_4^-}[OH^-]\gamma_{OH^-}}{[SO_4^{2-}]\gamma_{SO_4^{2-}}}$$

$$\Rightarrow K'_{base} = K_{base}\frac{\gamma_{SO_4^{2-}}}{\gamma_{HSO_4^-}\gamma_{OH^-}} = \frac{[HSO_4^-][OH^-]}{[SO_4^{2-}]} \tag{J}$$

$$K_w = [H^+]\gamma_{H^+}[OH^-]\gamma_{OH^-} \Rightarrow K'_w = \frac{K_w}{\gamma_{H^+}\gamma_{OH^-}} = [H^+][OH^-] \tag{K}$$

Substitute equilibrium expressions into mass balance (G):

$$[Ca^{2+}] + [CaOH^+] = [SO_4^{2-}] + [HSO_4^-]$$

$$[Ca^{2+}] + \frac{K'_{acid}[Ca^{2+}]}{[H^+]} = [SO_4^{2-}] + \frac{K'_{base}[SO_4^{2-}]}{[OH^-]} = [SO_4^{2-}]\left(1 + \frac{K'_{base}}{[OH^-]}\right) \tag{L}$$

Substitute expression for $[SO_4^{2-}]$ from (H) into right side of (L):

$$[Ca^{2+}] + \frac{K'_{acid}[Ca^{2+}]}{[H^+]} = \frac{K'_{sp}}{[Ca^{2+}]}\left(1 + \frac{K'_{base}}{[OH^-]}\right)$$

Express $[OH^-]$ in terms of $[H^+]$:

$$[Ca^{2+}] + \frac{K'_{acid}[Ca^{2+}]}{[H^+]} = \frac{K'_{sp}}{[Ca^{2+}]}\left(1 + \frac{K'_{base}[H^+]}{K'_w}\right)$$

Multiply both sides by $[Ca^{2+}]$ and solve for $[Ca^{2+}]$:

$$[Ca^{2+}]^2\left(1 + \frac{K'_{acid}}{[H^+]}\right) = K'_{sp}\left(1 + \frac{K'_{base}[H^+]}{K'_w}\right) \tag{X}$$

$$[Ca^{2+}]^2 = K'_{sp}\left(1 + \frac{K'_{base}[H^+]}{K'_w}\right) \bigg/ \left(1 + \frac{K'_{acid}}{[H^+]}\right) \tag{M}$$

In the following spreadsheet, we begin with a guess of pH = 7 in cell A14 and an
ionic strength of 0 in cell C16. The spreadsheet computes the concentrations in
row 14 and the sum of charges in cell H14.

	A	B	C	D	E	F	G	H
1	Saturated CaSO$_4$							
2								
3	K$_{sp}$ =	2.4E-05	K$_{sp}$' =	2.40E-05				
4	K$_{ip}$ =	5.0E-03	K$_{ip}$' =	5.00E-03				
5	K$_{acid}$ =	2.0E-13	K$_{acid}$' =	2.00E-13				
6	K$_{base}$ =	9.8E-13	K$_{base}$' =	9.80E-13				
7	K$_w$ =	1.0E-14	K$_w$' =	1.00E-14				
8								
9	Davies activity coefficients:							
10	Ca^{2+}	1.00		SO$_4^{2-}$	1.00		OH$^-$	1.00
11	CaOH$^+$	1.00		HSO$_4^-$	1.00		H$^+$	1.00
12								
13	pH	[H$^+$]	[OH$^-$]	[Ca^{2+}]	[CaOH$^+$]	[SO$_4^{2-}$]	[HSO$_4^-$]	Net charge
14	7.0000	1.00E-07	1.00E-07	4.90E-03	9.80E-09	4.90E-03	4.80E-08	3.82E-08
15								
16	Ionic strength =		0.00000	← Initial value is 0				
17	New ionic strength =		0.00000					
18								
19	Activity coefficients = 10^(-0.51*charge^2*(SQRT($C$16)/(1+SQRT($C$16))-0.3*$C$16))							
20	K$_{sp}$' = B3/(B10*E10)			[H$^+$] = 10^-A14/H11				
21	K$_{ip}$' = B4			[OH$^-$] = D7/B14				
22	K$_{acid}$' = B5*B10/(B11*H11)			[Ca^{2+}] = SQRT(D3*(1+D6*B14/D7)/(1+D5/B14))				
23	K$_{base}$' = B6*E10/(E11*H10)			[CaOH$^+$] =D5*D14/B14				
24	K$_w$' = B7/(H11*H10)			[SO$_4^{2-}$] = D3/D14				
25				[HSO$_4^-$] =D6*F14/C14				
26	Net charge = 2*D14+E14+B14-2*F14-G14-C14							
27	New ionic strength = 0.5*(B14+C14+4*D14+E14+4*F14+G14)							

We then use Solver or Goal Seek to vary pH in cell A14 until the net charge in cell H14 is close to 0. We set a limit of 1E-18 for this calculation. The spreadsheet computes a pH and new set of concentrations and an ionic strength in cell C17. We transcribe the ionic strength from C17 into cell C16 and repeat the cycle several times until ionic strength no longer changes. Final results are displayed in the following spreadsheet. The pH is 7.06 and the ionic strength is 0.041 M.

	A	B	C	D	E	F	G	H
1	Saturated CaSO$_4$							
2								
3	K_{sp} =	2.4E-05	K_{sp}' =	1.04E-04				
4	K_{ip} =	5.0E-03	K_{ip}' =	5.00E-03				
5	K_{add} =	2.0E-13	K_{acid}' =	1.39E-13				
6	K_{base} =	9.8E-13	K_{base}' =	6.80E-13				
7	K_w =	1.0E-14	K_w' =	1.44E-14				
8								
9	Davies activity coefficients:							
10	Ca^{2+}	0.48		SO$_4^{2-}$	0.48		OH$^-$	0.83
11	CaOH$^+$	0.83		HSO$_4^-$	0.83		H$^+$	0.83
12								
13	pH	[H$^+$]	[OH$^-$]	[Ca^{2+}]	[CaOH$^+$]	[SO$_4^{2-}$]	[HSO$_4^-$]	Net charge
14	7.0648	1.03E-07	1.39E-07	1.02E-02	1.37E-08	1.02E-02	4.96E-08	8.20E-19
15								
16	Ionic strength =		0.04071	← Initial value is 0				
17	New ionic strength =		0.04071					

12-14.

$$AgCN(s) \rightleftharpoons Ag^+ + CN^- \qquad [Ag^+] = K_{sp}'/[CN^-]$$

$$HCN(aq) \rightleftharpoons CN^- + H^+ \qquad [HCN(aq)] = [CN^-][H^+]/K_{HCN}'$$

$$Ag^+ + H_2O \rightleftharpoons AgOH(aq) + H^+ \qquad [AgOH] = K_{Ag}'[Ag^+]/[H^+]$$

$$Ag^+ + CN^- + OH^- \rightleftharpoons Ag(OH)(CN)^-$$

$$[Ag(OH)(CN)^-] = K_{AgOHCN}'[Ag^+][CN^-][OH^-] = K_{AgOHCN}'K_w'[Ag^+][CN^-]/[H^+]$$

$$Ag^+ + 2CN^- \rightleftharpoons Ag(CN)_2^- \qquad [Ag(CN)_2^-] = \beta_2'[Ag^+][CN^-]^2$$

$$Ag^+ + 3CN^- \rightleftharpoons Ag(CN)_3^{2-} \qquad [Ag(CN)_3^{2-}] = \beta_3'[Ag^+][CN^-]^3$$

$$H_2O \rightleftharpoons H^+ + OH^- \qquad [OH^-] = K_w'/[H^+]$$

$$K_{sp}' = \frac{K_{sp}}{\gamma_{Ag^+}\gamma_{CN^-}} \qquad K_{HCN}' = \frac{K_{HCN}}{\gamma_{CN^-}\gamma_{H^+}} \qquad K_{Ag}' = K_{Ag}\frac{\gamma_{Ag^+}}{\gamma_{H^+}}$$

$$K_{AgOHCN}' = K_{AgOHCN}\frac{\gamma_{Ag^+}\gamma_{CN^-}\gamma_{OH^-}}{\gamma_{Ag(OH)(CN)^-}} \qquad K_w' = \frac{K_w}{\gamma_{H^+}\gamma_{OH^-}}$$

$$\beta_2' = \beta_2\frac{\gamma_{Ag^+}\gamma_{CN^-}^2}{\gamma_{Ag(CN)_2^-}} \qquad \beta_3' = \beta_3\frac{\gamma_{Ag^+}\gamma_{CN^-}^3}{\gamma_{Ag(CN)_3^{2-}}}$$

$$K_{sp} = 10^{-15.66} \qquad K_{HCN} = 10^{-9.21} \qquad K_{Ag} = 10^{-12.0}$$

$$K_{AgOHCN} = 10^{13.22} \qquad \beta_2 = 10^{20.48} \qquad \beta_3 = 10^{21.7}$$

Charge balance:

$$[OH^-] + [CN^-] + [Ag(OH)(CN)^-] + [Ag(CN)_2^-] + 2[Ag(CN)_3^{2-}]$$

$$= [H^+] + [Ag^+] + [K^+] + [Na^+] \qquad \text{(A)}$$

Mass balances:

$[K^+] = 0.10$ M   (B)

{total silver } + $[K^+]$ = {total cyanide}

$$[Ag^+] + [AgOH] + [Ag(OH)(CN)^-] + [Ag(CN)_2^-] + [Ag(CN)_3^{2-}] + [K^+]$$
$$= [CN^-] + [HCN] + [Ag(OH)(CN)^-] + 2[Ag(CN)_2^-] + 3[Ag(CN)_3^{2-}]$$

which simplifies to

$$[Ag^+] + [AgOH] - [Ag(CN)_2^-] - 2[Ag(CN)_3^{2-}] + [K^+] - [CN^-] - [HCN] = 0 \quad (C)$$

In the following spreadsheet, the initial ionic strength was guessed to be 0.10 M in cell C23. $[CN^-]$ was then guessed in cell A17. $[H^+]$ and $[OH^-]$ in cells B17 and B18 were computed from the known pH. $[Ag^+]$ in cell D17 was calculated from $[Ag^+] = K'_{sp}/[CN^-]$. $[AgOH]$, $[Ag(OH)(CN)^-]$, $[Ag(CN)_2^-]$, and $[Ag(CN)_3^{2-}]$ were computed from $[CN^-]$, $[H^+]$, $[Ag^+]$, and $[OH^-]$ with equilibrium expressions. The mass balance Equation C is evaluated in cell H21. In the key step, the value in cell H21 is set equal to 0 by using Solver to vary $[CN^-]$ in cell A17. When the mass balance is satisfied, all concentrations must be correct. $[Na^+]$ is then computed in cell C22 from the charge balance Equation A and the new ionic strength is found in cell C24. The new ionic strength is then entered in cell C23 and the process is repeated until the ionic strength no longer changes. The spreadsheet shows the final concentrations when the mass balance is satisfied.

	A	B	C	D	E	F	G	H
1	Species in silver-cyanide solution							
2								
3	log K_{sp} =	-15.66	K_{sp} =	2.19E-16	K_{sp}' =	3.65E-16	$[K^+]$ =	0.10
4	log K_{HCN} =	-9.21	K_{HCN} =	6.17E-10	K_{HCN}' =	1.03E-09	pH =	12.00
5	log K_{Ag} =	-12.0	K_{Ag} =	1.00E-12	K_{Ag}' =	1.00E-12		
6	log K_{AgOHCN} =	13.22	K_{AgOHCN} =	1.66E+13	K_{AgOHCN}' =	9.95E+12		
7	log β_2 =	20.48	β_2 =	3.02E+20	β_2' =	1.81E+20		
8	log β_3 =	21.7	β_3 =	5.01E+21	β_3' =	5.01E+21		
9	log K_w =	-14.00	K_w =	1.00E-14	K_w' =	1.67E-14		
10								
11	Davies activity coefficients:							
12	Ag^+	0.77		$Ag(OH)(CN)^-$	0.77		OH^-	0.77
13	$Ag(CN)_2^-$	0.77		CN^-	0.77		H^+	0.77
14	$Ag(CN)_3^{2-}$	0.36						
15								
16	$[CN^-]$	$[H^+]$	$[OH^-]$	$[Ag^+]$	$[AgOH]$	$[Ag(OH)(CN)^-]$		
17	1.513E-06	1.29E-12	1.29E-02	2.410E-10	1.867E-10	4.688E-05		
18				$[Ag(CN)_2^-]$	$[Ag(CN)_3^{2-}]$	$[HCN]$		
19				0.09999	4.187E-06	1.901E-09		
20								
21	Mass balance: $[Ag^+] + [AgOH] - [Ag(CN)_2^-] - 2[Ag(CN)_3^{2-}] + [K^+] - [CN^-] - [HCN]$ =							-1.5E-17
22	$[Na^+]$ =		0.01296	(from charge balance)				
23	Ionic strength =		0.113	$\leftarrow$ Initial value is 0.1				
24	New ionic strength =		0.113	$\leftarrow$ Substitute this value into C23 for next iteration				
25	Total Ag + K^+ =		0.20004	check				
26	Total CN =		0.20004	check				
27								
28	Activity coefficients = 10^(-0.51*charge^2*(SQRT($C$23)/(1+SQRT($C$23))-0.3*$C$23))							
29	K_{sp}' = D3/(B12*E13)		K_{HCN}' = D4/(E13*H13)			K_{Ag}' = D5*B12/H13		
30	K_{AgOHCN}' = D6*B12*E13*H12/E12					β_2' = D7*B12*E13^2/B13		
31	K_w' = D9/(H12*H13)					β_3' = D8*B12*E13^3/B14		
32	$[H^+]$ = 10^-H4/H13		$[OH^-]$ = F9/B17			$[AgOH]$ = F5*D17/B17		
33	$[Ag^+]$ = F3/A17							
34	$[Ag(OH)(CN)^-]$ = F6*D17*A17*C17					$[Ag(CN)_2^-]$ = F7*D17*A17^2		
35	$[Ag(CN)_3^{2-}]$ = F8*D17*A17^3					$[HCN]$ = A17*B17/F4		
36	Mass balance = D17+E17-D19-2*E19+H3-A17-F19							
37	$[Na^+]$ = C17+A17+F17+D19+2*E19-B17-D17-H3							
38	New ionic strength = 0.5*(A17+B17+C17+D17+F17+D19+4*E19+H3+C22)							

12-15.

$$Fe^{2+} + G^- \rightleftharpoons FeG^+ \qquad\qquad [FeG^+] = \beta_1'[Fe^{2+}][G^-]$$

$$Fe^{2+} + 2G^- \rightleftharpoons FeG_2(aq) \qquad\qquad [FeG_2(aq)] = \beta_2'[Fe^{2+}][G^-]^2$$

$$Fe^{2+} + 3G^- \rightleftharpoons FeG_3^- \qquad\qquad [FeG_3^-] = \beta_3'[Fe^{2+}][G^-]^3$$

$$Fe^{2+} + H_2O \rightleftharpoons FeOH^+ + H^+ \qquad\qquad [FeOH^+] = K_a'[Fe^{2+}]/[H^+]$$

$$HG \rightleftharpoons G^- + H^+ \qquad\qquad [HG] = [G^-][H^+]/K_2'$$

$$H_2G^+ \rightleftharpoons HG + H^+ \qquad\qquad [H_2G^+] = [HG][H^+]/K_1' = [G^-][H^+]^2/(K_2'K_1')$$

$$H_2O \rightleftharpoons H^+ + OH^- \qquad\qquad [OH^-] = K_w'/[H^+]$$

$$\beta_1' = \beta_1 \frac{\gamma_{Fe^{2+}}\gamma_{G^-}}{\gamma_{FeG^+}} \qquad \beta_2' = \beta_2 \frac{\gamma_{Fe^{2+}}\gamma_{G^-}^2}{\gamma_{FeG_2}} \qquad \beta_3' = \beta_3 \frac{\gamma_{Fe^{2+}}\gamma_{G^-}^3}{\gamma_{FeG_3^-}}$$

$$K_a' = K_a \frac{\gamma_{Fe^{2+}}}{\gamma_{H^+}} \qquad K_2' = K_2\frac{\gamma_{HG}}{\gamma_{G^-}\gamma_{H^+}} \qquad K_1' = K_1\frac{\gamma_{H_2G^+}}{\gamma_{HG}\gamma_{H^+}} \qquad K_w' = \frac{K_w}{\gamma_{H^+}\gamma_{OH^-}}$$

$$\beta_1 = 10^{4.31} \qquad\qquad \beta_2 = 10^{7.65} \qquad \beta_3 = 10^{8.87}$$

$$K_a = 10^{-9.4} \qquad\qquad K_1 = 10^{-2.350} \qquad K_2 = 10^{-9.778}$$

Charge balance:

$$[OH^-] + [G^-] + [FeG_3^-] + [Cl^-] =$$
$$[H^+] + 2[Fe^{2+}] + [FeOH^+] + [FeG^+] + [H_2G^+] \qquad (A)$$

Mass balances:

$$F_{Fe} = 0.050\ M = [Fe^{2+}] + [FeG^+] + [FeG_2] + [FeG_3^-] + [FeOH^+] \qquad (B)$$

$$F_G = 0.100\ M = [G^-] + [HG] + [H_2G^+] + [FeG^+] + 2[FeG_2] + 3[FeG_3^-] \qquad (C)$$

Substitute equilibrium expressions into the mass balance B to express $[Fe^{2+}]$ in terms of $[G^-]$ and $[H^+]$:

$$F_{Fe} = [Fe^{2+}] + \beta_1'[Fe^{2+}][G^-] + \beta_2'[Fe^{2+}][G^-]^2 + \beta_3'[Fe^{2+}][G^-]^3 + K_a'[Fe^{2+}]/[H^+]$$

$$[Fe^{2+}] = \frac{F_{Fe}}{1 + \beta_1'[G^-] + \beta_2'[G^-]^2 + \beta_3'[G^-]^3 + K_a'/[H^+]} \qquad (D)$$

In the spreadsheet on the next page, we guess a value for $[G^-]$ in cell A17 and compute $[H^+] = 10^{-pH}/\gamma_{H^+}$ in cell B17. Equation D is used to find $[Fe^{2+}]$ in cell D17. Other concentrations in rows 17 and 19 are computed from $[G^-]$, $[H^+]$, and $[Fe^{2+}]$. Cell H24 checks that the sum of iron species equals $F_{Fe} = 0.050$ M. By using the mass balance for iron to find Fe^{2+} in cell D17, cell H24 must give F_{Fe} if we have not made a mistake. We use the mass balance for glycine for the first time in cell H25, where we add all the glycine species. Then use Solver to vary $[G^-]$ in cell A17 to satisfy the mass balance for glycine in cell H25. In Solver, Set Target Cell H25 Equal To Value of 0.1 By Changing Cells A17. $[Cl^-]$ in cell C20 is then found from the charge balance equation A. The new ionic strength is computed in cell C22. Enter the value from cell C22 into cell C21 and repeat the whole process several times until the ionic strength is constant.

	A	B	C	D	E	F	G	H
1	Composition of Fe^{2+}-glycine solution							
2								
3	log β_1 =	4.31	β_1 =	2.04E+04	β_1' =	1.16E+04	pH =	8.50
4	log β_2 =	7.65	β_2 =	4.47E+07	β_2' =	1.91E+07	F_{Fe} =	0.050
5	log β_3 =	8.87	β_3 =	7.41E+08	β_3' =	3.17E+08	F_G =	0.100
6	log K_a =	-9.4	K_a =	3.98E-10	$K_{a'}$ =	2.60E-10		
7	log K_1 =	-2.350	K_1 =	4.47E-03	K_1' =	4.47E-03		
8	log K_2 =	-9.778	K_2 =	1.67E-10	K_2' =	2.21E-10		
9	log K_w =	-14.00	K_w =	1.00E-14	K_w' =	1.33E-14		
10								
11	Davies activity coefficients:							
12	Fe^{2+}	0.57		$FeOH^+$	0.87		OH^-	0.87
13	FeG^+	0.87		G^-	0.87		H^+	0.87
14	FeG_3^-	0.87		H_2G^+	0.87			
15								
16	[G^-]	[H^+]	[OH^-]	[Fe^{2+}]	[$FeOH^+$]	[FeG^-]	[FeG_2]	[FeG_3^-]
17	1.10E-03	3.64E-09	3.64E-06	1.33E-03	9.50E-05	1.70E-02	3.10E-02	5.68E-04
18	Guess			[HG]	[H_2G^+]			
19				1.82E-02	1.48E-08			
20			[Cl^-] =	0.0181	(from charge balance)			
21		Ionic strength =	0.0211	← Initial value is 0.01				
22		New ionic strength =	0.0211	← Substitute this value into cell C21 for next iteration				
23								
24		Mass balance for Fe: [Fe^{2+}] + [FeG^+] + [FeG_2] + [FeG_3^-] + [$FeOH^+$] =						0.0500
25		Mass balance for G: [G^-] + [FeG^+] + 2[FeG_2] + 3[FeG_3^-] + [H_2G^+] + [HG] =						0.1000
26								
27	Activity coefficients = 10^(-0.51*charge^2*(SQRT($C$21)/(1+SQRT($C$21))-0.3*$C$21))							
28	β_1' = D3/(B12*E13)		β_2' = D4/(E13*H13)		β_3' = D5*B12/H13		K_w' = D9/(H12*H13)	
29	$K_{a'}$ = D6*B12*E13*H12/E12			K_1' = D7*B12*E13^2/B13			K_2' = D8*B12*E13^3/B14	
30	[H^+] = 10^-H3/H13		[OH^-] = F9/B17			[$FeOH^+$] = F6*D17/B17		
31	[Fe^{2+}] = H4/(1+F3*A17+F4*A17^2+F5*A17^3+F6/B17)							
32	[FeG^-] = F3*D17*A17			[FeG_2] = F4*D17*A17^2			[FeG_3^-] = F5*D17*A17^3	
33	[HG] = A17*B17/F8			[H_2G^+] = A17*B17^2/(F7*F8)				
34	[Cl^-] = B17+2*D17+E17+F17+E19-A17-C17-H17							
35	New ionic strength = 0.5*(A17+B17+C17+4*D17+E17+F17+H17+E19+C20)							

Fraction of Fe in each form:

$[Fe^{2+}]$, 2.7%; $[FeG^+]$, 34.0%; $[FeG_2]$, 62.0%; $[FeG_3^-]$, 1.1%; $[FeOH^+]$, 0.2%

Fraction of glycine in each form:

$[G^-]$, 1.1%; $[HG]$, 18.2%; $[H_2G^+]$, 0.000 015%; $[FeG^+]$, 17.0%; $2[FeG_2]$, 62.0%; $3[FeG_3^-]$, 1.7%

The chemistry: We dissolved FeG_2 and found that the principal species are FeG^+, FeG_2, and HG. The chemistry that requires HCl to be added to obtain pH 8.5 is $FeG_2 \rightleftharpoons FeG^+ + G^-$ followed by $G^- + H^+ \rightleftharpoons HG$. The base G^- is released when FeG_2 dissolves, so HCl is required to neutralize the base.

12-16. (a) A range of initial values of pK_1 and pK_2, such as 6 and 6 or 10 and 10, converge to the correct solution. Even choosing a ridiculous value for pK_w', such as 10, converges to the correct value after executing Solver more than once.

(b) Fixing pK_w' at 13.797 and using Solver gives the optimized values of pK_1 and $pK_2 = 2.312$ and 9.630, which are hardly different from the values obtained when pK_w' is allowed to vary. However, inspection of the following curves shows that $\bar{n}_H$(measured) deviates systematically from $\bar{n}_H$(theoretical) at the end of the titration when $\bar{n}_H$(measured) should approach 0.

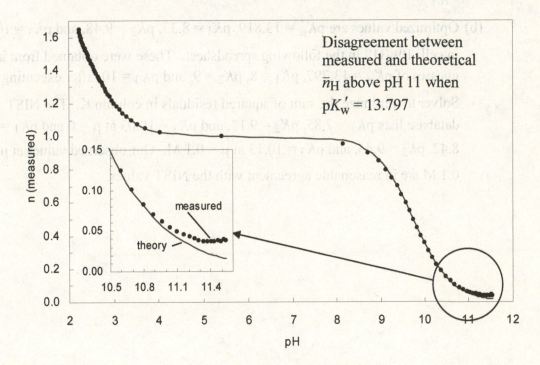

Disagreement between measured and theoretical $\bar{n}_H$ above pH 11 when $pK_w' = 13.797$

12-17. (a) $\bar{n}_H = \dfrac{\text{moles of bound } H^+}{\text{total moles of weak acid}} = \dfrac{3[H_3A^{3+}] + 2[H_2A^{2+}] + [HA^+]}{[H_3A^{3+}] + [H_2A^{2+}] + [HA^+] + [A]}$

or $\bar{n}_H F_{H_3A} = 3[H_3A^{3+}] + 2[H_2A^{2+}] + [HA^+]$ (A)

where $F_{H_3A} = [H_3A^{3+}] + [H_2A^{2+}] + [HA^+] + [A]$

charge balance:

$[H^+] + [Na^+] + \underbrace{3[H_3A^{3+}] + 2[H_2A^{2+}] + [HA^+]}_{= \bar{n}_H F_{H_3A} \text{ from Eq. (A)}} = [OH^-] + [Cl^-]_{HCl} + \underbrace{[Cl^-]_{H_3A}}_{= 3F_{H_3A}}$ (B)

where $[Cl^-]_{HCl}$ is from HCl and $[Cl^-]_{H_3A}$ is from H_3A^{3+}.

Each mol of H_3A^{3+} brings $3Cl^-$, so $[Cl^-]_{H_3A} = 3F_{H_3A}$

We can rearrange Eq. (B) to solve for $\bar{n}_H$:

$[H^+] + [Na^+] + \bar{n}_H F_{H_3A} = [OH^-] + [Cl^-]_{HCl} + 3F_{H_3A}$

$\bar{n}_H F_{H_3A} = 3F_{H_3A} + [OH^-] + [Cl^-]_{HCl} - [H^+] - [Na^+]$

$\bar{n}_H = 3 + \dfrac{[OH^-] + [Cl^-]_{HCl} - [H^+] - [Na^+]}{F_{H_3A}}$ (C)

Expression (C) is the same equation derived in the text, with $n = 3$.

The expression for $\bar{n}_H$(theoretical) is $\bar{n}_H$(theoretical) $= 3\alpha_{H_3A} + 2\alpha_{H_2A} + \alpha_{HA}$.

(b) Optimized values are $pK'_w = 13.819$, $pK_1 = 8.33$, $pK_2 = 9.48$, and $pK_3 = 10.19$ in cells B9:B12 in the following spreadsheet. These were obtained from initial guesses of $pK'_w = 13.797$, $pK_1 = 8$, $pK_2 = 9$, and $pK_3 = 10$, after executing Solver to minimize the sum of squared residuals in column K. The NIST database lists $pK_1 = 7.85$, $pK_2 = 9.13$, and $pK_3 = 10.03$ at $\mu = 0$ and $pK_1 = 8.42$, $pK_2 = 9.43$, and $pK_3 = 10.13$ at $\mu = 0.1$ M. Our observed values at $\mu = 0.1$ M are in reasonable agreement with the NIST values.

	A	B	C	D	E	F	G	H	I	J	K
1	Difference plot for tris(2-aminoethyl)amine										
2			C17 = 10^-B17/$B$8				D17 = 10^-$B$9/C17				
3	Titrant NaOH =	0.4905	C_b (M)	E17 = $B$7+($B$6-$B$3*A17-(C17-D17)*($B$4+A17))/$B$5							
4	Initial volume =	40	V_o (mL)	denom = $C17^3+$C17^2*$E$10+$C17*$E$10*$E$11+$E$10*$E$11*$E$12							
5	H_3A =	0.139	L (mmol)	F17 = $C17^3/denom							
6	HCl added =	0.115	A (mmol)	G17 = $C17^2*$E$10/denom)							
7	Number of H^+=	3	n	H17 = $C17*$E$10*$E$11/denom							
8	Activity coeff =	0.78	γ_H	I17 = $E$10*$E$11*$E$12/denom							
9	pK_w' =	13.819		J17 = 3*F17+2*G17+H17							
10	pK_1 =	8.334		K_1 =	4.636E-09	= 10^-B10					
11	pK_2 =	9.483		K_2 =	3.289E-10	= 10^-B11					
12	pK_3 =	10.188		K_3 =	6.485E-11	= 10^-B12					
13	$\Sigma(resid)^2$ =	0.0510	= sum of column K								
14											
15	v	pH	$[H^+]$ =	$[OH^-]$ =	Measured	α_{H3A}	α_{H2A}	α_{HA}	α_A	Theoretical	$(residuals)^2$ =
16	mL NaOH		$(10^{-pH})/\gamma_H$	$(10^{-pKw})/[H^+]$	n_H					n_H	$(n_{meas} - n_{thoer})^2$
17	0.00	2.709	2.51E-03	6.06E-12	3.106	1.000	0.000	0.000	0.000	3.000	0.011303
18	0.02	2.743	2.32E-03	6.55E-12	3.090	1.000	0.000	0.000	0.000	3.000	0.008046
19	:										
20	0.34	8.158	8.91E-09	1.70E-06	2.628	0.649	0.338	0.012	0.000	2.637	0.000077
21	0.36	8.283	6.68E-09	2.27E-06	2.558	0.579	0.401	0.020	0.000	2.558	0.000001
22	:										
23	0.54	9.087	1.05E-09	1.45E-05	1.926	0.145	0.641	0.201	0.012	1.919	0.000045
24	0.56	9.158	8.91E-10	1.70E-05	1.856	0.121	0.630	0.232	0.017	1.855	0.000002
25	:										
26	0.78	9.864	1.75E-10	8.66E-05	1.100	0.010	0.277	0.520	0.192	1.106	0.000031
27	0.80	9.926	1.52E-10	9.98E-05	1.034	0.008	0.243	0.525	0.224	1.035	0.000001
28	:										
29	1.00	10.545	3.66E-11	4.15E-04	0.421	0.000	0.039	0.346	0.615	0.424	0.000011
30	1.02	10.615	3.11E-11	4.88E-04	0.372	0.000	0.030	0.314	0.656	0.375	0.000007
31	:										
32	1.38	11.496	4.09E-12	3.71E-03	0.062	0.000	0.001	0.059	0.940	0.061	0.000002
33	1.40	11.521	3.86E-12	3.93E-03	0.057	0.000	0.001	0.056	0.943	0.058	0.000000

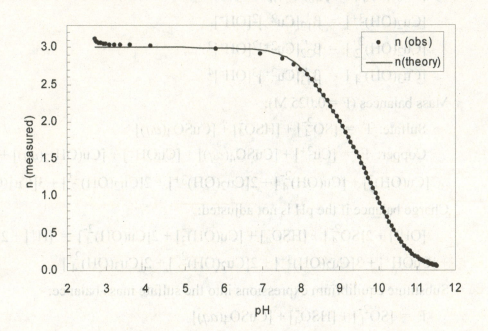

(c) The fractions of each species are computed in columns F, G, H, and I beginning in row 17 in the spreadsheet. Results are shown following graph.

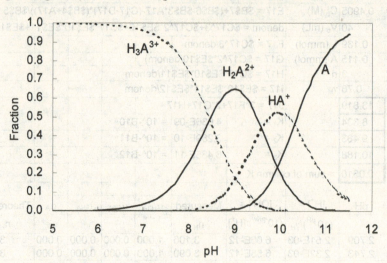

12-18. (a) From the equilibrium expressions, we write

$$[CuSO_4(aq)] = K'_{ip}[Cu^{2+}][SO_4^{2-}]$$

$$[HSO_4^-] = [SO_4^{2-}][H^+]/K'_a$$

$$[CuOH^+] = \beta'_1[Cu^{2+}][OH^-]$$

$$[Cu(OH)_2(aq)] = \beta'_2[Cu^{2+}][OH^-]^2$$

$$[Cu(OH)_3^-] = \beta'_3[Cu^{2+}][OH^-]^3$$

$$[Cu(OH)_4^{2-}] = \beta'_4[Cu^{2+}][OH^-]^4$$

$$[Cu_2(OH)^{3+}] = \beta'_{12}[Cu^{2+}]^2[OH^-]$$

$$[Cu_2(OH)_2^{2+}] = \beta'_{22}[Cu^{2+}]^2[OH^-]^2$$

$$[Cu_3(OH)_4^{2+}] = \beta'_{43}[Cu^{2+}]^3[OH^-]^4$$

Mass balances (F = 0.025 M):

Sulfate: $F = [SO_4^{2-}] + [HSO_4^-] + [CuSO_4(aq)]$

Copper: $F = [Cu^{2+}] + [CuSO_4(aq)] + [CuOH^+] + [Cu(OH)_2(aq)] +$

$[Cu(OH)_3^-] + [Cu(OH)_4^{2-}] + 2[Cu_2(OH)^{3+}] + 2[Cu_2(OH)_2^{2+}] + 3[Cu_3(OH)_4^{2+}]$

Charge balance if the pH is not adjusted:

$[OH^-] + 2[SO_4^{2-}] + [HSO_4^-] + [Cu(OH)_3^-] + 2[Cu(OH)_4^{2-}] = [H^+] + 2[Cu^{2+}]$

$+ [CuOH^+] + 3[Cu_2(OH)^{3+}] + 2[Cu_2(OH)_2^{2+}] + 2[Cu_3(OH)_4^{2+}]$

Substitute equilibrium expressions into the sulfate mass balance:

$$F = [SO_4^{2-}] + [HSO_4^-] + [CuSO_4(aq)]$$

$$= [SO_4^{2-}] + [SO_4^{2-}][H^+]/K_a' + K_{ip}'[Cu^{2+}][SO_4^{2-}]$$

$$\text{or} \quad [SO_4^{2-}] = \frac{F}{1 + [H^+]/K_a' + K_{ip}'[Cu^{2+}]} \tag{A}$$

Therefore, we can find $[SO_4^{2-}]$ if we know $[Cu^{2+}]$ and $[H^+]$. We can also find $[HSO_4^-]$ from the equilibrium relationship $[HSO_4^-] = [SO_4^{2-}][H^+]/K_a'$. We find the concentrations of copper species from their equilibrium relationships.

In the spreadsheet on the next page, pH is input in column A beginning at cell A14. $[H^+]$ and $[OH^-]$ are computed from pH in columns B and C. The initial value of $[Cu^{2+}]$ in each row of column D is a *guess*. We use Solver later to find the correct value of $[Cu^{2+}]$. $[SO_4^{2-}]$ in column E is computed with Equation A. The remaining concentrations in columns F through N are computed from the equilibrium relationships. The sum of copper species is tallied in column O. The key step is to use Solver to vary $[Cu^{2+}]$ in column A, so the total copper in column O equals 0.025 M. Solver must be applied separately to each row of the spreadsheet.

(b) Column P gives the sum of all charges. This sum will be zero at the pH of 0.025 M CuSO₄, to which no acid or base has been added. Trial-and-error variation of the pH shows that the charge is closest to zero at pH 4.61. Column Q was not necessary for the problem, but it shows that the ionic strength of the solution is near 0.075 M at most pH values, if no solids precipitate.

(c) To find out if the solubility of any salt has been exceeded, we evaluate the reaction quotient $[Cu^{2+}][OH^-]^2$ for $Cu(OH)_2(s)$ and $CuO(s)$ and the reaction quotient $[Cu^{2+}][OH^-]^{3/2}[SO_4^{2-}]^{1/4}$ for $Cu(OH)_{1.5}(SO_4)_{0.25}(s)$. The solubility of $Cu(OH)_{1.5}(SO_4)_{0.25}(s)$ is exceeded above pH $\approx$ 4.5. The solubility of $CuO(s)$ is exceeded above pH $\approx$ 5. The solubility of $Cu(OH)_2(s)$ is exceeded above pH $\approx$ 5.5. We predict that 0.025 M CuSO₄ is not a stable solution. At the calculated pH of 4.61, some $Cu(OH)_{1.5}(SO_4)_{0.25}(s)$ will precipitate.

	A	B	C	D	E	F	G	H	I
1	Copper-sulfate-hydroxide system								
2	F =	0.025	M						
3	log K_{ip}' =	1.26	K_{ip}' =	1.8E+01	log K_{sp}' (OH-SO$_4$) =		-16.68		
4	log K_a' =	-1.54	K_a' =	2.9E-02	log K_{sp}' (OH) =		-18.7		
5	log β_1' =	6.1	β_1' =	1.E+06	log K_{sp}' (O) =		-19.7		
6	log β_2' =	11.2	β_2' =	2.E+11	K_{sp}' (OH-SO$_4$) =		2.1E-17		
7	log β_3' =	14.5	β_3' =	3.E+14	K_{sp}' (OH) =		2.E-19		
8	log β_4' =	15.6	β_4' =	4.E+15	K_{sp}' (O) =		2.E-20		
9	log β_{12}' =	8.2	β_{12}' =	2.E+08	log K_w' =		-13.79		
10	log β_{22}' =	16.8	β_{22}' =	6.E+16	K_w' =		1.62E-14		
11	log β_{43}' =	33.5	β_{43}' =	3.E+33	γ_{H^+} =		0.78		
12									
13	pH	[H$^+$]	[OH$^-$]	[Cu^{2+}]	[SO$_4^{2-}$]	[HSO$_4^-$]	[CuSO$_4$]	[CuOH$^+$]	[Cu(OH)$_2$]
14	3	1.3E-03	1.3E-11	0.018824	1.8E-02	8.0E-04	6.2E-03	3.0E-07	4.8E-13
15	4	1.3E-04	1.3E-10	0.018665	1.9E-02	8.3E-05	6.3E-03	3.0E-06	4.7E-11
16	4.61	3.1E-05	5.2E-10	0.018602	1.9E-02	2.0E-05	6.3E-03	1.2E-05	7.8E-10
17	5	1.3E-05	1.3E-09	0.018476	1.9E-02	8.3E-06	6.3E-03	2.9E-05	4.7E-09
18	6	1.3E-06	1.3E-08	0.014136	2.0E-02	8.8E-07	5.1E-03	2.3E-04	3.6E-07
19	8	1.3E-08	1.3E-06	9.8E-05	2.5E-02	1.1E-08	4.4E-05	1.6E-04	2.5E-05
20	12	1.3E-12	1.3E-02	3.3E-11	2.5E-02	1.1E-12	1.5E-11	5.2E-07	8.3E-04

	J	K	L	M	N	O	P	Q
12						Total	Charge	Ionic
13	[Cu(OH)$_3^-$]	[Cu(OH)$_4^{2-}$]	[Cu$_2$(OH)$^{3+}$]	[Cu$_2$(OH)$_2^{2+}$]	[Cu$_3$(OH)$_4^{2+}$]	copper	balance	strength
14	1.2E-20	1.9E-30	7.1E-07	3.6E-09	5.4E-16	0.0250000	2.1E-03	0.075
15	1.2E-17	1.9E-26	7.0E-06	3.5E-07	5.3E-12	0.0250000	2.0E-04	0.075
16	8.1E-16	5.2E-24	2.8E-05	5.8E-06	1.4E-09	0.0250000	-1.0E-07	0.075
17	1.2E-14	1.9E-22	6.8E-05	3.4E-05	5.1E-08	0.0250000	-1.5E-04	0.075
18	9.0E-12	1.4E-18	4.0E-04	2.0E-03	2.3E-04	0.0250000	-5.6E-03	0.074
19	6.3E-08	1.0E-12	1.9E-06	9.7E-04	7.6E-03	0.0250000	-3.2E-02	0.067
20	2.1E-02	3.3E-03	2.1E-15	1.1E-08	2.8E-06	0.0250000	-9.0E-02	0.073

	A	B	C	D	E	F	G	H	I
22	[H$^+$]	B14 = 10^-A14/$G$11				[OH$^-$]	C14 = $G$10/B14		
23	[Cu^{2+}]	Enter guess and find value with SOLVER to make total copper in column 0 = 0.025							
24	[SO$_4^{2-}$]	E14 = $B$2/(1+B14/$D$4+$D$3*D14)				[HSO$_4^-$]	F14 = E14*B14/$D$4		
25	[CuSO$_4$]	G14 = $D$3*D14*E14				[CuOH$^+$]	H14 = $D$5*D14*C14		
26	[Cu(OH)$_2$]	I14 = $D$6*D14*C14^2				[Cu(OH)$_3^-$]	J14 = $D$7*D14*C14^3		
27	[Cu(OH)$_4^{2-}$]	K14 = $D$8*D14*C14^4				[Cu$_2$(OH)$^{3+}$]	L14 = $D$9*D14^2*C14		
28	[Cu$_2$(OH)$_2^{2+}$]	M14 = $D$10*D14^2*C14^2				[Cu$_3$(OH)$_4^{2+}$]	N14 = $D$11*D14^3*C14^4		
29	Total copper	O14 = D14+G14+H14+I14+J14+K14+2*L14+2*M14+3*N14							
30	Charge balance	P14 = B14-C14+2*D14-2*E14-F14+H14-J14-2*K14+3*L14+2*M14+2*N14							
31	Ionic strength	Q14 = 0.5*(B14+C14+4*D14+4*E14+F14+H14+J14+4*K14+9*L14+4*M14+4*N14)							

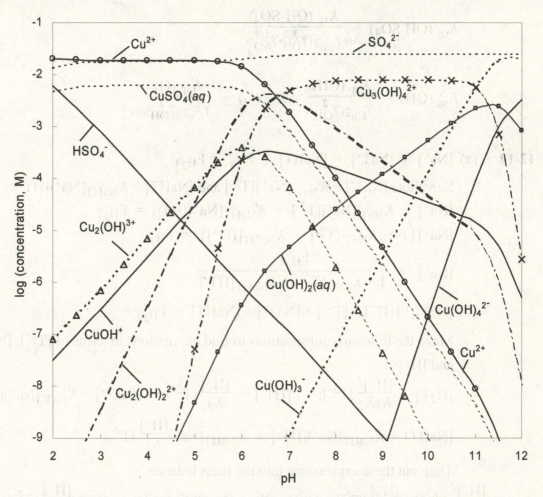

The following formulas were used when writing this problem to compute
equilibrium constants K' for $\mu = 0.1$ M from equilibrium constants K for $\mu = 0$.
If K' was reported for another ionic strength, such as 1 M, it was first converted
to K for $\mu = 0$ and then converted from K to K' for $\mu = 0.1$ M. Davies activity
coefficients were used.

$$K'_{ip} = K_{ip}\frac{\gamma_{Cu^{2+}}\gamma_{SO_4^{2-}}}{\gamma_{CuSO_4}} \qquad K'_a = K_a\frac{\gamma_{HSO_4^-}}{\gamma_{SO_4^{2-}}\gamma_{H^+}} \qquad \beta'_1 = \beta_1\frac{\gamma_{Cu^{2+}}\gamma_{OH^-}}{\gamma_{CuOH^+}}$$

$$\beta'_2 = \beta_2\frac{\gamma_{Cu^{2+}}\gamma_{OH^-}^2}{\gamma_{Cu(OH)_2}} \qquad \beta'_3 = \beta_3\frac{\gamma_{Cu^{2+}}\gamma_{OH^-}^3}{\gamma_{Cu(OH)_3^-}} \qquad \beta'_4 = \beta_4\frac{\gamma_{Cu^{2+}}\gamma_{OH^-}^4}{\gamma_{Cu(OH)_4^{2-}}}$$

$$\beta'_{12} = \beta_{12}\frac{\gamma_{Cu^{2+}}^2\gamma_{OH^-}}{\gamma_{Cu_2(OH)^{3+}}} \qquad \beta'_{22} = \beta_{22}\frac{\gamma_{Cu^{2+}}^2\gamma_{OH^-}^2}{\gamma_{Cu_2(OH)_2^{2+}}} \qquad \beta'_{43} = \beta_{43}\frac{\gamma_{Cu^{2+}}^3\gamma_{OH^-}^4}{\gamma_{Cu_3(OH)_4^{2+}}}$$

$$K'_{sp}(\text{OH-SO}_4) = \frac{K_{sp}(\text{OH-SO}_4)}{\gamma_{\text{Cu}^{2+}}\gamma_{\text{OH}^-}^{3/2}\gamma_{\text{SO}_4^{2-}}^{1/4}}$$

$$K'_{sp}(\text{OH}) = \frac{K_{sp}(\text{OH})}{\gamma_{\text{Cu}^{2+}}\gamma_{\text{OH}^-}^2} \qquad K'_{sp}(\text{O}) = \frac{K_{sp}(\text{O})}{\gamma_{\text{Cu}^{2+}}\gamma_{\text{OH}^-}^2}$$

12-19. (a) $[\text{Na}^+] + [\text{NaT}^-] + [\text{NaHT}] = F_{\text{Na}} = F_{\text{H}_2\text{T}}$

Substitute $[\text{NaT}^-] = K_{\text{NaT}^-}[\text{Na}^+][\text{T}^{2-}]$ and $[\text{NaHT}] = K_{\text{NaHT}}[\text{Na}^+][\text{HT}^-]$:

$[\text{Na}^+] + K_{\text{NaT}^-}[\text{Na}^+][\text{T}^{2-}] + K_{\text{NaHT}}[\text{Na}^+][\text{HT}^-] = F_{\text{H}_2\text{T}}$

$[\text{Na}^+]\{1 + K_{\text{NaT}^-}[\text{T}^{2-}] + K_{\text{NaHT}}[\text{HT}^-]\} = F_{\text{H}_2\text{T}}$

$$[\text{Na}^+] = \frac{F_{\text{H}_2\text{T}}}{1 + K_{\text{NaT}^-}[\text{T}^{2-}] + K_{\text{NaHT}}[\text{HT}^-]} \tag{A}$$

(b) $[\text{H}_2\text{T}] + [\text{HT}^-] + [\text{T}^{2-}] + [\text{NaT}^-] + [\text{NaHT}] = F_{\text{H}_2\text{T}}$

Make the following substitutions to find expressions in terms of $[\text{T}^{2-}]$, $[\text{Na}^+]$, and $[\text{H}^+]$:

$$[\text{H}_2\text{T}] = \frac{[\text{H}^+]^2}{K_1K_2}[\text{T}^{2-}]; \qquad [\text{HT}^-] = \frac{[\text{H}^+]}{K_2}[\text{T}^{2-}]; \qquad [\text{NaT}^-] = K_{\text{NaT}^-}[\text{Na}^+][\text{T}^{2-}]$$

$$[\text{NaHT}] = K_{\text{NaHT}}[\text{Na}^+][\text{HT}^-] = K_{\text{NaHT}}[\text{Na}^+]\frac{[\text{H}^+]}{K_2}[\text{T}^{2-}]$$

Then put these expressions into the mass balance:

$$\frac{[\text{H}^+]^2}{K_1K_2}[\text{T}^{2-}] + \frac{[\text{H}^+]}{K_2}[\text{T}^{2-}] + [\text{T}^{2-}] + K_{\text{NaT}^-}[\text{Na}^+][\text{T}^{2-}] + K_{\text{NaHT}}[\text{Na}^+]\frac{[\text{H}^+]}{K_2}[\text{T}^{2-}] = F_{\text{H}_2\text{T}}$$

and solve for $[\text{T}^{2-}]$:

$$[\text{T}^{2-}] = \frac{F_{\text{H}_2\text{T}}}{\frac{[\text{H}^+]^2}{K_1K_2} + \frac{[\text{H}^+]}{K_2} + 1 + K_{\text{NaT}^-}[\text{Na}^+] + K_{\text{NaHT}}[\text{Na}^+]\frac{[\text{H}^+]}{K_2}} \tag{B}$$

(c) To find $[\text{HT}^-]$, make the following substitutions in the mass balance for H_2T:

$$[\text{H}_2\text{T}] = \frac{[\text{H}^+]}{K_1}[\text{HT}^-]; \qquad [\text{T}^{2-}] = \frac{K_2}{[\text{H}^+]}[\text{HT}^-]; \qquad [\text{NaHT}] = K_{\text{NaHT}}[\text{Na}^+][\text{HT}^-]$$

$$[\text{NaT}^-] = K_{\text{NaT}^-}[\text{Na}^+][\text{T}^{2-}] = K_{\text{NaT}^-}[\text{Na}^+]\frac{K_2}{[\text{H}^+]}[\text{HT}^-]$$

Then put these expressions into the mass balance:

$$\frac{[\text{H}^+]}{K_1}[\text{HT}^-] + [\text{HT}^-] + \frac{K_2}{[\text{H}^+]}[\text{HT}^-] + K_{\text{NaT}^-}[\text{Na}^+]\frac{K_2}{[\text{H}^+]}[\text{HT}^-] + K_{\text{NaHT}}[\text{Na}^+][\text{HT}^-] = F_{\text{H}_2\text{T}}$$

and solve for $[\text{HT}^-]$:

$$[HT^-] = \cfrac{F_{H_2T}}{\cfrac{[H^+]}{K_1} + 1 + \cfrac{K_2}{[H^+]} + K_{NaT^-}[Na^+]\cfrac{K_2}{[H^+]} + K_{NaHT}[Na^+]} \tag{C}$$

To find $[H_2T]$, make the following substitutions in the mass balance for H_2T:

$$[HT^-] = \frac{K_1}{[H^+]}[H_2T]; \qquad [T^{2-}] = \frac{K_1K_2}{[H^+]^2}[H_2T]$$

$$[NaT^-] = K_{NaT^-}[Na^+][T^{2-}] = K_{NaT^-}[Na^+]\frac{K_1K_2}{[H^+]^2}[H_2T]$$

$$[NaHT] = K_{NaHT}[Na^+][HT^-] = K_{NaHT}[Na^+]\frac{K_1}{[H^+]}[H_2T]$$

Then put these expressions into the mass balance:

$$[H_2T] + \frac{K_1}{[H^+]}[H_2T] + \frac{K_1K_2}{[H^+]^2}[H_2T] + K_{NaT^-}[Na^+]\frac{K_1K_2}{[H^+]^2}[H_2T] + K_{NaHT}[Na^+]\frac{K_1}{[H^+]}[H_2T]$$
$$= F_{H_2T}$$

and solve for $[H_2T]$:

$$[H_2T] = \cfrac{F_{H_2T}}{1 + \cfrac{K_1}{[H^+]} + \cfrac{K_1K_2}{[H^+]^2} + K_{NaT^-}[Na^+]\cfrac{K_1K_2}{[H^+]^2} + K_{NaHT}[Na^+]\cfrac{K_1}{[H^+]}} \tag{D}$$

(d) Insert equations A, B, C, and D into the following spreadsheet to compute $[Na^+]$, $[H_2T]$, $[HT^-]$, and $[T^{2-}]$ in cells B12, H10, E11, and E12. Excel indicates a circular reference problem with these new formulas. In Excel 2007, click the Microsoft Office button at the upper left of the spreadsheet. Click on Excel Options at the bottom of the window. Select Formulas. In Calculation options, check Enable iterative calculation and set Maximum Change to 1e-16. Click OK. (In earlier versions of Excel, go to the Tools menu and choose Options. Select Calculation and choose Iteration. Set the maximum change to 1e-16 and click OK.) Guess a pH (such as 6) in cell H13. Select Solver and Set Target Cell E15 Equal To Value of 0 By Changing Cells H13. Click Solve and your spreadsheet should find the concentrations in the spreadsheet on the next page.

	A	B	C	D	E	F	G	H	I
1	Mixture of 0.020 M Na⁺HT⁻, 0.015 M PyH⁺Cl⁻, and 0.010 M KOH								
2	With some Na⁺ ion pairs								
3	F_{H2T} =	0.020		F_{PyH+} =	0.015		[K⁺] =	0.010	
4	pK_1 =	3.036		pK_a =	5.20		K_w =	1.00E-14	
5	pK_2 =	4.366		K_a =	6.31E-06		K_{NaT-} =	8	
6	K_1 =	9.20E-04					K_{NaHT} =	1.6	
7	K_2 =	4.31E-05							
8									
9	Species in charge balance:						Other concentrations:		
10	[H⁺] =	5.45E-05		[OH⁻] =	1.84E-10		[H₂T] =	5.93E-04	
11	[PyH⁺] =	1.34E-02		[HT⁻] =	1.00E-02		[Py] =	1.56E-03	
12	[Na⁺] =	0.0185		[T²⁻] =	7.92E-03		[NaHT] =	2.97E-04	
13	[K⁺] =	0.0100		[Cl⁻] =	0.0150		pH =	4.264	← initial value
14				[NaT⁻] =	1.17E-03				is a guess
15	Positive charge minus negative charge =				-3.69E-18				
16					E15 = B10+B11+B12+B13-E10-E11-2*E12-E13-E14				
17	Check: [PH⁺] + [P] =				0.01500				
18	Check: [H₂T]+[HT⁻]+[T²⁻]+[NaT⁻]+[NaHT] =				0.02000				
19	Check: [Na⁺]+[NaT⁻]+[NaHT] =				0.02000				
20	Formulas:								
21	B6 = 10^-B4			B7 = 10^-B5		E5 = 10^-E4		E10 = H4/B10	
22	B10 = 10^-H13					B13 = H3		E13 = E3	
23	E11 = B3/(B10/B6 + 1 + B7/B10 + H5*B12*B7/B10 + H6*B12)								
24	E12 = B3/((B10^2/(B6*B7))+(B10/B7)+1+H5*B12+H6*B12*B10/B7)								
25	H10 = B3/(1 + B6/B10 + B6*B7/B10^2 + H5*B12*B6*B7/B10^2 + H6*B12*B6/B10)								
26	B12 = B3/(1 + H5*E12 + H6*E11)						B11 = B10*E3/(B10+E5)		
27	E14 = H5*B12*E12						H11 = E5*E3/(B10+E5)		
28	H12 = H6*B12*E11								
29									
30									
31	Na distribution (%)			H₂T distribution (%)					
32	[Na⁺] =	92.6		[H₂T] =	3.0				
33	[NaT⁻] =	5.9		[HT⁻] =	50.1				
34	[NaHT] =	1.5		[T²⁻] =	39.6				
35				[NaT⁻] =	5.9				
36				[NaHT] =	1.5				

CHAPTER 13
FUNDAMENTALS OF ELECTROCHEMISTRY

13-1. Electric charge (coulombs) refers to the quantity of positive or negative particles. Current (amperes) is the quantity of charge moving past a point in a circuit each second. Electric potential (volts) measures the work that can be done by (or must be done to) each coulomb of charge as it moves from one point to another.

13-2. (a) $1/1.602\,176\,53 \times 10^{-19}$ C/electron = $6.241\,509\,48 \times 10^{18}$ electrons/C

    (b) $F = 96\,485.338\,3$ C/mol

13-3. (a) I = coulombs/s. Every mol of O_2 accepts 4 mol of e^-. 16 mol O_2/day

        = 64 mol e^-/day = 7.41×10^{-4} mol e^-/s = 71.5 C/s = $71._5$ A

    (b) I = Power/E = 500 W/115 V = 4.35 A. The resting human uses 16 times as much current as the refrigerator.

    (c) Power = $E \cdot I$ = (1.1 V) ($71._5$ A) = 79 W

13-4. (a) $I = \dfrac{6.00 \text{ V}}{2.0 \times 10^3 \text{ W}}$ = 3.00 mA = 3.00×10^{-3} C/s

        $\left(\dfrac{3.00 \times 10^{-3} \text{ C/s}}{9.649 \times 10^4 \text{ C/mol}}\right)(6.022 \times 10^{23}$ e^-/mole$) = 1.87 \times 10^{16}$ e^-/s

    (b) $P = E \cdot I$ = (6.00 V) (3.00×10^{-3} A) = 1.80×10^{-2} W

        $\Rightarrow \dfrac{1.80 \times 10^{-2} \text{ J/s}}{1.87 \times 10^{16} \text{ } e^-\text{/s}} = 9.63 \times 10^{-19}$ J/e^-

    (c) $(1.87 \times 10^{16}$ e^-/s$)(1\,800$ s$) = 3.37 \times 10^{19}$ electrons = 5.60×10^{-5} mol

    (d) $P = EI = E(E/R) = E^2/R \Rightarrow E = \sqrt{PR} = \sqrt{(100 \text{ W})(2.00 \times 10^3 \text{ W})}$ = 447 V

13-5. (a) $I_2 + 2e^- \rightleftharpoons 2I^-$ (I_2 is the oxidant)

    (b) $2S_2O_3^{2-} \rightleftharpoons S_4O_6^{2-} + 2e^-$ ($S_2O_3^{2-}$ is the reductant)

    (c) 1.00 g $S_2O_3^{2-}$/ (112.13 g/mol) = 8.92 mmol $S_2O_3^{2-}$ = 8.92 mmol e^-

        $(8.92 \times 10^{-3}$ mol$)(9.649 \times 10^4$ C/mol$)$ = 861 C

    (d) Current (A) = coulombs/s = 861 C/60 s = 14.3 A

13-6. (a) Oxidation numbers of reactants: N (in NH_4^+)   Cl (in ClO_4^-)   Al

                                                             –3   +7   0

        Oxidation numbers of products: N (in N_2)   Cl (in HCl)   Al (in Al_2O_3)

                                                                0   –1   +3

        NH_4^+ and Al are reducing agents and ClO_4^- is the oxidizing agent.

      (b) Formula mass of reactants = 6(FM NH_4ClO_4) + 10(FM Al) = 974.75

            Heat released per gram = 9 334 kJ/974.75 g = 9.576 kJ/g

13-7. In a galvanic cell, two half-reactions are physically separated from each other. At the anode, oxidation generates electrons that can flow through the electric circuit to reach the cathode, where a reduction occurs. The favorable free energy change for the net reaction provides the driving force for electrons to flow through the circuit. There must be a connector (such as a salt bridge) between the two half-cells to allow ions to flow to maintain electroneutrality.

13-8. (a) Fe(s) | FeO(s) | KOH(aq) | Ag_2O(s) | Ag(s)

           FeO(s) + H_2O + 2e$^-$ $\rightleftharpoons$ Fe(s) + 2OH$^-$

           Ag_2O + H_2O + 2e$^-$ $\rightleftharpoons$ 2Ag(s) + 2OH$^-$

      (b) Pb(s) | $PbSO_4$(s) | K_2SO_4(aq) || H_2SO_4(aq) | $PbSO_4$(s) | PbO_2(s) | Pb(s)

           $PbSO_4$(s) + 2e$^-$ $\rightleftharpoons$ Pb(s) + SO_4^{2-}

           PbO_2 + 4H$^+$ + SO_4^{2-} + 2e$^-$ $\rightleftharpoons$ $PbSO_4$(s) + 2H_2O

13-9. Fe^{3+} + e$^-$ $\rightleftharpoons$ Fe^{2+}

      $Cr_2O_7^{2-}$ + 14H$^+$ + 6e$^-$ $\rightleftharpoons$ 2Cr^{3+} + 7H_2O

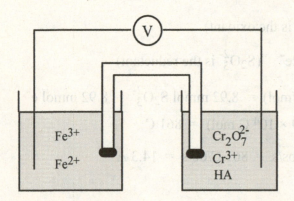

13-10. (a) $Zn^{2+} + 2e^- \rightleftharpoons Zn(s)$ $E° = -0.762$ V

$Cl_2(l) + 2e^- \rightleftharpoons 2Cl^-$ $E° \approx 1.4$ V

The Appendix lists standard reduction potentials for $Cl_2(g)$ and $Cl_2(aq)$, but not for $Cl_2(l)$. Both listed potentials are close to 1.4 V, so the potential for $Cl_2(l)$ is probably also close to 1.4 V. Electrons flow from the more negative electrode (Zn) through the circuit to the more positive electrode (C).

(b)   One mol of Cl_2 requires 2 mol of e^-.

Moles of Cl_2 consumed in 1.00 hr $= \frac{1}{2}$ (mol of e^-/hr) $=$

$\left[\frac{1}{2}\left(1.00 \times 10^3 \frac{C}{s}\right)/(9.64 \times 10^4 \text{ C/mol})\right](3\ 600 \text{ s/hr}) = 18.7$ mol of $Cl_2 = 1.32$ kg.

13-11. (a)   anode: $C_6Li \rightleftharpoons C_6 + Li^+ + e^-$

cathode: $CoO_2 + Li^+ + e^- \rightleftharpoons LiCoO_2$

(b) 1 mA $= 1 \times 10^{-3} \frac{C}{s}$. 1 h $= 3\ 600$ s

1 mA·h $= \left(1 \times 10^{-3} \frac{C}{s}\right)(3\ 600 \text{ s}) = 3.6$ C

(c) $\frac{1 \text{ g LiCoO}_2}{97.87 \text{ g/mol}} = 1.02_{17} \times 10^{-2}$ mol $LiCoO_2$ which holds $1.02_{17} \times 10^{-2}$ mol Li^+ and $1.02_{17} \times 10^{-2}$ mol e^-. $(1.02_{17} \times 10^{-2}$ mol $e^-)\left(9.649 \times 10^5 \frac{C}{\text{mol } e^-}\right) = 985._8$ C

Charge capacity $= \left(985._8 \frac{C}{\text{g LiCoO}_2}\right)\left(\frac{1 \text{ mA·h}}{3.6 \text{ C}}\right) = 273._8 \frac{\text{mA·h}}{\text{g LiCoO}_2}$

(d)   Fraction of Li available $= \dfrac{140 \text{ mA·h/g}}{273._8 \text{ mA·h/g}} = 0.51$

(e)   Energy stored per unit mass $=$ work that can be done / mass $= E·q$ / mass

$= (3.7 \text{ V})\left(140 \frac{\text{mA·h}}{\text{g LiCoO}_2}\right) = (3.7 \text{ V})\left(0.140 \frac{\text{A·h}}{\text{g LiCoO}_2}\right) = 0.52 \frac{\text{W·h}}{\text{g LiCoO}_2}$

13-12. Cl_2 is strongest because it has the most positive reduction potential.

13-13. (a)   Since it becomes harder to reduce Fe(III) to Fe(II) in the presence of CN^-, Fe(III) is stabilized more than Fe(II).

(b)   Since it becomes easier to reduce Fe(III) to Fe(II) in the presence of phenanthroline, Fe(II) is stabilized more than Fe(III).

13-14. $E°$ applies when activities of reactants and products are unity. E applies to whatever activities exist. At equilibrium, E goes to zero. $E°$ is constant.

13-15. (a) $Zn(s) \mid Zn^{2+}(0.1\ M) \parallel Cu^{2+}(0.1\ M) \mid Cu(s)$

right half-cell: $Cu^{2+} + 2e^- \rightleftharpoons Cu(s)$ $E°_+ = 0.339\ V$

left half-cell: $Zn^{2+} + 2e^- \rightleftharpoons Zn(s)$ $E°_- = -0.762\ V$

$$E = \left\{ 0.339 - \frac{0.059\,16}{2} \log \frac{1}{0.1} \right\} - \left\{ -0.762 - \frac{0.059\,16}{2} \log \frac{1}{0.1} \right\} = 1.101\ V$$

Since the voltage is positive, electrons are transferred from Zn to Cu. The net reaction is $Cu^{2+} + Zn(s) \rightleftharpoons Cu(s) + Zn^{2+}$.

(b)   Since Cu^{2+} ions are consumed in the right half-cell, Zn^{2+} ions must migrate from the left half-cell into the salt bridge to help balance charge. I hope you like Zn^{2+}, because that is what your body will take up.

13-16. (a) $E = -0.238 - \dfrac{0.059\,16}{3} \log \dfrac{P_{AsH_3}}{[H^+]^3}$

(b) $E = -0.238 - \dfrac{0.059\,16}{3} \log \dfrac{1.0 \times 10^{-3}}{(10^{-3.00})^3} = -0.356\ V$

13-17. (a) $Pt(s) \mid Br_2(l) \mid HBr(aq, 0.10\ M) \parallel Al(NO_3)_3(aq, 0.010\ M) \mid Al(s)$

(b)   right half-cell: $Al^{3+} + 3e^- \rightleftharpoons Al(s)$ $E°_+ = -1.677\ V$

left half-cell: $Br_2(l) + 2e^- \rightleftharpoons 2Br^-$ $E°_- = 1.078\ V$

right half-cell: $E_+ = \left\{ -1.677 - \dfrac{0.059\,16}{3} \log \dfrac{1}{[0.010]} \right\} = -1.716_4\ V$

left half-cell: $E_- = \left\{ 1.078 - \dfrac{0.059\,16}{2} \log [0.10]^2 \right\} = 1.137_2\ V$

$E = E_+ - E_- = -1.716_4 - 1.137_2 = -2.854\ V$. The right electrode is more negative, so electrons flow from Al to Pt. Reduction occurs at the left-hand electrode. The spontaneous reaction is

$$\tfrac{3}{2} Br_2(l) + Al(s) \rightleftharpoons 3Br^- + Al^{3+}$$

(c)   14.3 mL of Br_2 = 44.6 g = 0.279 mol of Br_2. 12.0 g of Al = 0.445 mol of Al. The reaction requires 3/2 mol of Br_2 for every mol of Al. The Br_2 will be used up first.

(d)   0.231 mL of Br_2 = 0.721 g of Br_2 = 4.51×10^{-3} mol Br_2 = 9.02×10^{-3} mol e^- = 870 C.   Work = $E \cdot q$ = (1.50)(870) = 1.31 kJ.

(e) $I = \sqrt{P/R} = \sqrt{(1.00 \times 10^{-4})/(1.20 \times 10^3)} = 2.89 \times 10^{-4}$ A

$= 2.99 \times 10^{-9}$ mol e^-/s $= 9.97 \times 10^{-10}$ mol Al/s $= 2.69 \times 10^{-8}$ g/s

13-18. The activities of the solid reagents do not change until they are used up. The only aqueous species, OH^-, is created at the cathode and consumed in equal amounts at the anode, so its concentration remains constant in the cell. Therefore, none of the activities change during the life cycle of the cell until something is used up.

13-19. (a) right half-cell: $E_+ = \left\{ 0.222 - \dfrac{0.059\,16}{2} \log [Cl^-]^2 \right\} = 0.281_2$ V

left half-cell: $E_- = \left\{ -0.350 - \dfrac{0.059\,16}{2} \log [F^-]^2 \right\} = -0.290_8$ V

$E = E_+ - E_- = 0.281_2 - (-0.290_8) = 0.572$ V

(b) Electrons flow from the left half-cell ($E = -0.290_8$ V) to the right half-cell ($E = 0.281_2$ V).

(c) $[Pb^{2+}] = K_{sp}$ (for PbF_2) $/ [F^-]^2 = (3.6 \times 10^{-8}) / (0.10)^2 = 3.6 \times 10^{-6}$ M

$[Ag^+] = K_{sp}$ (for $AgCl$) $/ [Cl^-] = (1.8 \times 10^{-10}) / (0.10) = 1.8 \times 10^{-9}$ M

right half-cell: $E_+ = \left\{ 0.799 - \dfrac{0.059\,16}{2} \log \dfrac{1}{[Ag^+]^2} \right\} = 0.281_2$ V

left half-cell: $E_- = \left\{ -0.126 - \dfrac{0.059\,16}{2} \log \dfrac{1}{[Pb^{2+}]} \right\} = -0.287_0$ V

$E = E_+ - E_- = 0.281_2 - (-0.287_0) = 0.568$ V

The agreement between the two calculations is reasonable.

13-20. A hydrogen pressure of 727.2 Torr corresponds to (727.2 Torr)/(760 Torr/atm) = 0.956 8 atm.   (0.956 8 atm)(1.013 25 bar/atm) = 0.969 5 bar.

$0.798\,3 = E^\circ_{Ag^+ \mid Ag} - 0.059\,16 \log \dfrac{[0.010\,00] (0.914)}{(0.969\,5)^{1/2} [0.010\,00](0.898)}$

$\Rightarrow E^\circ_{Ag^+ \mid Ag} = 0.799\,2$ V

13-21. Balanced reaction: $HOBr + 2e^- + H^+ \rightleftharpoons Br^- + H_2O$

$HOBr \rightarrow \frac{1}{2} Br_2 \qquad\qquad \Delta G^\circ_1 = -1F(1.584)$

$\frac{1}{2} Br_2 \rightarrow Br^- \qquad\qquad \Delta G^\circ_2 = -1F(1.098)$

$\overline{\qquad\qquad\qquad\qquad\qquad\qquad\qquad\qquad}$

$HOBr \rightarrow Br^- \qquad\qquad \Delta G^\circ_3 = \Delta G^\circ_1 + \Delta G^\circ_2 = -2FE^\circ_3$

$E^\circ_3 = \dfrac{-1F(1.584) - 1F(1.098)}{-2F} = 1.341$ V

13-22.

$$2X^+ + 2e^- \rightleftharpoons 2X(s) \qquad\qquad E_+^\circ = E_2^\circ$$
$$-\underline{X^{3+} + 2e^- \rightleftharpoons X^+} \qquad\qquad \underline{E_-^\circ = E_1^\circ}$$
$$3X^+ \rightleftharpoons X^{3+} + 2X(s) \qquad\qquad E_3^\circ = E_2^\circ - E_1^\circ$$

If $E_2^\circ > E_1^\circ$, then $E_3^\circ \geq 0$ and disproportionation is spontaneous.

13-23.

right half-cell: $Cu^{2+} + 2e^- \rightleftharpoons Cu(s)$ $\qquad\qquad E_+^\circ = 0.339$ V

left half-cell: $Ni^{2+} + 2e^- \rightleftharpoons Ni(s)$ $\qquad\qquad E_-^\circ = -0.236$ V

The ionic strength of the right half-cell is $0.009\,0$ M, and the ionic strength of the left half-cell is $0.008\,0$ M. At $\mu = 0.009\,0$ M, $\gamma_{Cu^{2+}} = 0.690$.

At $\mu = 0.008\,0$ M, $\gamma_{Ni^{2+}} = 0.705$.

$$E_+ = E_+^\circ - \frac{0.059\,16}{2} \log \frac{1}{[Cu^{2+}]\gamma_{Cu^{2+}}}$$

$$= 0.339 - \frac{0.059\,16}{2} \log \frac{1}{(0.003\,0)(0.690)} = 0.259_6 \text{ V}$$

$$E_- = E_-^\circ - \frac{0.059\,16}{2} \log \frac{1}{[Ni^{2+}]\gamma_{Ni^{2+}}}$$

$$= -0.236 - \frac{0.059\,16}{2} \log \frac{1}{(0.002\,0)(0.705)} = -0.320_3 \text{ V}$$

$$E = E_+ - E_- = 0.580 \text{ V}$$

Electrons flow from Ni ($E = -0.320_3$ V) to Cu ($E = 0.259_6$ V).

13-24. (a) $\quad E^\circ = \dfrac{-\Delta G^\circ}{nF} = \dfrac{(+257 \times 10^3 \text{ J/mol})}{(2)(9.648\,5 \times 10^4 \text{ C/mol})} = 1.33$ V

(b) $\quad K = 10^{nE^\circ/0.059\,16} = 1 \times 10^{45}$

13-25. (a)

$$4[Co^{3+} + e^- \rightleftharpoons Co^{2+}] \qquad\qquad E_+^\circ = 1.92 \text{ V}$$
$$-2[\tfrac{1}{2}O_2 + 2H^+ + 2e^- \rightleftharpoons H_2O] \qquad\qquad \underline{E_-^\circ = 1.229 \text{ V}}$$
$$4Co^{3+} + 2H_2O \rightleftharpoons 4Co^{2+} + O_2 + 4H^+ \qquad\qquad E^\circ = 0.69_1 \text{ V}$$

$$\Delta G^\circ = -4FE^\circ = -2.7 \times 10^5 \text{ J} \qquad K = 10^{4E^\circ/0.059\,16} = 10^{47}$$

(b)

$$Ag(S_2O_3)_2^{3-} + e^- \rightleftharpoons Ag(s) + 2S_2O_3^{2-} \qquad\qquad E_+^\circ = 0.017 \text{ V}$$
$$-\underline{Fe(CN)_6^{3-} + e^- \rightleftharpoons Fe(CN)_6^{4-}} \qquad\qquad \underline{E_-^\circ = 0.356 \text{ V}}$$
$$Ag(S_2O_3)_2^{3-} + Fe(CN)_6^{4-} \rightleftharpoons Ag(s) + 2S_2O_3^{2-} + Fe(CN)_6^{3-} \quad E^\circ = -0.339 \text{ V}$$

$$\Delta G^\circ = -1FE^\circ = 32.7 \text{ kJ} \qquad K = 10^{1E^\circ/0.059\,16} = 1.9 \times 10^{-6}$$

13-26. (a) $5Ce^{4+} + 5e^- \rightleftharpoons 5Ce^{3+}$ $E_+^\circ = 1.70$ V

$-\ MnO_4^- + 8H^+ + 5e^- \rightleftharpoons Mn^{2+} + 4H_2O$ $E_-^\circ = 1.507$ V

$\overline{5Ce^{4+} + Mn^{2+} + 4H_2O \rightleftharpoons 5Ce^{3+} + MnO_4^- + 8H^+}$ $E^\circ = 0.19_3$ V

(b) $\Delta G^\circ = -5FE^\circ = -93._1$ kJ $K = 10^{5E^\circ/0.059\ 16} = 2 \times 10^{16}$

(c) $E = \left\{ 1.70 - \dfrac{0.059\ 16}{5} \log \dfrac{[Ce^{3+}]^5}{[Ce^{4+}]^5} \right\} - \left\{ 1.507 - \dfrac{0.059\ 16}{5} \log \dfrac{[Mn^{2+}]}{[MnO_4^-][H^+]^8} \right\}$

$= 1.52_{23} - 1.54_{25} = -0.02_0$ V

(d) $\Delta G = -5FE = +10$ kJ

(e) At equilibrium, $E = 0 \Rightarrow E^\circ = \dfrac{0.059\ 16}{5} \log \dfrac{[Ce^{3+}]^5[MnO_4^-][H^+]^8}{[Ce^{4+}]^5[Mn^{2+}]}$

$\Rightarrow [H^+] = 0.62 \Rightarrow pH = 0.21$

13-27. right half-cell: $Sn^{4+} + 2e^- \rightleftharpoons Sn^{2+}$

left half-cell: $-\ 2VO^{2+} + 4H^+ + 2e^- \rightleftharpoons 2V^{3+} + 2H_2O$

net reaction: $\overline{Sn^{4+} + 2V^{3+} + 2H_2O \rightleftharpoons Sn^{2+} + 2VO^{2+} + 4H^+}$

$E = E^\circ - \dfrac{0.059\ 16}{2} \log \dfrac{[VO^{2+}]^2[H^+]^4[Sn^{2+}]}{[V^{3+}]^2[Sn^{4+}]}$

$-0.289 = E^\circ - \dfrac{0.059\ 16}{2} \log \dfrac{(0.116)^2(1.57)^4(0.031\ 8)}{(0.116)^2(0.031\ 8)} \Rightarrow E^\circ = -0.266$ V

$\Rightarrow K = 10^{2E^\circ/0.059\ 16} = 1.0 \times 10^{-9}$

13-28. $Pd(OH)_2(s) + 2e^- \rightleftharpoons Pd(s) + 2OH^-$ E_+°

$-\ Pd^{2+} + 2e^- \rightleftharpoons Pd(s)$ $E_-^\circ = 0.915$ V

$\overset{K_{sp}}{\overline{Pd(OH)_2 \rightleftharpoons Pd^{2+} + 2OH^-}}$ $E^\circ = E_+^\circ - 0.915$

But $K_{sp} = 3 \times 10^{-28} \Rightarrow E^\circ = \dfrac{0.059\ 16}{2} \log K_{sp} = -0.814$

$-0.814 = E_+^\circ - 0.915 \Rightarrow E_+^\circ = 0.101$ V

13-29. $Br_2(l) + 2e^- \rightleftharpoons 2Br^-$ $E_+^\circ = 1.078$ V

$-\ Br_2(aq) + 2e^- \rightleftharpoons 2Br^-$ $E_-^\circ = 1.098$ V

$\overset{K}{\overline{Br_2(l) \rightleftharpoons Br_2(aq)}}$ $E^\circ = -0.020$ V

At equilibrium, $E = 0$. Therefore, $0 = -0.020 - \dfrac{0.059\ 16}{2} \log \dfrac{[Br_2(aq)]}{[Br_2(l)]}$

$$\Rightarrow K = \frac{[Br_2(aq)]}{[Br_2(l)]} = 0.21_1 \text{ M.}$$

That is, the solubility of Br_2 in water is 0.21_1 M $= 34$ g/L.

13-30.

$$FeY^- + e^- \rightleftharpoons FeY^{2-} \qquad\qquad E_+^\circ$$

$$\underline{-\ FeY^- + e^- \rightleftharpoons Fe^{2+} + Y^{4-}} \qquad\qquad \underline{E_-^\circ = -0.730 \text{ V}}$$

$$Fe^{2+} + Y^{4-} \rightleftharpoons FeY^{2-} \qquad\qquad E^\circ = E_+^\circ + 0.730$$

But $E^\circ = 0.059\,16 \ \log [K_f \ (\text{for } FeY^{2-})] = 0.846 \text{ V} \Rightarrow E_+^\circ = E^\circ - E_-^\circ = 0.116$ V.

13-31. $\quad E^\circ(T) = E^\circ + \dfrac{dE^\circ}{dT}\Delta T$

$$E^\circ(50^\circ \text{ C}) = -1.677 \text{ V} + (0.533 \times 10^{-3} \text{ V/K})(25 \text{ K}) = -1.664 \text{ V}$$

13-32.

1. $2Cu(s) + 2I^- \rightleftharpoons 2CuI(s) + 2e^- \qquad \Delta G_1^\circ = +2F(-0.185) = -35.7_0 \text{ kJ}$

2. $2Cu^{2+} + 4e^- \rightleftharpoons 2Cu(s) \qquad\qquad \Delta G_2^\circ = -4F(0.339) = -130._{83} \text{ kJ}$

3. $\text{hydro} \rightleftharpoons \text{quinone} + 2H^+ + 2e^- \qquad \underline{\Delta G_3^\circ = +2F(0.700) = 135._{08} \text{ kJ}}$

4. $(1)+(2)+(3)$: $2Cu^{2+} + 2I^- + \text{hydro} \rightleftharpoons \qquad \Delta G_4^\circ = \Delta G_1^\circ + \Delta G_2^\circ + \Delta G_3^\circ$

$\qquad\qquad 2CuI(s) + \text{quinone} + 2H^+ \qquad\qquad = -31._4 \text{ kJ}$

Since $2e^-$ are transferred in the net reaction, $E_4^\circ = \dfrac{-\Delta G_4^\circ}{2F} = +0.16_3$ V

$$K = 10^{2(0.163)/0.059\,16} = 3._2 \times 10^5.$$

13-33. (a) $\quad E(\text{right}) = E^\circ(\text{right}) - \dfrac{RT}{2F} \ln \dfrac{\mathcal{A}_{F^-}^2(\text{right})}{P_{O_2}^{1/2}(\text{right})}$

$$E(\text{left}) = E^\circ(\text{left}) - \frac{RT}{2F} \ln \frac{\mathcal{A}_{F^-}^2(\text{left})}{P_{O_2}^{1/2}(\text{left})}$$

Net reaction: reverse left half-reaction and add it to right half-reaction:

$$MgF_2(s) + Al_2O_3(s) + \tfrac{1}{2} O_2(g) + 2e^- \rightleftharpoons MgAl_2O_4(s) + 2F^-$$

$$MgO(s) + 2F^- \rightleftharpoons MgF_2(s) + \tfrac{1}{2} O_2(g) + 2e^-$$

net reaction: $Al_2O_3(s) + MgO(s) \rightleftharpoons MgAl_2O_4(s)$

Nernst equation for net reaction:

$$E(\text{right}) - E(\text{left}) = E^\circ(\text{right}) - E^\circ(\text{left}) - \frac{RT}{2F} \ln \frac{\mathcal{A}_{F^-}^2(\text{right})}{P_{O_2}^{1/2}(\text{right})} + \frac{RT}{2F} \ln \frac{\mathcal{A}_{F^-}^2(\text{left})}{P_{O_2}^{1/2}(\text{left})}$$

The activities of F⁻ are the same on both sides and the activities of O_2 are also the same on both sides, so the ln terms cancel, leaving $E(\text{cell}) = E°(\text{right}) - E°(\text{left}) = E°(\text{cell})$.

(b) $\Delta G° = -nFE° = -(2)(9.648\ 5 \times 10^4\ \text{C/mol})(0.152\ 9\ \text{J/C}) = -29.51\ \text{kJ/mol}$, where we made use of the fact that a volt is equivalent to one joule/coulomb.

(c) $\Delta G° = \Delta H° - T\Delta S°$

$-nFE° = \Delta H° - T\Delta S° = -nF(0.122\ 3 + 3.06 \times 10^{-5}\ T)$

$\qquad\qquad = -nF(0.122\ 3\ \text{V}) - nF(3.06 \times 10^{-5}\ T)$

$\qquad\qquad = \underbrace{-nF(0.122\ 3\ \text{V})}_{\Delta H°} - \underbrace{T\,\{nF(3.06 \times 10^{-5}\ \text{V/K})\}}_{\Delta S°}$

$\Delta H° = -nF(0.122\ 3\ \text{V}) = -(2)(9.648\ 5 \times 10^4\ \text{C/mol})(0.122\ 3\ \text{J/C})$

$\qquad = -23.60\ \text{kJ/mol}$

$\Delta S° = nF(3.06 \times 10^{-5}\ \text{V/K})$

$\qquad = (2)(9.648\ 5 \times 10^4\ \text{C/mol})(3.06 \times 10^{-5}\ \text{V/K})$

$\qquad = 5.90\ \text{C·V/(K·mol)} = 5.90\ \text{J/(K·mol)}$, where we made use of the conversion coulomb·volt = joule.

13-34. In the right half-cell, the reaction $Hg^{2+} + Y^{4-} \rightleftharpoons HgY^{2-}$ is at equilibrium, even though the net cell reaction $Hg^{2+} + H_2 \rightleftharpoons Hg(l) + 2H^+$ is not at equilibrium.

13-35. (a)

$$AgCl(s) + e^- \rightleftharpoons Ag(s) + Cl^- \qquad\qquad E_+° = 0.222\ \text{V}$$

$$-\quad H^+ + e^- \rightleftharpoons \tfrac{1}{2}\,H_2(g) \qquad\qquad\qquad E_-° = 0\ \text{V}$$

$$\overline{AgCl(s) + \tfrac{1}{2}H_2(g) \rightleftharpoons Ag(s) + H^+ + Cl^- \qquad E° = 0.222\ \text{V}}$$

$$E = 0.222 - 0.059\ 16 \log\frac{[H^+][Cl^-]}{\sqrt{P_{H_2}}}$$

(b) $0.485 = 0.222 - 0.059\ 16 \log\dfrac{(10^{-3.60})[Cl^-]}{\sqrt{1.00}} \Rightarrow [Cl^-] = 0.14_3\ \text{M}$

13-36. (a) Left: quinone $+ 2H^+ + 2e^- \rightleftharpoons$ hydroquinone $\qquad E_-° = 0.700\ \text{V}$

Right: $Hg_2Cl_2 + 2e^- \rightleftharpoons 2Hg(l) + 2Cl^- \qquad E_+° = 0.268\ \text{V}$

$$E(\text{left}) = 0.700 - \frac{0.059\ 16}{2}\log\frac{[\text{hydroquinone}]}{[\text{quinone}][H^+]^2}$$

$$E(\text{right}) = 0.268 - \frac{0.059\ 16}{2}\log[Cl^-]^2$$

(b) $E(\text{cell}) = E(\text{right}) - E(\text{left})$

$$= \left\{ 0.268 - \frac{0.059\,16}{2} \log\,[\text{Cl}^-]^2 \right\} - \left\{ 0.700 - \frac{0.059\,16}{2} \log \frac{[\text{hydroquinone}]}{[\text{quinone}][\text{H}^+]^2} \right\}$$

$$E(\text{cell}) = -0.432 - \frac{0.059\,16}{2} \log \frac{[\text{quinone}][\text{H}^+]^2[\text{Cl}^-]^2}{[\text{hydroquinone}]}$$

Setting $[\text{Cl}^-] = 0.50$ M and noting $[\text{quinone}] = [\text{hydroquinone}]$, we find

$$E(\text{cell}) = -0.432 - 0.059\,16 \log\,(0.50) - 0.059\,16 \log\,[\text{H}^+]$$

$$E(\text{cell}) = -0.414 + 0.059\,16 \text{ pH} \quad (A = -0.414, \, B = 0.059\,16 \text{ V per pH unit})$$

(c) $E(\text{cell}) = -0.414 + 0.059\,16\,(4.50) = -0.148$

Since $E < 0$, electrons flow from right to left (Hg $\rightarrow$ Pt) through the meter.

13-37. $\text{H}^+\,(1.00\text{ M}) + \text{e}^- \rightleftharpoons \frac{1}{2}\,\text{H}_2\,(g, 1.00\text{ bar}) \qquad\qquad E_+^\circ = 0\text{ V}$

$\underline{-\quad \text{H}^+\,(x\text{ M}) + \text{e}^- \quad \rightleftharpoons \quad \frac{1}{2}\,\text{H}_2\,(g, 1.00\text{ bar}) \qquad\qquad E_-^\circ = 0\text{ V}}$

$\text{H}^+\,(1.00\text{ M}) \qquad \rightleftharpoons \qquad \text{H}^+\,(x\text{ M}) \qquad\qquad\qquad\qquad E^\circ = 0\text{ V}$

$$E = 0.490 = 0 - 0.059\,16 \log\,[\text{H}^+] \Rightarrow [\text{H}^+] = 5.2 \times 10^{-9}\text{ M}$$

$$K_b = \frac{[\text{RNH}_3^+][\text{OH}^-]}{[\text{RNH}_2]} = \frac{(0.050)(K_w/[\text{H}^+])}{0.10} = 9.6 \times 10^{-7}$$

13-38. $\text{M}^{2+} + 2\text{e}^- \rightleftharpoons \text{M}(s) \qquad\qquad\qquad\qquad\qquad\qquad E_+^\circ = -0.266\text{ V}$

$\underline{-\quad 2\text{H}^+ + 2\text{e}^- \quad \rightleftharpoons \quad \text{H}_2\,(g, 0.50\text{ bar}) \qquad\qquad\qquad E_-^\circ = 0\text{ V}}$

$\text{H}_2(g) + \text{M}^{2+} \rightleftharpoons 2\text{H}^+ + \text{M}(s) \qquad\qquad\qquad\qquad E^\circ = -0.266\text{ V}$

$$E = -0.266 - \frac{0.059\,16}{2} \log \frac{[\text{H}^+]^2}{P_{\text{H}_2}\,[\text{M}^{2+}]}$$

$[\text{H}^+]$ in the left half-cell is found by considering the titration of 28.0 mL of the tetraprotic pyrophosphoric acid (abbreviated H_4P) with 72.0 mL of KOH.

28.0 mL of 0.010 0 M $\text{H}_4\text{P} = 0.280$ mmol

72.0 mL of 0.010 0 M KOH $= 0.720$ mmol

First, 0.280 mmol OH^- consumes 0.280 mmol of H_4P, giving 0.280 mmol of H_3P^- and $(0.720 - 0.280 =)$ 0.440 mmol of OH^-. Then 0.280 mmol OH^- consumes 0.280 mmol of H_3P^-, giving 0.280 mmol of H_2P^{2-} and $(0.440 - 0.280 =)$ 0.160 mmol of OH^-. Finally, 0.160 mmol of OH^- reacts with 0.280 mmol of H_2P^{2-} to create 0.160 mmol of HP^{3-}, leaving 0.120 mmol of unreacted H_2P^{2-}.

$$\begin{array}{ccccccc} \text{H}_4\text{P} & + & \text{OH}^- & \rightarrow & \text{H}_2\text{P}^{2-} & + & \text{HP}^{3-} \\ 0.280\text{ mmol} & & 0.720\text{ mmol} & & 0.120\text{ mmol} & & 0.160\text{ mmol} \end{array}$$

$$\text{pH} = \text{p}K_3 + \log \frac{[\text{HP}^{3-}]}{[\text{H}_2\text{P}^{2-}]} = 6.70 + \log \frac{0.160}{0.120} = 6.82 \Rightarrow [\text{H}^+] = 1.5_0 \times 10^{-7}\text{ M}$$

Putting the known values of $[H^+]$ and P_{H_2} into the Nernst equation gives

$$-0.246 = -0.266 - \frac{0.059\,16}{2} \log \frac{[H^+]^2}{P_{H_2}[M^{2+}]}$$

$$= -0.266 - \frac{0.059\,16}{2} \log \frac{[1.5_0 \times 10^{-7}]^2}{0.50\,[M^{2+}]} \Rightarrow [M^{2+}] = 2.1_3 \times 10^{-13}\ M$$

In the right half-cell we have the equilibrium

$$M^{2+} \quad + \quad F_{EDTA} \quad \rightleftharpoons \quad MY^{2-}$$

initial mmol/mL $\frac{0.280}{100}$ $\frac{0.720}{100}$ —

final mmol/mL   small $\frac{0.440}{100}$ $\frac{0.280}{100}$

$$K_f = \frac{[MY^{2-}]}{[M^{2+}]\,\alpha_{Y4^-}\,F_{EDTA}} = \frac{0.280/100}{(2.1_3 \times 10^{-13})(0.004\,2)(0.440/100)} = 7.1 \times 10^{14}$$

13-39. right half-cell: $Pb^{2+}(\text{right}) + 2e^- \rightleftharpoons Pb(s)$ $E_+^\circ = -0.126\ V$

left half-cell: $^-\ Pb^{2+}(\text{left}) + 2e^- \rightleftharpoons Pb(s)$ $E_-^\circ = -0.126\ V$

$$Pb^{2+}(\text{right}) \rightleftharpoons Pb^{2+}(\text{left})\qquad\qquad E^\circ = 0$$

Nernst equation for net cell reaction:

$$-0.001\,8 = -\frac{0.059\,16}{2} \log \frac{[Pb^{2+}(\text{left})]}{[Pb^{2+}(\text{right})]} \Rightarrow \frac{[Pb^{2+}(\text{left})]}{[Pb^{2+}(\text{right})]} = 1.15$$

For each half-cell, we can write $[CO_3^{2-}] = K_{sp}$ (for $PbCO_3$) / $[Pb^{2+}]$

$$\frac{[CO_3^{2-}(\text{left})]}{[CO_3^{2-}(\text{right})]} = \frac{K_{sp}\ (\text{for } PbCO_3)/[Pb^{2+}(\text{left})]}{K_{sp}\ (\text{for } PbCO_3)/[Pb^{2+}\ (\text{right})]} = \frac{1}{1.15} = 0.87$$

In each compartment the Ca^{2+} concentration is equal to the total concentration of all carbonate species (since $PbCO_3$ is much less soluble than $CaCO_3$). Let the fraction of all carbonate species in the form CO_3^{2-} be $\alpha_{CO_3^{2-}}$ (i.e., $[CO_3^{2-}] = \alpha_{CO_3^{2-}}$ [total carbonate]). We can say that $[Ca^{2+}] = $ [total carbonate] $= [CO_3^{2-}] / \alpha_{CO_3^{2-}}$. The value of $\alpha_{CO_3^{2-}}$ is the same in both compartments, since the pH is the same. Now we can write

$$\frac{K_{sp}(\text{calcite})}{K_{sp}(\text{aragonite})} = \frac{[Ca^{2+}(\text{left})][CO_3^{2-}(\text{left})]}{[Ca^{2+}(\text{right})][CO_3^{2-}(\text{right})]} = \frac{[CO_3^{2-}(\text{left})]^2 / \alpha_{CO_3^{2-}}}{[CO_3^{2-}(\text{right})]^2 / \alpha_{CO_3^{2-}}}$$

$$= (0.87)^2 = 0.76.$$

13-40.

$$Cu^{2+} + 2e^- \rightleftharpoons Cu(s) \qquad\qquad E^\circ_+ = 0.339 \text{ V}$$

$$- \underline{Ni^{2+} + 2e^- \rightleftharpoons Ni(s)} \qquad\qquad \underline{E^\circ_- = -0.236 \text{ V}}$$

$$Cu^{2+} + Ni(s) \rightleftharpoons Cu(s) + Ni^{2+} \qquad E^\circ = 0.575 \text{ V}$$

The ionic strength of the right half-cell is 0.10 M and $\gamma_{Cu^{2+}} = 0.405$.

The ionic strength of the left half-cell is 0.010 M and $\gamma_{Ni^{2+}} = 0.675$.

$$0.512 = 0.575 - \frac{0.059\,16}{2} \log \frac{(0.002\,5)(0.675)}{[Cu^{2+}](0.405)} \Rightarrow [Cu^{2+}] = 3.09 \times 10^{-5} \text{ M}$$

$$K_{sp} = [Cu^{2+}]\gamma_{Cu^{2+}}[IO_3^-]^2\gamma_{IO_3^-}^2) = (3.09 \times 10^{-5})(0.405)(0.10)^2(0.775)^2 = 7.5 \times 10^{-8}$$

13-41. $E^{\circ\prime}$ is the effective reduction potential for a half-reaction at pH 7, instead of pH 0. Since living systems tend to have a pH much closer to 7 than to 0, $E^{\circ\prime}$ provides a better indication of redox behavior in an organism.

13-42. (a) $E = 0.731 - \dfrac{0.059\,16}{2} \log \dfrac{P_{C_2H_4}}{P_{C_2H_2}[H^+]^2}$

(b) $E = \underbrace{0.731 + 0.059\,16 \log [H^+]}_{\text{This is } E^{\circ\prime} \text{ when pH} = 7} - \dfrac{0.059\,16}{2} \log \dfrac{P_{C_2H_4}}{P_{C_2H_2}}$

(c) $E^{\circ\prime} = 0.731 + 0.059\,16 \log (10^{-7.00}) = 0.317 \text{ V}$

13-43. $E = E^\circ - \dfrac{0.059\,16}{2} \log \dfrac{[HCN]^2}{P_{(CN)_2}[H^+]^2}$

Substituting $[HCN] = \dfrac{[H^+] F_{HCN}}{[H^+] + K_a}$ into the Nernst equation gives

$$E = 0.373 - \frac{0.059\,16}{2} \log \frac{[H^+]^2 F_{HCN}^2}{([H^+] + K_a)^2 P_{(CN)_2}[H^+]^2}$$

$$E = 0.373 + \underbrace{0.059\,16 \log ([H^+] + K_a)}_{\text{This is } E^{\circ\prime} \text{ when pH} = 7} - \frac{0.059\,16}{2} \log \frac{F_{HCN}^2}{P_{(CN)_2}}.$$

Inserting $K_a = 6.2 \times 10^{-10}$ for HCN and $[H^+] = 10^{-7.00}$ gives

$$E^{\circ\prime} = 0.373 + 0.059\,16 \log (10^{-7.00} + 6.2 \times 10^{-10}) = -0.041 \text{ V}.$$

13-44. $H_2C_2O_4 + 2H^+ + 2e^- \rightleftharpoons 2HCO_2H$

$$E = 0.204 - \frac{0.059\,16}{2} \log \frac{[HCO_2H]^2}{[H_2C_2O_4][H^+]^2}$$

But $[HCO_2H] = \dfrac{[H^+]\,F_{HCO_2H}}{[H^+] + K_a}$ and $[H_2C_2O_4] = \dfrac{[H^+]^2\,F_{H_2C_2O_4}}{[H^+]^2 + K_1[H^+] + K_1K_2}.$

Putting these expressions into the Nernst equation gives

$$E = 0.204 - \frac{0.059\,16}{2} \log \frac{[H^+]^2\,F_{HCO_2H}^2\,([H^+]^2 + K_1[H^+] + K_1K_2)}{([H^+] + K_a)^2\,[H^+]^2\,F_{H_2C_2O_4}\,[H^+]^2}$$

$$E = 0.204 - \underbrace{\frac{0.059\,16}{2} \log \frac{[H^+]^2 + K_1[H^+] + K_1K_2}{([H^+] + K_a)^2\,[H^+]^2}}_{\text{This is } E^{\circ\prime} \text{ when pH} = 7} - \frac{0.059\,16}{2} \log \frac{F_{HCO_2H}^2}{F_{H_2C_2O_4}}.$$

Putting in $[H^+] = 10^{-7.00}$ M, $K_a = 1.80 \times 10^{-4}$, $K_1 = 5.62 \times 10^{-2}$, and $K_2 = 5.42 \times 10^{-5}$ gives $E^{\circ\prime} = -0.268$ V.

13-45. $E = E^\circ - 0.059\,16 \log \dfrac{[H_2Red^-]}{[HOx]}$

But $[HOx] = \dfrac{[H^+]\,F_{HOx}}{[H^+] + K_a}$ and $[H_2Red^-] = \dfrac{[H^+]^2\,F_{H_2Red^-}}{[H^+]^2 + [H^+]K_1 + K_1K_2}.$

Putting these values into the Nernst equation gives

$$E = E^\circ - 0.059\,16 \log \frac{[H^+]^2\,F_{H_2Red^-}\,([H^+] + K_a)}{[H^+]\,F_{HOx}\,([H^+]^2 + [H^+]K_1 + K_1K_2)}$$

$$E = E^\circ - \underbrace{0.059\,16 \log \frac{[H^+]\,([H^+] + K_a)}{[H^+]^2 + [H^+]K_1 + K_1K_2}}_{E^{\circ\prime}} - 0.059\,16 \log \frac{F_{H_2Red^-}}{F_{HOx}}.$$

Since $E^{\circ\prime} = 0.062$ V, we find $E^\circ = -0.036$ V.

13-46. $E = E^\circ - \dfrac{0.059\,16}{2} \log \dfrac{[HNO_2]}{[NO_3^-][H^+]^3}$

But $[HNO_2] = \dfrac{[H^+]F_{HNO_2}}{[H^+] + K_a}$ and $[NO_3^-] = F_{NO_3^-}.$

Putting these values into the Nernst equation gives

$$E = E^\circ - \underbrace{\frac{0.059\,16}{2} \log \frac{1}{([H^+] + K_a)[H^+]^2}}_{E^{\circ\prime}} - \frac{0.059\,16}{2} \log \frac{F_{HNO_2}}{F_{NO_3^-}}$$

$$E^{\circ\prime} = 0.433 = 0.940 - \frac{0.059\,16}{2} \log \frac{1}{(10^{-7} + K_a)(10^{-7})^2} \Rightarrow K_a = 7.2 \times 10^{-4}.$$

13-47.

$$\begin{array}{lr} H_2PO_4^- + H^+ + 2e^- \rightleftharpoons HPO_3^{2-} + H_2O & E_+^\circ \\[4pt] - HPO_4^{2-} + 2H^+ + 2e^- \rightleftharpoons HPO_3^{2-} + H_2O & E_-^\circ = -0.234 \text{ V} \\[2pt] \hline H_2PO_4^- \rightleftharpoons HPO_4^{2-} + H^+ & E^\circ = E_+^\circ - E_-^\circ \end{array}$$

$$E^\circ = \frac{0.059\,16}{2} \log K_{a2} \text{ (for } H_3PO_4\text{)} = \frac{0.059\,16}{2} \log(6.34 \times 10^{-8}) = -0.213 \text{ V}$$

$$E_+^\circ = E^\circ - 0.234 \text{ V} = -0.447 \text{ V}$$

13-48. (a) $A = 0.500 =$

$(1.12 \times 10^4 \text{ M}^{-1}\text{cm}^{-1})[\text{Ox}](1.00 \text{ cm}) + (3.82 \times 10^3 \text{ M}^{-1}\text{cm}^{-1})[\text{Red}](1.00 \text{ cm})$

But $[\text{Ox}] = 5.70 \times 10^{-5} - [\text{Red}]$. Combining these two equations gives

$[\text{Ox}] = 3.82 \times 10^{-5}$ M and $[\text{Red}] = 1.88 \times 10^{-5}$ M.

(b) $[\text{S}^-] = [\text{Ox}] = 3.82 \times 10^{-5}$ M and $[\text{S}] = \text{Red} = 1.88 \times 10^{-5}$ M.

(c)

$$\begin{array}{lr} S + e^- \rightleftharpoons S^- & E_+^{\circ\prime} = ? \\[4pt] - Ox + e^- \rightleftharpoons Red & E_-^{\circ\prime} = -0.128 \text{ V} \\[2pt] \hline Red + S \rightleftharpoons Ox + S^- & E^{\circ\prime} = 0.059\,16 \log \dfrac{[\text{Ox}][\text{S}^-]}{[\text{Red}][\text{S}]} = 0.036 \text{ V} \end{array}$$

$$E_+^{\circ\prime} = E^{\circ\prime} + E_-^{\circ\prime} = -0.092 \text{ V}$$

CHAPTER 14
ELECTRODES AND POTENTIOMETRY

14-1. (a) $AgCl(s) + e^- \rightleftharpoons Ag(s) + Cl^-$

$Hg_2Cl_2(s) + 2e^- \rightleftharpoons 2Hg(l) + 2Cl^-$

(b) $E = E_+ - E_- = 0.241 - 0.197 = 0.044$ V

14-2. (a) 0.326 V (b) 0.086 V (c) 0.019 V (d) –0.021 V (e) 0.021 V

14-3. $E = E_+ - E_-$

$E = \left\{ 0.771 - 0.059\,16 \ \log \frac{[Fe^{2+}]}{[Fe^{3+}]} \right\} - (0.241) = 0.684$ V

14-4. $E = E° - 0.059\,16 \log \mathcal{A}_{Cl^-}$

$0.280 = 0.268 - 0.059\,16 \log \mathcal{A}_{Cl^-} \Rightarrow \mathcal{A}_{Cl^-} = 0.627$

14-5. For the saturated Ag-AgCl electrode, we can write: $E = E° - 0.059\,16 \log \mathcal{A}_{Cl^-}$.
Putting in $E = 0.197$ and $E° = 0.222$ V gives $\mathcal{A}_{Cl^-} = 2.6_5$. For the S.C.E., we
can write: $E = E° - 0.059\,16 \log \mathcal{A}_{Cl^-} = 0.268$ V $- 0.059\,16 \log 2.6_5$
$= 0.243$ V.

14-6. (a) $Cu^{2+} + 2e^- \rightleftharpoons Cu(s)$ $\qquad E° = 0.339$ V

(b) $E_+ = 0.339 - \dfrac{0.059\,16}{2} \log \dfrac{1}{[Cu^{2+}]} = 0.309$ V

(c) $E = E_+ - E_- = 0.309 - 0.241 = 0.068$ V

14-7. A silver electrode serves as an indicator for Ag^+ by virtue of the equilibrium
$Ag^+ + e^- \rightleftharpoons Ag(s)$ that occurs at its surface. If the solution is saturated with
silver halide, then $[Ag^+]$ is affected by changes in halide concentration.
Therefore, the electrode is also an indicator for halide.

14-8. $V_e = 20.0$ mL. $Ag^+ + e^- \rightleftharpoons Ag(s) \Rightarrow E_+ = 0.799 - 0.059\,16 \log \dfrac{1}{[Ag^+]}$

0.1 mL: $\qquad [Ag^+] = \underbrace{\left(\dfrac{19.9}{20.0} \right)}_{\substack{\text{Fraction} \\ \text{remaining}}} \underbrace{(0.050\,0 \ \text{M})}_{\substack{\text{Original} \\ \text{concentration}}} \underbrace{\left(\dfrac{10.0}{10.1} \right)}_{\substack{\text{Dilution} \\ \text{factor}}} = 0.049\,3$ M

$E = E_+ - E_- = \left\{ 0.799 - 0.059\,16 \log \dfrac{1}{0.049\,3} \right\} - 0.241 = 0.481$ V

189

30.0 mL: This is 10.0 mL past $V_e \Rightarrow [Br^-] = \left(\frac{10.0}{40.0}\right)(0.025\,0\text{ M}) = 0.006\,25\text{ M}$

$$[Ag^+] = K_{sp}/[Br^-] = (5.0 \times 10^{-13})/0.006\,25 = 8.0 \times 10^{-11}\text{ M}$$

$$E = E_+ - E_- = \left\{0.799 - 0.059\,16\log\frac{1}{8.0 \times 10^{-11}}\right\} - 0.241 = -0.039\text{ V}.$$

14-9. The reaction in the right half-cell is $Hg^{2+} + 2e^- \rightleftharpoons Hg(l)$

$$E = E_+ - E_-$$

$$-0.027 = 0.852 - \frac{0.059\,16}{2}\log\frac{1}{[Hg^{2+}]} - (0.241)$$

$$\Rightarrow [Hg^{2+}] = 2.7 \times 10^{-22}\text{ M}.$$

The cell contains 5.00 mmol EDTA (in all forms) and 1.00 mmol Hg(II) in 100 mL. 1.00 mmol EDTA reacts with 1.00 mmol Hg(II), leaving 4.00 mmol EDTA.

$$K_f = \frac{[HgY^{2-}]}{[Hg^{2+}][Y^{4-}]} = \frac{[HgY^{2-}]}{[Hg^{2+}]\alpha_{Y4}\cdot[EDTA]}$$

$$K_f = \frac{(1.00\text{ mmol}/100\text{ mL})}{(2.7 \times 10^{-22})(0.30)(4.00\text{ mmol}/100\text{ mL})} = 3._1 \times 10^{21}$$

14-10. (a) $Fe^{3+} + e^- \rightleftharpoons Fe^{2+}$ $\qquad\qquad E° = 0.771\text{ V}$

(b) $E = E_+ - E_-$

$$-0.126 = 0.771 - 0.059\,16\log\frac{[Fe^{2+}]}{[Fe^{3+}]} - 0.241 \Rightarrow \frac{[Fe^{2+}]}{[Fe^{3+}]} = 1._2 \times 10^{11}$$

(c) $\dfrac{K_f(FeEDTA^-)}{K_f(FeEDTA^{2-})} = \dfrac{[FeEDTA^-]}{[Fe^{3+}][EDTA^{4-}]} \div \dfrac{[FeEDTA^{2-}]}{[Fe^{2+}][EDTA^{4-}]}$

$$= \frac{[FeEDTA^-]}{[FeEDTA^{2-}]}\cdot\frac{[Fe^{2+}]}{[Fe^{3+}]} = \left(\frac{1.00 \times 10^{-3}}{2.00 \times 10^{-3}}\right)(1._2 \times 10^{11}) = 6 \times 10^{10}$$

14-11. $E = E_+ - E_- = -0.429 - 0.059\,16\log\dfrac{[CN^-]^2}{[Cu(CN)_2^-]} - 0.197$

Putting in $E = -0.440$ V and $[Cu(CN)_2^-] = 1.00$ mM gives $[CN^-] = 0.847$ mM.

$$pH = pK_a(HCN) + \log\frac{[CN^-]}{[HCN]} = 9.21 + \log\frac{8.47 \times 10^{-4}}{1.00 \times 10^{-3} - 8.47 \times 10^{-4}} = 9.95_4$$

Now we use the pH to see how much HA reacted with KOH:

	HA	+	OH⁻	→	A⁻	+	H₂O
initial mmol	10.0		x		—		
final mmol	10.0 − x		—		x		

$$pH = pK_a(HA) + \log \frac{[A^-]}{[HA]}$$

$$9.95_4 = 9.50 + \log \frac{x}{10.0 - x} \Rightarrow x = 7.4_0 \text{ mmol of OH}^-$$

$$[KOH] = \frac{7.4_0 \text{ mmol}}{25.0 \text{ mL}} = 0.29_6 \text{ M}$$

14-12. Junction potential arises because different ions diffuse at different rates across a liquid junction, leading to a separation of charge. The resulting electric field retards fast-moving ions and accelerates the slow-moving ions until a steady-state junction potential is reached. This limits the accuracy of a potentiometric measurement, because we do not know what part of a measured cell voltage is due to the process of interest and what is due to the junction potential. The cell in Figure 13-4 has no junction potential because there are no liquid junctions.

14-13. H^+ has greater mobility than K^+. The HCl side of the $HCl\,|\,KCl$ junction will be negative because H^+ diffuses into the KCl region faster than K^+ diffuses into the HCl region. K^+ has a greater mobility than Na^+, so this junction has the opposite sign. The $HCl\,|\,KCl$ voltage is larger, because the difference in mobility between H^+ and K^+ is greater than the difference in mobility between K^+ and Na^+.

14-14. Relative mobilities:

$$K^+ \rightarrow 7.62 \qquad NO_3^- \rightarrow 7.40$$

$$5.19 \leftarrow Na^+ \qquad 7.91 \leftarrow Cl^-$$

Both the cation and anion diffusion cause negative charge to build up on the <u>left</u>.

14-15. Velocity = mobility × field = $(36.30 \times 10^{-8} \text{ m}^2/(\text{s·V})) \times (7\,800 \text{ V/m}) = 2.83 \times 10^{-3}$ m s^{-1} for H^+ and $(7.40 \times 10^{-8})(7\,800) = 5.77 \times 10^{-4}$ m s^{-1} for NO_3^-. To cover 0.120 m will require $(0.120 \text{ m})/(2.83 \times 10^{-3} \text{ m s}^{-1}) = 42.4$ s for H^+ and $(0.120)/(5.77 \times 10^{-4}) = 208$ s for NO_3^-.

14-16. (a) $E° = 0.799 \text{ V} \Rightarrow K = 10^{0.799/0.059\,16} = 3._{20} \times 10^{13}$

(b) $K' = 10^{0.801/0.059\,16} = 3._{46} \times 10^{13}$. $K'/K = 1.08$. The increase is 8%.

(c) $K = 10^{0.100/0.059\,16} = 49.0$. $K' = 10^{0.102/0.059\,16} = 53.0$

$K'/K = 1.08$. The change is still 8%.

14-17. Both half-cell reactions are the same ($AgCl + e^- \rightleftharpoons Ag + Cl^-$) and the concentration of Cl^- is the same on both sides. In principle, the voltage of the cell would be 0 if there were no junction potential. The measured voltage can be attributed to the junction potential. In practice, if both sides contained 0.1 M HCl (or 0.1 M KCl), the two electrodes would probably produce a small voltage because no two real cells are identical. This voltage can be measured and subtracted from the voltage measured with the HCl | KCl junction.

14-18. (a) In phase α, we have 0.1 M H^+ ($u = 36.3 \times 10^{-8}\ m^2\ s^{-1}\ V^{-1}$) and 0.1 M Cl^- ($u = 7.91 \times 10^{-8}\ m^2\ s^{-1}\ V^{-1}$). In phase β, we have 0.1 M K^+ ($u = 7.62 \times 10^{-8}\ m^2\ s^{-1}\ V^{-1}$) and 0.1 M Cl^- ($u = 7.91 \times 10^{-8}\ m^2\ s^{-1}\ V^{-1}$). Substituting into the Henderson equation gives

$$E_j = \frac{(36.3 \times 10^{-8})[0 - 0.1] + (7.62 \times 10^{-8})[0.1 - 0] - (7.91 \times 10^{-8})[0.1 - 0.1]}{(36.3 \times 10^{-8})[0 - 0.1] + (7.62 \times 10^{-8})[0.1 - 0] + (7.91 \times 10^{-8})[0.1 - 0.1]} \times$$

$$0.059\,16 \log \frac{(36.3 \times 10^{-8})(0.1) + (7.91 \times 10^{-8})(0.1)}{(7.62 \times 10^{-8})(0.1) + (7.91 \times 10^{-8})(0.1)} = 26.9\ mV$$

(b)

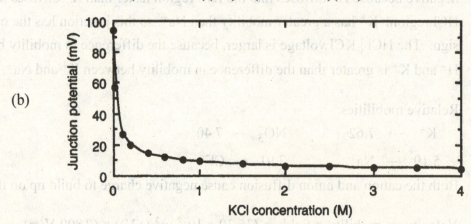

(c)

[HCl]	y M HCl | 1mM KCl	y M HCl | 4 M KCl
10^{-4} M	9.1 mV	4.6 mV
10^{-3} M	26.9 mV	3.6 mV
10^{-2} M	57.3 mV	3.0 mV
10^{-1} M	93.6 mV	4.7 mV

14-19. Ideally, the electrode should be calibrated at 37° using two buffers bracketing the pH of the blood. It would be reasonable to use the MOPSO and HEPES buffers in Table 14-3 that are recommended for use with physiologic fluids. The pH of these standards at 37°C is 6.695 and 7.370. The standards should be thermostatted to 37° during calibration, and the blood should also be at 37° during the measurement.

14-20. Uncertainty in pH of standard buffers, junction potential, junction potential drift, sodium or acid errors at extreme pH values, equilibration time, hydration of glass, temperature of measurement and calibration, and cleaning of electrode

14-21. The error measured in the graph is –0.33 pH units. The electrode will indicate $11.00 - 0.33 = 10.67$.

14-22. Saturated potassium hydrogen tartrate and 0.05 m potassium hydrogen phthalate

14-23. If the alkaline solution has a high concentration of Na^+ (as in NaOH), the Na^+ cation competes with H^+ for cation exchange sites on the glass surface. The glass responds as if H^+ were present, and the apparent pH is lower than the actual pH.

14-24. The junction potential changes from –6.4 mV to –0.2 mV. A change of $6.4 - 0.2 = +6.2$ mV appears to be a pH change of $+6.2/59.16 = +0.10$ pH units.

14-25. (a) $(4.63)(59.16 \text{ mV}) = 274$ mV. The factor 59.16 mV is the value of $(RT \ln 10)/F$ at 298.15 K.

(b) At 310.15 K (37°C), $(RT \ln 10)/F$

$= (8.3145 \text{ J mol}^{-1} \text{ K}^{-1})(310.15 \text{ K})(\ln 10)/(96\,485 \text{ C mol}^{-1}) = 61.54$ mV.

$(4.63)(61.54 \text{ mV}) = 285$ mV.

14-26. pH of 0.025 m KH_2PO_4/0.025 m Na_2HPO_4 at 20°C = 6.881

pH of 0.05 m potassium hydrogen phthalate at 20°C = 4.002

$$\frac{E_{unknown} - E_{S1}}{pH_{unknown} - pH_{S1}} = \frac{E_{S2} - E_{S1}}{pH_{S2} - pH_{S1}}$$

$$\frac{E_{unknown} - (-18.3 \text{ mV})}{pH_{unknown} - 6.881} = \frac{(+146.3 \text{ mV}) - (-18.3 \text{ mV})}{4.002 - 6.881} = -57.17_3 \text{ mV/pH unit}$$

$$pH_{unknown} = \frac{E_{unknown} - (-18.3 \text{ mV})}{-57.17_3 \text{ mV/pH unit}} + 6.881 = 5.686$$

Observed slope = -57.17_3 mV/pH unit

Theoretical slope $= -\dfrac{RT \ln 10}{F}$

$$= -\frac{[8.314\,47 \text{ V·C/(mol·K)}][293.15 \text{ K}] \ln 10}{9.648\,53 \times 10^4 \text{ C/mol}} = -0.058\,17 \text{ V}$$

$$\beta = \frac{\text{observed slope}}{\text{theoretical slope}} = \frac{-57.17_3 \text{ mV}}{-58.17 \text{ mV}} = 0.983$$

14-27. (a) There is negligible change in the concentrations of the buffer species when we mix the acid $H_2PO_4^-$ with its conjugate base, HPO_4^{2-}. The ionic strength of 0.025 0 m KH_2PO_4 (a 1:1 electrolyte) is 0.025 0 m. The ionic strength of 0.025 0 m Na_2HPO_4 (a 2:1 electrolyte) is 0.075 0 m. The total ionic strength is 0.100 m.

(b) $K_2 = \dfrac{[H^+]\gamma_{H^+}[HPO_4^{2-}]\gamma_{HPO_4^{2-}}}{[H_2PO_4^-]\gamma_{H_2PO_4^-}}$

But $K_2 = 10^{-7.198}$ and $[H^+]\gamma_{H^+} = 10^{-pH} = 10^{-6.865}$

Therefore, $\dfrac{\gamma_{HPO_4^{2-}}}{\gamma_{H_2PO_4^-}} = \dfrac{K_2[H_2PO_4^-]}{[H^+]\gamma_{H^+}[HPO_4^{2-}]} = \dfrac{10^{-7.198}[0.025\ 0]}{10^{-6.865}[0.025\ 0]} = 0.464_5$

(We can use molality or any other units for concentrations because they cancel in the numerator and denominator.)

(c) To get a pH of 7.000, we need to increase the concentration of base (HPO_4^{2-}) and decrease the concentration of acid ($H_2PO_4^-$). To maintain a constant ionic strength, we must decrease KH_2PO_4 three times as much as we increase Na_2HPO_4, because Na_2HPO_4 contributes three times as much as KH_2PO_4 to the ionic strength. So let's increase Na_2HPO_4 by x and decrease KH_2PO_4 by $3x$.

$$K_2 = \frac{[H^+]\gamma_{H^+}[HPO_4^{2-}]\gamma_{HPO_4^{2-}}}{[H_2PO_4^-]\gamma_{H_2PO_4^-}}$$

$$\Rightarrow 10^{-7.198} = \frac{10^{-7.000}[0.025\ 0 + x]}{[0.025\ 0 - 3x]}(0.464_5) \Rightarrow x = 0.001\ 8 \ m.$$

The new concentrations should be Na_2HPO_4 = 0.026 8 m and KH_2PO_4 = 0.019 6 m.

14-28. Analyte ions equilibrate with ion-exchange sites at the outer surface of the ion-selective membrane. Diffusion of analyte ions out of the membrane creates a slight charge imbalance (an electric potential difference) across the interface between the membrane and the analyte solution. Changes in analyte ion concentration in the solution change the potential difference across the outer boundary of the ion-selective membrane.

A compound electrode contains a second chemically active membrane outside the ion-selective membrane. The second membrane may be semipermeable and only allow the species of interest to pass through. Alternatively, the second membrane may contain a substance (such as an enzyme) that reacts with analyte to generate the species to which the ion-selective membrane responds.

14-29. The selectivity coefficient $K_{A,X}^{Pot}$ tells us the relative response of an ion-selective electrode to the ion of interest (A) and an interfering ion (X). The smaller $K_{A,X}^{Pot}$, the more selective is the electrode (smaller response to the interfering ion).

14-30. A mobile molecule dissolved in the membrane liquid phase binds tightly to the ion of interest and weakly to interfering ions.

14-31. A metal ion buffer maintains the desired (small) concentration of metal ion from a large reservoir of metal complex (ML) and free ligand (L). If you just tried to dissolve 10^{-8} M metal ion in most solutions or containers, the metal would probably bind to the container wall or to an impurity in the solution and be lost.

14-32. Electrodes respond to *activity*. If the ionic strength is constant, the activity coefficient of analyte will be constant in all standards and unknowns. In this case, the calibration curve can be written directly in terms of concentration.

14-33. (a) $-0.230 = \text{constant} - 0.059\,16 \log (1.00 \times 10^{-3}) \Rightarrow \text{constant} = -0.407$ V

(b) $-0.300 = -0.407 - 0.059\,16 \log x \Rightarrow x = 1.5_5 \times 10^{-2}$ M

(c) $-0.230 = \text{constant} - 0.059\,16 \log (1.00 \times 10^{-3})$

$\underline{-0.300 = \text{constant} - 0.059\,16 \log\ x}$

subtract: $0.070 = -0.059\,16 \log \dfrac{1.00 \times 10^{-3}}{x} \Rightarrow x = 1.5_2 \times 10^{-2}$ M

14-34. $E_1 = \text{constant} + \dfrac{0.059\,16}{2} \log [1.00 \times 10^{-4}]$

$E_2 = \text{constant} + \dfrac{0.059\,16}{2} \log [1.00 \times 10^{-3}]$

$\Delta E = E_2 - E_1 = \dfrac{0.059\,16}{2} \log \dfrac{1.00 \times 10^{-3}}{1.00 \times 10^{-4}} = +0.029\,6$ V

14-35. $[F^-]_{\text{Providence}} = 1.00$ mg $F^-/L = 5.26 \times 10^{-5}$ M

$E_{\text{Providence}} = \text{constant} - 0.059\,16 \log [5.26 \times 10^{-5}]$

$E_{\text{Foxboro}} = \text{constant} - 0.059\,16 \log [F^-]_{\text{Foxboro}}$

$$\Delta E = E_{Foxboro} - E_{Providence} = 0.040\ 0\ V$$
$$= -0.059\ 16 \log \frac{[F^-]_{Foxboro}}{5.26 \times 10^{-5}} \Rightarrow [F^-]_{Foxboro} = 1.11 \times 10^{-5}\ M = 0.211\ mg/L$$

14-36. K^+ has the largest selectivity coefficient of Group 1 ions and therefore interferes the most. Sr^{2+} and Ba^{2+} are the worst of the Group 2 ions. Since $\log K_{Li^+,K^+}^{Pot} \approx$ –2, there must be 100 times more K^+ than Li^+ to give equal response.

14-37. $\dfrac{[ML]}{[M][L]} = 4.0 \times 10^8 = \dfrac{0.030\ M}{[M](0.020\ M)} \Rightarrow [M] = 3.8 \times 10^{-9}\ M$

14-38. (a) The least squares parameters are
$$E = 51.10\ (\pm0.24) + 28.14\ (\pm0.08_5) \log [Ca^{2+}] \quad (s_y = 0.2_7)$$

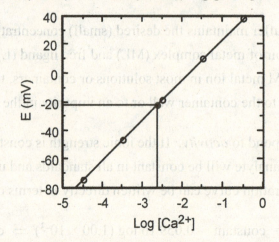

(b) The slope is $0.028\ 14\ V = \beta(0.059\ 16\ V)/2 \Rightarrow \beta = 0.951$.

(c) If we use Equation 4-27 in a spreadsheet, we find $\log [Ca^{2+}] = -2.615_3(\pm0.007_2)$ using $s_y = 0.3$ and $k = 4$.

From Table 3-1, we can write that if $F = 10^x$, $e_F/F = (\ln 10)e_x$.

In this problem, $F = [Ca^{2+}] = 10^{-2.615_3(\pm0.007_2)}$

$e_F/F = (\ln 10)(0.007_2) = 0.016_6$

$e_F = (0.016_6)\ F = 4.0 \times 10^{-5} \Rightarrow F = 2.43(\pm0.04) \times 10^{-3}\ M$.

14-39. At pH 7.2 the effect of H^+ will be negligible because $[H^+] \ll [Li^+]$:

$-0.333\ V = $ constant $+ 0.059\ 16 \log [3.44 \times 10^{-4}] \Rightarrow$ constant $= -0.128\ V$.

At pH 1.1 ($[H^+] = 0.079\ M$), we must include interference by H^+:

$E = -0.128 + 0.059\ 16 \log [3.44 \times 10^{-4} + (4 \times 10^{-4})(0.079)] = -0.331\ V$.

14-40. The function to plot on the y-axis is $(V_0 + V_s) 10^{E/S}$, where $S = -(\beta R T \ln 10)/nF$. (The minus sign comes from the equation for the response of the electrode, which has a minus sign in front of the log term.) Putting in $\beta = 0.933$, $R = 8.314\ 5$ J/(mol·K), $F = 96\ 485$ C/mol, $T = 303.15$ K, and $n = 2$ gives $S = -0.028\ 06$ J/C $= 0.028\ 06$ V. (You can get the relation of J/C = V from the equation $\Delta G = -nFE$, in which the units are J = (mol)(C/mol)(V).)

V_s (mL)	E (V)	y
0	0.079 0	0.084 1
0.100	0.072 4	0.144 9
0.200	0.065 3	0.259 9
0.300	0.058 8	0.443 8
0.800	0.050 9	0.856 5

The graph has a slope of $m = 0.989$ and an intercept of $b = 0.080\ 9$, giving an x-intercept of $-b/m = -0.081\ 8$ mL. The concentration of original unknown is

$$c_X = -\frac{(x\text{-intercept})\,c_S}{V_0} = -\frac{(-0.081\ 8\ \text{mL})(0.020\ 0\ \text{M})}{55.0\ \text{mL}} = 3.0 \times 10^{-5}\ \text{M}.$$

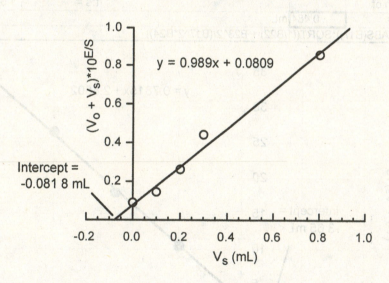

14-41. (a) The following spreadsheet and graph are shown. The x-intercept is at -3.65 mL with a standard deviation in cell B26 of $s = 0.484$ mL. The intercept gives us the concentration of ammonia nitrogen in the volume $V_0 = 101.0$ mL:

$$x\text{-intercept} = -3.65\ \text{mL} = -\frac{V_0 c_X}{c_S}$$

$$\Rightarrow c_X = \frac{(x\text{-intercept})c_S}{V_0} = \frac{(3.65\ \text{mL})(10.0\ \text{ppm})}{101.0\ \text{mL}} = 0.361_4\ \text{ppm}$$

	A	B	C	D	E	F	G	H
1	Standard Addition: Ammonia in Seawater							
2								
3	$V_0 =$	101	mL					
4	$c_S =$	10	ppm					
5	$s =$	0.0566	V					
6								
7	x			y				
8	Added standard	$E =$ cell	$V_0 + V_s$	$(V_0 + V_s)10^{E/s}$				
9	(mL)	voltage (V)	(mL)	(mL)				
10	0.00	-0.0844	101.0	3.26				
11	10.00	-0.0581	111.0	10.44				
12	20.00	-0.0469	121.0	17.95				
13	30.00	-0.0394	131.0	26.37				
14	40.00	-0.0347	141.0	34.37				
15								
16		LINEST output:			Highlight cells B17:D19			
17	m	0.7815	2.8502	b	Type '=LINEST(D10:D14,A10:A14,TRUE,TRUE)			
18	s_m	0.0137	0.3364	s_b	Press CTRL+SHIFT+ENTER (on PC)			
19	R^2	0.9991	0.4343	s_y	Press COMMAND+RETURN (on Mac)			
20								
21	x-intercept = -b/m =	-3.647	mL		To find 95% confidence interval			
22	n =	5	B22 = COUNT(A10:A14)		we need Student's t for			
23	Mean y =	18.479	B23 = AVERAGE(D10:D14)		3 degrees of freedom			
24	$\Sigma(x_i - \text{mean } x)^2 =$	1000	B24 = DEVSQ(A10:A14)		TINV(0.05,3) =	3.182446		
25	Std deviation of				t*s =	1.541109		
26	x-intercept =	0.484	mL					
27	B26 = (C19/ABS(B17))*SQRT((1/B22) + B23^2/(B17^2*B24))							

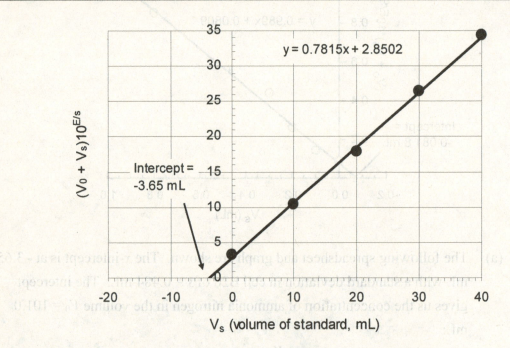

$$y = 0.7815x + 2.8502$$

Intercept =
-3.65 mL

V_s (volume of standard, mL)

The concentration of ammonia nitrogen in the original 100.0 mL of seawater, which had been diluted from 100.0 to 101.0 mL, is $\frac{101.0 \text{ mL}}{100.0 \text{ mL}}$ (0.3614 ppm) = 0.365 ppm. The 95% confidence interval is equal to Student's t times the standard deviation:

$$95\% \text{ confidence interval} = \pm t \cdot s = \pm (3.18)(0.484 \text{ mL}) = \pm 1.54 \text{ mL}$$

where t is for 95% confidence and $5 - 2 = 3$ degrees of freedom because there are 5 data points on the standard addition curve. You can find t in the table of Student's t or you can compute it with the statement "=TINV(0.05,3)" in cell G24 in the spreadsheet. The confidence interval of ± 1.54 mL corresponds to a relative uncertainty of $100 \times \frac{1.54 \text{ mL}}{3.65 \text{ mL}} = 42\%$.

The absolute uncertainty is $(0.042)(0.365 \text{ ppm}) = 0.15$ ppm. The concentration of ammonia nitrogen in the seawater can be expressed as 0.36 ± 0.15 ppm.

(b)   Added standards should increase analytical signal by a factor of 1.5 to 3. In this experiment, analytical signal is $(V_0 + V_S)10^{E/S} = 3.26$ mL for unknown and 34.37 mL for final standard. The final signal is 10 times as great as the initial signal, which is ~3 times more than the recommended limit.

14-42. For the first line of data, with A = Na^+ and X = Mg^{2+}

$$\log K_{A,X}^{Pot} = \frac{z_A F(E_X - E_A)}{RT \ln 10} + \log \left(\frac{\mathcal{A}_A}{\mathcal{A}_X^{z_A/z_X}} \right)$$

$$= \frac{(+1)(96\,485 \text{ C/mol})(-0.385 \text{ V})}{(8.3145 \text{ V·C/[mol·K]})(294.65 \text{ K}) \ln 10} + \log \left(\frac{10^{-3}}{(10^{-3})^{1/2}} \right) = -8.09.$$

For the second line of data, $\log K_{A,X}^{Pot} = -8.15$.

The first and second lines should give the same selectivity coefficient. The difference is experimental error.

For the third line of data, with A = Na^+ and X = K^+:

$$\log K_{A,X}^{Pot} = \frac{(+1)(96\,485 \text{ C/mol})(-0.285 \text{ V})}{(8.3145 \text{ V·C/[mol·K]})(294.65 \text{ K}) \ln 10} + \log \left(\frac{10^{-3}}{(10^{-3})^{1/1}} \right) = -4.87.$$

For the fourth line of data, $\log K_{A,X}^{Pot} = -4.87$.

14-43. For Na^+: error in $\mathcal{A}_{H^+}(\%) = \frac{(10^{-8.6})^{1/1}(10^{-2.0})}{(10^{-8.0})^{1/1}} \times 100 = 0.25\%$

For Ca^{2+}: error in $\mathcal{A}_{H^+}(\%) = \frac{(10^{-7.8})^{2/1}(10^{-2.0})}{(10^{-8.0})^{2/1}} \times 100 = 2.5\%$

14-44. $E = \text{constant} + \dfrac{\beta(0.059\,16)}{2}\;\log\,([Ca^{2+}] + K^{Pot}_{Ca^{2+},Mg^{2+}}\,[Mg^{2+}])$

$\phantom{E = \text{constant} + }\underset{A}{}\qquad\underset{B}{}$

For the first two solutions we can write

$-52.6\text{ mV} = A + B\,\log\,(1.00 \times 10^{-6}) = A - 6\,B$

$+16.1\text{ mV} = A + B\,\log\,(2.43 \times 10^{-4}) = A - 3.614\,B.$

Subtraction gives $68.7\text{ mV} = 2.386\,B \Rightarrow B = 28.80\text{ mV}.$

Putting this value of B back into the first equation gives $A = 120.2\text{ mV}.$

The third set of data now gives the selectivity coefficient:

$-38.0\text{ mV} = 120.2 + 28.80\,\log\,[10^{-6} + K^{Pot}_{Ca^{2+},Mg^{2+}}\,(3.68 \times 10^{-3})]$

$\Rightarrow K^{Pot}_{Ca^{2+},Mg^{2+}} = 6.0 \times 10^{-4}$

$E = 120.2 + 28.80\,\log\,([Ca^{2+}] + 6.0 \times 10^{-4}\,[Mg^{2+}]).$

14-45.   There is a large excess of EDTA in the buffer. We expect essentially all lead to be in the form PbY^{2-} (where Y = EDTA).

$[PbY^{2-}] = \dfrac{1.0}{101.0}\,(0.10\text{ M}) = 9.9 \times 10^{-4}\text{ M}$

$\text{Total EDTA} = \dfrac{100.0}{101.0}\,(0.050\text{ M}) = 0.049_5\text{ M}$

$\text{Free EDTA} = 0.049_5\text{ M} - 9.9 \times 10^{-4}\text{ M} = 0.048_5\text{ M}$

$$\underset{\substack{\text{EDTA bound}\\\text{to }Pb^{2+}}}{}$$

$Pb^{2+} + Y^{4-} \rightleftharpoons PbY^{2-}\qquad K_f' = \alpha_{Y^{4-}}\,K_f = (1.46 \times 10^{-8})(10^{18.0}) = 1.4_6 \times 10^{10}$

$K_f' = \dfrac{[PbY^{2-}]}{[Pb^{2+}][EDTA]}$

$\Rightarrow [Pb^{2+}] = \dfrac{[PbY^{2-}]}{K_f'[EDTA]} = \dfrac{9.9 \times 10^{-4}}{(1.4_6 \times 10^{10})(0.048_5)} = 1.4 \times 10^{-12}\text{ M}$

14-46. $[Hg^{2+}]$ in the buffers is computed from equilibrium constants for the solubility of HgX_2 and formation of complex ions such as HgX_3^-. Since the data for $HgCl_2$ are not in line with the data for $Hg(NO_3)_2$ and $HgBr_2$, equilibrium constants used for the $HgCl_2$ system could be in error. Whenever we make a buffer by mixing *calculated* quantities of reagents, we are at the mercy of the quality of tabulated equilibrium constants.

14-47.   (a)   slope $= 29.58 \text{ mV} = \dfrac{E_2 - E_1}{\log \mathcal{A}_2 - \log \mathcal{A}_1} = \dfrac{(-25.90) - 2.06}{\log \mathcal{A}_2 - (-3.000)}$

$\Rightarrow \mathcal{A}_2 = 1.13 \times 10^{-4}$

(b)

$$\underset{5.00 \times 10^{-4} - x}{Ca^{2+}} \quad + \quad \underset{(0.998)[(5.00 \times 10^{-4}) - x]}{A^{3-}} \quad \rightleftharpoons \quad \underset{x}{CaA^-}$$

But $\mathcal{A}_{Ca^{2+}} = 1.13 \times 10^{-4} = (5.00 \times 10^{-4} - x)\,(0.405)$

$\uparrow$

γ from Table 7-1

$\Rightarrow \quad x = 2.2 \times 10^{-4} \text{ M}$

$K_f = \dfrac{[CaA^-]\gamma_{CaA^-}}{[Ca^{2+}]\gamma_{Ca^{2+}}[A^{3-}]\gamma_{A^{3-}}}$

$K_f = \dfrac{(2.20 \times 10^{-4})(0.79)}{(1.13 \times 10^{-4})[(0.998)(5 \times 10^{-4} - 2.20 \times 10^{-4})](0.115)}$

$K_f = 4.8 \times 10^4$

14-48.   Analyte adsorbed on the surface of the gate changes the electric potential of the gate. This, in turn, changes the current between the source and drain. The potential that must be applied by the external circuit to restore the current to its initial value is a measure of the change in gate potential. Following the Nernst equation, there is close to a 59 mV change in gate potential for each factor-of-10 change in activity of univalent analyte at 25°C. The key to ion-specific response is to have a chemical on the gate that selectively binds one analyte.

CHAPTER 15
REDOX TITRATIONS

15-1. (a) $Ce^{4+} + Fe^{2+} \rightarrow Ce^{3+} + Fe^{3+}$

(b) $Fe^{3+} + e^- \rightleftharpoons Fe^{2+}$ $E° = 0.767$ V

$Ce^{4+} + e^- \rightleftharpoons Ce^{3+}$ $E° = 1.70$ V

(c) $E = \left\{0.767 - 0.059\,16 \log \dfrac{[Fe^{2+}]}{[Fe^{3+}]}\right\} - \left\{0.241\right\}$ (A)

$E = \left\{1.70 - 0.059\,16 \log \dfrac{[Ce^{3+}]}{[Ce^{4+}]}\right\} - \left\{0.241\right\}$ (B)

(d) <u>10.0 mL</u>: Use Eq. (A) with $[Fe^{2+}]/[Fe^{3+}] = 40.0/10.0$, since

$V_e = 50.0$ mL $\Rightarrow E = 0.490$ V.

<u>25.0 mL</u>: $[Fe^{2+}]/[Fe^{3+}] = 25.0/25.0 \Rightarrow E = 0.526$ V

<u>49.0 mL</u>: $[Fe^{2+}]/[Fe^{3+}] = 1.0/49.0 \Rightarrow E = 0.626$ V

<u>50.0 mL</u>: This is V_e, where $[Ce^{3+}] = [Fe^{3+}]$ and $[Ce^{4+}] = [Fe^{2+}]$.

Eq. 15-11 gives $E_+ = 1.23$ V and $E = 0.99$ V.

<u>51.0 mL</u>: Use Eq. (B) with $[Ce^{3+}]/[Ce^{4+}] = 50.0/1.0 \Rightarrow E = 1.36$ V.

<u>60.0 mL</u>: $[Ce^{3+}]/[Ce^{4+}] = 50.0/10.0 \Rightarrow E = 1.42$ V

<u>100.0 mL</u>: $[Ce^{3+}]/[Ce^{4+}] = 50.0/50.0 \Rightarrow E = 1.46$ V

15-2. (a) $Ce^{4+} + Cu^+ \rightarrow Ce^{3+} + Cu^{2+}$

(b) $Ce^{4+} + e^- \rightleftharpoons Ce^{3+}$ $E° = 1.70$ V

$Cu^{2+} + e^- \rightleftharpoons Cu^+$ $E° = 0.161$ V

(c) $E = \left\{1.70 - 0.059\,16 \log \dfrac{[Ce^{3+}]}{[Ce^{4+}]}\right\} - \left\{0.197\right\}$ (A)

$E = \left\{0.161 - 0.059\,16 \log \dfrac{[Cu^+]}{[Cu^{2+}]}\right\} - \left\{(0.197)\right\}$ (B)

(d) <u>1.00 mL</u>: Use Eq. (A) with $[Ce^{3+}]/[Ce^{4+}] = 1.00/24.0$, since

$V_e = 25.0$ mL $\Rightarrow E = 1.58$ V.

<u>12.5 mL</u>: $[Ce^{3+}]/[Ce^{4+}] = 12.5/12.5 \Rightarrow E = 1.50$ V

<u>24.5 mL</u>: $[Ce^{3+}]/[Ce^{4+}] = 24.5/0.5 \Rightarrow E = 1.40$ V

<u>25.0 mL</u>: $E_+ = 1.70 - 0.059\,16 \log \dfrac{[Ce^{3+}]}{[Ce^{4+}]}$

$E_+ = 0.161 - 0.059\,16 \log \dfrac{[Cu^+]}{[Cu^{2+}]}$

$2E_+ = 1.86_1 - 0.059\,16 \log \dfrac{[Ce^{3+}][Cu^+]}{[Ce^{4+}][Cu^{2+}]}$

At the equivalence point, $[Ce^{3+}] = [Cu^{2+}]$ and $[Ce^{4+}] = [Cu^+]$.
Therefore, the log term above is zero and $E_+ = 1.86_1/2 = 0.930$ V.

$$E = 0.930 - 0.197 = 0.733 \text{ V}$$

<u>25.5 mL</u>:   Use Eq. (B) with $[Cu^+]/[Cu^{2+}] = 0.5/25.0 \Rightarrow E = 0.065$ V.

<u>30.0 mL</u>: $[Cu^+]/[Cu^{2+}] = 5.0/25.0 \Rightarrow E = 0.005$ V

<u>50.0 mL</u>: $[Cu^+]/[Cu^{2+}] = 25.0/25.0 \Rightarrow E = -0.036$ V

15-3.   (a) $Sn^{2+} + Tl^{3+} \rightarrow Sn^{4+} + Tl^+$

(b) $Sn^{4+} + 2e^- \rightleftharpoons Sn^{2+} \quad E° = 0.139$ V

$Tl^{3+} + 2e^- \rightleftharpoons Tl^+ \quad E° = 0.77$ V

(c) $E = \left\{ 0.139 - \dfrac{0.059\,16}{2} \log \dfrac{[Sn^{2+}]}{[Sn^{4+}]} \right\} - \left\{ 0.241 \right\}$ (A)

$E = \left\{ 0.77 - \dfrac{0.059\,16}{2} \log \dfrac{[Tl^+]}{[Tl^{3+}]} \right\} - \left\{ 0.241 \right\}$ (B)

(d)   <u>1.00 mL</u>:   Use Eq. (A) with $[Sn^{2+}]/[Sn^{4+}] = 4.00/1.00$, since
$V_e = 5.00$ mL $\Rightarrow E = -0.120$ V.

<u>2.50 mL</u>: $[Sn^{2+}]/[Sn^{4+}] = 2.50/2.50 \Rightarrow E = -0.102$ V

<u>4.90 mL</u>: $[Sn^{2+}]/[Sn^{4+}] = 0.10/4.90 \Rightarrow E = -0.052$ V

<u>5.00 mL</u>: $E_+ = 0.139 - \dfrac{0.059\,16}{2} \log \dfrac{[Sn^{2+}]}{[Sn^{4+}]}$

$E_+ = 0.77 - \dfrac{0.059\,16}{2} \log \dfrac{[Tl^+]}{[Tl^{3+}]}$

$$\overline{2E_+ = 0.90_9 - \dfrac{0.059\,16}{2} \log \dfrac{[Sn^{2+}][Tl^+]}{[Sn^{4+}][Tl^{3+}]}}$$

At the equivalence point, $[Sn^{4+}] = [Tl^+]$ and $[Sn^{2+}] = [Tl^{3+}]$.
Therefore, the log term above is zero and $E_+ = 0.90_9/2 = 0.45_4$ V.

$$E = 0.45_4 - 0.241 = 0.21 \text{ V}$$

<u>5.10 mL</u>:   Use Eq. (B) with $[Tl^+]/[Tl^{3+}] = 5.00/0.10 \Rightarrow E = 0.48$ V

<u>10.0 mL</u>:   Use Eq. (B) with $[Tl^+]/[Tl^{3+}] = 5.00/5.00 \Rightarrow E = 0.53$ V

15-4.   (a) $2Fe^{3+} + \text{ascorbic acid} + H_2O \rightarrow 2Fe^{2+} + \text{dehydroascorbic acid} + 2H^+$

(b)   The equivalence volume is 10.0 mL.
At <u>5.0 mL</u>, half of the Fe^{3+} is titrated and the ratio $[Fe^{2+}]/[Fe^{3+}]$ is 5.0/5.0:
$Fe^{3+} + e^- \rightleftharpoons Fe^{2+} \qquad\qquad E° = 0.767$ V

$E = E_+ - E_- = \left\{ 0.767 - 0.059\,16 \log \dfrac{[Fe^{2+}]}{[Fe^{3+}]} \right\} - \left\{ 0.197 \right\}$

$$= \left\{0.767 - 0.059\,16 \log \frac{5.0}{5.0}\right\} - \left\{0.197\right\} = 0.570$$

<u>10.0 mL</u> is the equivalence point. We multiply the ascorbic acid Nernst equation by 2 and add it to the iron Nernst equation:

$$E_+ = 0.767 - 0.059\,16 \log \frac{[Fe^{2+}]}{[Fe^{3+}]}$$

$$2E_+ = 2\left(0.390 - \frac{0.059\,16}{2} \log \frac{[\text{ascorbic acid}]}{[\text{dehydro}][H^+]^2}\right)$$

$$3E_+ = 1.547 - 0.059\,16 \log \frac{[Fe^{2+}][\text{ascorbic acid}]}{[Fe^{3+}][\text{dehydro}][H^+]^2}$$

At the equivalence point, the stoichiometry of the titration reaction tells us that $[Fe^{2+}] = 2[\text{dehydroascorbic acid}]$ and $[Fe^{3+}] = 2[\text{ascorbic acid}]$. Inserting these equalities into the log term just shown gives

$$3E_+ = 1.547 - 0.059\,16 \log \frac{2[\text{dehydro}][\text{ascorbic acid}]}{2[\text{ascorbic acid}][\text{dehydro}][H^+]^2}$$

$$3E_+ = 1.547 - 0.059\,16 \log \frac{1}{[H^+]^2}$$

Using $[H^+] = 10^{-0.30}$ gives $E_+ = 0.504$ V and $E = 0.504 - 0.197 = 0.307$ V.

At <u>15.0 mL</u>, the ratio [dehydro]/[ascorbic acid] is 10.0/5.0:

dehydroascorbic acid $+ 2H^+ + 2e^- \rightleftharpoons$ ascorbic acid $+ H_2O$ $E° = 0.390$ V

$$E = E_+ - E_- = \left\{0.390 - \frac{0.059\,16}{2} \log \frac{[\text{ascorbic acid}]}{[\text{dehydro}][H^+]^2}\right\} - \left\{0.197\right\}$$

$$= \left\{0.390 - \frac{0.059\,16}{2} \log \frac{[5.0]}{[10.0][10^{-0.30}]^2}\right\} - \left\{0.197\right\} = 0.184 \text{ V}$$

15-5.   (a)   Titration reaction: $Sn^{2+} + 2Fe^{3+} \rightarrow Sn^{4+} + 2Fe^{2+}$ $V_e = 25.0$ mL

   (b) $Fe^{3+} + e^- \rightleftharpoons Fe^{2+}$ $E° = 0.732$ V

   $Sn^{4+} + 2e^- \rightleftharpoons Sn^{2+}$ $E° = 0.139$ V

   (c) $E = \left\{0.732 - 0.059\,16 \log \frac{[Fe^{2+}]}{[Fe^{3+}]}\right\} - \left\{0.241\right\}$ (A)

   $E = \left\{0.139 - \frac{0.059\,16}{2} \log \frac{[Sn^{2+}]}{[Sn^{4+}]}\right\} - \left\{0.241\right\}$ (B)

   (d)   Representative calculations:

   <u>1.0 mL</u>: $E_+ = 0.139 - \frac{0.059\,16}{2} \log \frac{[Sn^{2+}]}{[Sn^{4+}]}$

   initial mol $Sn^{2+} = (25.0 \text{ mL})(0.050\,0 \frac{\text{mmol}}{\text{mL}}) = 1.25$ mmol

   mol Fe^{3+} added $= (1.0 \text{ mL})(0.100 \frac{\text{mmol}}{\text{mL}}) = 0.10$ mmol

$$[Sn^{4+}] = \frac{\frac{1}{2}\,(0.10\text{ mmol})}{26.0\text{ mL}} = 1.9_2 \times 10^{-3}\text{ M}$$

$$[Sn^{2+}] = \frac{1.25 - \frac{1}{2}\,(0.10)\text{ mmol}}{26.0\text{ mL}} = 4.62 \times 10^{-2}\text{ M}$$

$$E_+ = 0.139 - \frac{0.059\,16}{2}\log\frac{[Sn^{2+}]}{[Sn^{4+}]}$$

$$E_+ = 0.139 - \frac{0.059\,16}{2}\log\frac{4.62 \times 10^{-2}]}{[1.9_2 \times 10^{-3}]} = 0.098\text{ V}$$

$$E = E_+ - E_- = 0.098 - 0.241 = -0.143\text{ V}$$

<u>25.0 mL</u>: At the equivalence point, we add the two indicator electrode Nernst equations. To make the factor in front of the log term the same in both equations, we can multiply the $Sn^{4+}\,|\,Sn^{2+}$ equation by 2:

$$E_+ = 0.139 - \frac{0.059\,16}{2}\log\frac{[Sn^{2+}]}{[Sn^{4+}]}$$

$$2E_+ = 0.278 - 0.059\,16\log\frac{[Sn^{2+}]}{[Sn^{4+}]}$$

$$E_+ = 0.732 - 0.059\,16\log\frac{[Fe^{2+}]}{[Fe^{3+}]}$$

Now add the last two equations to get

$$3E_+ = 1.010 - 0.059\,16\log\left(\frac{[Sn^{2+}][Fe^{2+}]}{[Sn^{4+}][Fe^{3+}]}\right)$$

But at the equivalence point, $2[Sn^{4+}] = [Fe^{2+}]$ and $2[Sn^{2+}] = [Fe^{3+}]$. Substituting these identities into the log term gives

$$3E_+ = 1.010 - 0.059\,16\log\left(\frac{[Sn^{2+}]2[Sn^{4+}]}{[Sn^{4+}]2[Sn^{2+}]}\right)$$

So the log quotient in the log term is 1 and the logarithm is 0. Therefore, $E_+ = 1.010/3 = 0.337$ V and $E = E_+ - E_- = 0.337 - 0.241 = 0.096$ V.

<u>26.0 mL</u>: $E_+ = 0.732 - 0.059\,16\log\dfrac{[Fe^{2+}]}{[Fe^{3+}]}$

There is 1.0 mL of Fe^{3+} beyond the equivalence point.

$$[Fe^{3+}] = \frac{(1.0\text{ mL})\,(0.100\text{ M})}{51.0\text{ mL}} = 1.9_6 \times 10^{-3}\text{ M}$$

The first 25.0 mL of Fe^{3+} were converted to Fe^{2+}, so

$$[Fe^{2+}] = \frac{(25.0\text{ mL})\,(0.100\text{ M})}{51.0\text{ mL}} = 4.90 \times 10^{-2}\text{ M}$$

$$E_+ = 0.732 - 0.059\,16\log\frac{[Fe^{2+}]}{[Fe^{3+}]}$$

$$E_+ = 0.732 - 0.059\,16\log\frac{[4.90 \times 10^{-2}]}{[1.9_6 \times 10^{-3}]} = 0.649\text{ V}$$

$$E = E_+ - E_- = 0.649 - 0.241 = 0.408 \text{ V}$$

mL	E (V)	mL	E (V)	mL	E (V)
1.0	−0.143	24.0	−0.061	26.0	0.408
12.5	−0.102	25.0	0.096	30.0	0.450

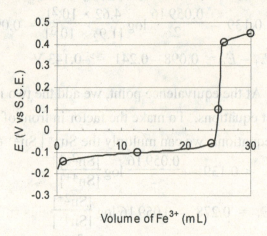

15-6. Diphenylamine sulfonic acid: colorless → red-violet

Diphenylbenzidine sulfonic acid: colorless → violet

tris (2,2'-bipyridine) iron: red → pale blue

Ferroin: red → pale blue

15-7. The reduction potentials are

$Sn^{4+} + 2e^- \rightleftharpoons Sn^{2+}$ $E° = 0.139 \text{ V}$

$Mn(EDTA)^- + e^- \rightleftharpoons Mn(EDTA)^{2-}$ $E° = 0.825 \text{ V}$

The end point will be between 0.139 and 0.825 V. Tris(2,2'-bipyridine) iron has too high a reduction potential (1.120 V) to be useful for this titration.

15-8. Preoxidation and prereduction refer to adjusting the oxidation state of analyte to a suitable value for a titration. The preoxidation or prereduction agent must be destroyed so it does not interfere with the titration by reacting with titrant.

15-9. $2S_2O_8^{2-} + 2H_2O \xrightarrow{\text{boiling}} 4SO_4^{2-} + O_2 + 4H^+$

$4Ag^{2+} + 2H_2O \xrightarrow{\text{boiling}} 4Ag^+ + O_2 + 4H^+$

$2H_2O_2 \xrightarrow{\text{boiling}} O_2 + 2H_2O$

15-10. A Jones reductor is a column packed with zinc granules coated with zinc amalgam. Prereduction is accomplished by passing analyte solution through the column.

15-11. Cr^{3+} and TiO^{2+} would interfere if they were reduced to Cr^{2+} and Ti^{3+}. In the Jones reductor, Zn is a strong enough reductant to react with Cr^{3+} and TiO^{2+}.

$E° = -0.764$ for the $Zn^{2+}|Zn$ couple

$E° = -0.42$ for the $Cr^{3+}|Cr^{2+}$ couple

$E° = 0.1$ for the $TiO^{2+}|Ti^{3+}$ couple

In the Walden reductor, Ag is not strong enough to reduce Cr^{3+} and TiO^{2+}:

$E° = 0.222$ for the $AgCl|Ag$ couple

15-12. A weighed amount of the solid mixture is added to a solution containing excess standard Fe^{2+} plus phosphoric acid. Each mol of $(NH_4)_2S_2O_8$ oxidizes 2 mol of Fe^{2+} to Fe^{3+}. Excess Fe^{2+} is then titrated with standard $KMnO_4$ to find out how much Fe^{2+} was consumed by the $(NH_4)_2S_2O_8$. The phosphoric acid masks the yellow color of Fe^{3+}, making the end point easier to see.

15-13. (a) $MnO_4^- + 8H^+ + 5e^- \rightleftharpoons Mn^{2+} + 4H_2O$

(b) $MnO_4^- + 4H^+ + 3e^- \rightleftharpoons MnO_2(s) + 2H_2O$

(c) $MnO_4^- + e^- \rightleftharpoons MnO_4^{2-}$

15-14. $3MnO_4^- + 5Mo^{3+} + 4H^+ \rightarrow 3Mn^{2+} + 5MoO_2^{2+} + 2H_2O$

$(16.43 - 0.04) = 16.39$ mL of $0.010\,33$ M $KMnO_4 = 0.169\,3$ mmol of MnO_4^-, which will react with $(5/3)(0.169\,3) = 0.282\,2$ mmol of Mo^{3+}.

$[Mo^{3+}] = 0.282\,2$ mmol/25.00 mL $= 0.011\,29$ M (= original $[MoO_4^{2-}]$).

15-15. $2MnO_4^- + 5H_2O_2 + 6H^+ \rightarrow 2Mn^{2+} + 5O_2 + 8H_2O$

$(27.66 - 0.04) = 27.62$ mL of $0.021\,23$ M $KMnO_4 = 0.586\,3_7$ mmol of MnO_4^-, which reacts with $(5/2)(0.586\,3_7) = 1.465\,9$ mmol of H_2O_2, which came from 25.00 mL of diluted solution $\Rightarrow$ $[H_2O_2] = 1.465\,9$ mmol/25.00 mL $= 0.058\,64$ M in the dilute solution. The original solution was ten times more concentrated $= 0.586\,4$ M.

15-16.   (a)   Scheme 1:

$$2\,[8H^+ + MnO_4^- + 5e^- \rightarrow Mn^{2+} + 4H_2O]$$
$$\;{+7}{+2}$$
$$5\,[H_2O_2 \rightarrow O_2 + 2e^- + 2H^+]$$
$${-1}0$$

$$6H^+ + 2MnO_4^- + 5H_2O_2 \rightarrow 2Mn^{2+} + 5O_2 + 8H_2O$$

Scheme 2:

$$2\,[MnO_4^- \rightarrow Mn^{2+} + 2O_2 + 3e^-]$$
$${+7}\;{-2}{+2}0$$
$$3\,[H_2O_2 + 2H^+ + 2e^- \rightarrow 2H_2O]$$
$${-1}{-2}$$

$$6H^+ + 2MnO_4^- + 3H_2O_2 \rightarrow 2Mn^{2+} + 4O_2 + 6H_2O$$

(b) $\dfrac{1.023 \text{ g NaBO}_3 \cdot 4H_2O}{153.86 \text{ g/mol}} = 6.649$ mmol NaBO$_3$

One tenth of this quantity was titrated $= 0.664\,9$ mmol NaBO$_3$, producing

$0.664\,9$ mmol H$_2$O$_2$ by the reaction BO$_3^- + 2H_2O \rightarrow$ H$_2$O$_2$ + H$_2$BO$_3^-$.

In Scheme 1, 2MnO$_4^-$ react with 5H$_2$O$_2$

$\Rightarrow\ 0.664\,9$ mmol H$_2$O$_2$ requires $\frac{2}{5}(0.664\,9) = 0.266\,0$ mmol MnO$_4^-$

$$\frac{0.266\,0 \text{ mmol MnO}_4^-}{0.010\,46 \text{ mmol KMnO}_4/\text{ml}} = 25.43 \text{ mL KMnO}_4 \text{ required}$$

In Scheme 2, 2MnO$_4^-$ react with 3H$_2$O$_2$

$\Rightarrow\ 0.664\,9$ mmol H$_2$O$_2$ requires $\frac{2}{3}(0.664\,9) = 0.443\,3$ mmol MnO$_4^-$

$$\frac{0.443\,3 \text{ mmol MnO}_4^-}{0.010\,46 \text{ mmol KMnO}_4/\text{ml}} = 42.38 \text{ mL KMnO}_4 \text{ required}$$

15-17. $2MnO_4^- + 5H_2C_2O_4 + 6H^+ \rightarrow 2Mn^{2+} + 10CO_2 + 8H_2O$

18.04 mL of 0.006 363 M KMnO$_4$ = 0.114 8 mmol of MnO$_4^-$, which reacts with

$(5/2)(0.114\,8) = 0.287\,0$ mmol of H$_2$C$_2$O$_4$, which came from $(2/3)(0.287\,0) =$

0.191 3 mmol of La^{3+}.   [La^{3+}] = 0.191 3 mmol/50.00 mL = 3.826 mM.

15-18. $C_3H_8O_3$ + $3H_2O$ $\rightleftharpoons$ $3HCO_2H$ + $8e^-$ + $8H^+$

                 glycerol   formic acid

        (average oxidation   (oxidation

        number of C = −2/3)   number of C = +2)

$$8Ce^{4+} + 8e^- \rightleftharpoons 8Ce^{3+}$$

$$C_3H_8O_3 + 8Ce^{4+} + 3H_2O \rightleftharpoons 3HCO_2H + 8Ce^{3+} + 8H^+$$

One mole of glycerol requires eight moles of Ce^{4+}.

50.0 mL of 0.083 7 M Ce^{4+} = 4.185 mmol

12.11 mL of 0.044 8 M Fe^{2+} = 0.543 mmol

Ce^{4+} reacting with glycerol = 3.642 mmol

glycerol = (1/8) (3.642) = 0.455$_2$ mmol = 41.9 mg $\Rightarrow$ original solution = 41.9 wt% glycerol

15-19.   50.00 mL of 0.118 6 M Ce^{4+} = 5.930 mmol Ce^{4+}

31.13 mL of 0.042 89 M Fe^{2+} = $\underline{1.335 \text{ mmol } Fe^{2+}}$

4.595 mmol Ce^{4+} consumed by NO_2^-

Since two moles of Ce^{4+} react with one mole of NO_2^-, there must have been

1/2 (4.595) = 2.298 mmol of $NaNO_2$ = 0.158 5 g in 25.0 mL. In 500.0 mL,

there would be $\left(\dfrac{500.0}{25.0}\right)$(0.158 5) = 3.170 g = 78.67% of the 4.030 g sample.

15-20.   Step 2 gives the total Cr content of the crystal, since each Cr^{x+} ion in any

oxidation state is oxidized and reacts with $3Fe^{2+}$.

Step 2: $\dfrac{(0.703 \text{ mL})(2.786 \text{ mM})}{0.156 6 \text{ g of crystal}} = \dfrac{12.51 \text{ μmol } Fe^{2+}}{\text{g of crystal}}$

$\dfrac{\frac{1}{3}(12.51) \text{ μmol Cr}}{\text{g of crystal}} = \dfrac{4.169 \text{ μmol Cr}}{\text{g of crystal}}$

Step 1 tells us how much Cr^{x+} is oxidized above the +3 state. Each Cr^{x+} reacts

with $(x - 3)$ Fe^{2+}.

Step 1: $\dfrac{(0.498 \text{ mL})(2.786 \text{ mM})}{0.437 5 \text{ g of crystal}} = \dfrac{3.171 \text{ μmol } Fe^{2+}}{\text{g of crystal}}$

Since one gram of crystal contains 4.169 μmol of Cr that reacts with 3.171 μm of

Fe^{2+}, the average oxidation state of Cr is $3 + \dfrac{3.171}{4.169} = +3.761$.

Total Cr (from Step 2) = 4.169 μmol Cr per gram = 217 μg per gram of crystal.

15-21.   (a) Theoretical molarity = (3.214 g/L)/(158.034 g/mol) = 0.020 33$_7$ M.

(b) 25.00 mL of 0.020 33$_7$ M $KMnO_4$ = 0.508 4$_3$ mmol. But two moles of MnO_4^-

react with five moles of H_3AsO_3, which comes from $\frac{5}{4}$ moles of As_4O_6.

The moles of As_4O_6 needed to react with 0.508 4$_3$ mmol of MnO_4^- =

$(^1/_2)(^5/_4)(0.508 4_3)$ = 0.317 7$_7$ mmol = 0.125 7$_4$ g of As_4O_6.

(c) $\dfrac{0.508\,4_3\ \text{mmol KMnO}_4}{0.125\,7_4\ \text{g As}_2\text{O}_3} = \dfrac{x\ \text{mmol KMnO}_4}{0.146\,8\ \text{g As}_4\text{O}_6} \Rightarrow x = 0.593\,6_1\ \text{mmol}$

KMnO$_4$ in $(29.98 - 0.03) = 29.95$ mL $\Rightarrow$ [KMnO$_4$] = 0.019 82 M.

15-22. I$^-$ reacts with I$_2$ to give I$_3^-$. This reaction increases the solubility of I$_2$ and decreases its volatility.

15-23. Standard triiodide can be prepared from a weighed amount of KIO$_3$ with acid plus excess iodide. Alternatively, triiodide solution can be standardized by reaction with standard $S_2O_3^{2-}$ prepared from anhydrous Na$_2$S$_2$O$_3$.

15-24. Starch is not added until just before the end point in iodometry, so it does not irreversibly bind to I$_2$ which is present during the whole titration.

15-25. (a) 50.00 mL contains exactly 1/10 of the KIO$_3$ = 0.102 2 g = 0.477 5$_7$ mmol KIO$_3$. Each mol of iodate makes 3 mol of triiodide, so I$_3^-$ = 3(0.477 5$_7$) = 1.43 2$_7$ mmol.

(b) Two moles of thiosulfate react with one mole of I$_3^-$. Therefore, there must have been 2(1.43 2$_7$) = 2.86 5$_4$ mmol of thiosulfate in 37.66 mL, so the concentration is (2.86 5$_4$ mmol)/(37.66 mL) = 0.076 08$_7$ M.

(c) 50.00 mL of KIO$_3$ make 1.43 2$_7$ mmol I$_3^-$. The unreacted I$_3^-$ requires 14.22 mL of sodium thiosulfate = (14.22 mL)(0.076 08$_7$ M) = 1.082$_0$ mmol, which reacts with $\frac{1}{2}$(1.082$_0$ mmol) = 0.541$_0$ mmol I$_3^-$. The ascorbic acid must have consumed the difference = 1.43 2$_7$ − 0.541$_0$ = 0.891$_7$ mmol I$_3^-$. Each mole of ascorbic acid consumes one mole of I$_3^-$, so mol ascorbic acid = 0.891$_7$ mmol, which has a mass of (0.891$_7$ × 10^{-3} mol)(176.13 g/mol) = 0.157$_1$ g. Ascorbic acid in the unknown = 100 × (0.157$_1$ g)/(1.223 g) = 12.8 wt%.

(d) Starch should not be added until just before the end point because I$_3^-$ is present throughout the titration and will irreversibly bind to starch if the starch is added too early.

15-26. $2\text{Cu}^{2+} + 5\text{I}^- \rightarrow 2\text{CuI}(s) + \text{I}_3^-$ $\text{I}_3^- + 2\text{S}_2\text{O}_3^{2-} \rightarrow 3\text{I}^- + \text{S}_4\text{O}_6^{2-}$

23.33 mL of 0.046 68 M Na$_2$S$_2$O$_3$ = 1.089$_0$ mmol S$_2$O$_3^{2-}$ = 0.544 5 mmol I$_3^-$, which came from 1.089$_0$ mmol Cu^{2+} = 69.20 mg Cu. This much Cu comes from

1/5 of the original solid, which therefore contained 346.0 mg Cu = 11.43 wt%. There is a great deal of I_3^- present at the start of the titration, so starch should not be added until just before the end point.

15-27. $H_2S + I_3^- \rightarrow S(s) + 3I^- + 2H^+$ $\qquad\qquad$ $I_3^- + 2S_2O_3^{2-} \rightarrow 3I^- + S_4O_6^{2-}$

25.00 mL of 0.010 44 M I_3^- = 0.261 0$_0$ mmol I_3^-

14.44 mL of 0.009 336 M $Na_2S_2O_3$ = 0.134 8$_1$ mmol $Na_2S_2O_3$, which would

have reacted with 0.067 40$_6$ mmol I_3^-. Therefore, the quantity of I_3^- that reacted

with H_2S was 0.261 0$_0$ – 0.067 40$_6$ = 0.193 5$_9$ mmol. Since 1 mol of I_3^- reacts

with 1 mol of H_2S, the H_2S concentration was 0.193 5$_9$ mmol/25.00 mL =

0.007 744 M. I_3^- is present at the start of the titration, so starch should not be

added until just before the end point.

15-28. (a) $I_2(aq) + 2e^- \rightleftharpoons 2I^-$ $\qquad\qquad\qquad\qquad\qquad$ $E° = 0.620$ V

$\qquad\qquad\qquad\qquad$ $3I^- \rightleftharpoons I_3^- + 2e^-$ $\qquad\qquad\qquad$ $E° = -0.535$ V

$\qquad\qquad$ $I_2(aq) + I^- \rightleftharpoons I_3^-$ $\qquad\qquad\qquad\qquad\qquad$ $E° = 0.085$ V

$\qquad\qquad$ $K = 10^{2(0.085)/0.059\,16} = 7 \times 10^2$

$\qquad$ (b) $I_2(s) + 2e^- \rightleftharpoons 2I^-$ $\qquad\qquad\qquad\qquad\qquad$ $E° = 0.535$ V

$\qquad\qquad\qquad\qquad$ $3I^- \rightleftharpoons I_3^- + 2e^-$ $\qquad\qquad\qquad$ $E° = -0.535$ V

$\qquad\qquad$ $I_2(s) + I^- \rightleftharpoons I_3^-$ $\qquad\qquad\qquad\qquad\qquad\quad$ $E° = 0.000$ V

$\qquad\qquad$ $K = 10^{2(-0.000)/0.059\,16} = 1.0$

$\qquad$ (c) $I_2(s) + 2e^- \rightleftharpoons 2I^-$ $\qquad\qquad\qquad\qquad\qquad$ $E° = 0.535$ V

$\qquad\qquad\qquad\qquad$ $2I^- \rightleftharpoons I_2(aq) + 2e^-$ $\qquad\qquad$ $E° = -0.620$ V

$\qquad\qquad$ $I_2(s) \rightleftharpoons I_2(aq)$ $\qquad\qquad\qquad\qquad\qquad\qquad$ $E° = -0.085$ V

$\qquad\qquad$ $K = [I_2(aq)] = 10^{2(-0.085)/0.059\,16} = 1.3 \times 10^{-3}$ M = 0.34 g of I_2/L

15-29. Each mole of NH_3 liberated in the Kjeldahl digestion reacts with 1 mole of H^+ in the standard H_2SO_4 solution. Six moles of H^+ left (3 moles of H_2SO_4) after reaction with NH_3 will react with 1 mole of iodate by Reaction 15-18 to release 3 moles of I_3^-. Two moles of thiosulfate react with 1 mole of I_3^- in Reaction 15-19. Therefore each mole of thiosulfate corresponds to $\frac{1}{2}$ mol of residual H_2SO_4.

$\qquad\qquad$ mol NH_3 = 2 (initial mol H_2SO_4 – final mol H_2SO_4)

$$\text{mol NH}_3 \; = \; 2 \left(\text{initial mol H}_2\text{SO}_4 - \tfrac{1}{2} \times \text{mol thiosulfate} \right)$$

15-30. (a) $IO_3^- + 8I^- + 6H^+ \rightleftharpoons 3I_3^- + 3H_2O$. The stock solution contained

$\{0.804\,3$ g KIO_3 (FM 214.00) + 5 g KI (FM 166.00)$\}$ / 100 mL, which

translates into 0.037 58 M KIO_3 plus 0.30 M KI, giving a mole ratio

$KI/KIO_3 = 18$, which is a good excess of the 8:1 ratio required in the

reaction. 5.00 mL of the stock solution contain 0.187 9$_2$ mmol KIO_3 plus

1.5 mmol KI. 1.0 mL of 6.0 M H_2SO_4 contains 6 mmol H_2SO_4, which is a

large excess for the reaction. Neither KI nor H_2SO_4 needs to be measured

accurately.

(b) $I_3^- + SO_3^{2-} + H_2O \rightarrow 3I^- + SO_4^{2-} + 2H^+$

(c) 0.187 9$_2$ mmol KIO_3 delivered to the wine generates $3 \times 0.187\,9_2 = 0.563\,7_6$

mmol I_3^-. The excess, unreacted I_3^- required 12.86 mL of 0.048 18 M

$Na_2S_2O_3 = 0.619\,5_9$ mmol $Na_2S_2O_3$. Each mole of unreacted I_3^- requires 2

moles of $Na_2S_2O_3$, so there must have been $(0.619\,5_9)/2 = 0.309\,8$ mmol I_3^-

left over from the reaction with sulfite. Therefore, the I_3^- that reacted with

sulfite was $(0.563\,7_6 - 0.309\,8) = 0.254\,0$ mmol I_3^-. One mole of I_3^- reacts

with 1 mole of sulfite, so there must have been 0.254 0 mmol SO_3^{2-} in 50.0

mL of wine. $[SO_3^{2-}] = 0.254\,0$ mmol/50.0 mL = $5.07\,9 \times 10^{-3}$ M. With a

formula mass of 80.06 for sulfite, the sulfite content is 406.6 mg/L.

(d) $s_{\text{pooled}} = \sqrt{\dfrac{2.2^2\,(3-1) + 2.1^2\,(3-1)}{3+3-2}} = 2.15$

$t_{\text{calculated}} = \dfrac{|277.7 - 273.2|}{2.15} \sqrt{\dfrac{3 \cdot 3}{3+3}} = 2.56$

$t_{\text{table}} = 2.776$ for 95% confidence and $3 + 3 - 2 = 4$ degrees of freedom

$t_{\text{calculated}} < t_{\text{table}}$, so the difference <u>is not significant</u> at 95% confidence level.

15-31. 25.00 mL of 0.020 00 M $KBrO_3 = 0.500\,0$ mmol of BrO_3^-, which generates 1.500

mmol of Br_2. One mole of excess Br_2 generates one mole of I_2 (from I^-) and one

mole of I_2 consumes 2 moles of $S_2O_3^{2-}$. Since mmol of $S_2O_3^{2-} = (8.83)(0.051\,13)$

$= 0.451\,5$ mmol, $I_2 = 0.225\,7$ mmol and Br_2 consumed by reaction with

8-hydroxyquinoline $= 1.500 - 0.225\,7 = 1.274$ mmol. But one mole of

8-hydroxyquinoline consumes 2 moles of Br_2, so 8-hydroxyquinoline $= 0.637\,1$ mmol and $Al^{3+} = 0.6371/3 = 0.212\,4$ mmol $= 5.730$ mg.

15-32. (a) $YBa_2Cu_3O_7$ contains 1 Cu^{3+} and 2 Cu^{2+}. $YBa_2Cu_3O_{6.5}$ contains no Cu^{3+} and 3 Cu^{2+}. The moles of Cu^{3+} in the formula $YBa_2Cu_3O_{7-z}$ are therefore $1 - 2z$. The moles of superconductor in 1 g of superconductor are $(1\ g)/[(666.246 - 15.999\,4\ z)g/mol]$. The difference between experiments B and A is $5.68 - 4.55 = 1.13$ mmol $S_2O_3^{2-}$/g of superconductor. Since 1 mol of thiosulfate is equivalent to 1 mol of Cu^{3+}, there are 1.13 mmol Cu^{3+}/g of superconductor.

$$\frac{mol\ Cu^{3+}}{mol\ superconductor} = 1 - 2z = \frac{1.13\ \ 10^{-3}\ mol\ Cu^{3+}}{\left(\dfrac{1\ g\ superconductor}{(666.246 - 15.999\,4\ z)\ g/mol}\right)}$$

Solving this equation gives $z = 0.125$. The formula is $YBa_2Cu_3O_{6.875}$.

(b) $1 - 2z = \dfrac{[5.68(\pm 0.05) - 4.55(\pm 0.10)]\ \ 10^{-3}}{\left(\dfrac{1}{666.246 - 15.999\,4\ z}\right)}$

$1 - 2z = \dfrac{1.13\ (\pm 0.112)\ \ 10^{-3}}{\left(\dfrac{1}{666.246 - 15.999\,4\ z}\right)}$

$1 - 2z = 0.752\,86\ (\pm 0.074\,49) - 0.018\,079\ (\pm 0.001\,789)\ z$

$0.247\,124\ (\pm 0.074\,488) = 1.981\,92\ (\pm 0.001\,79)\ z$

$z = 0.125 \pm 0.038.$ The formula is $YBa_2Cu_3O_{6.875\ \pm 0.038}$.

15-33. A superconductor containing unknown quantities of Cu(I), Cu(II), Cu(III), and peroxide (O_2^{2-}) is dissolved in a known excess of Cu(I) in oxygen-free HCl solution. Possible reactions are

$$Cu^{3+} + Cu^+ \rightarrow 2Cu^{2+}$$
$$H_2O_2 + 2Cu^+ + 2H^+ \rightarrow 2H_2O + 2Cu^{2+}$$

Unreacted Cu(I) is then measured by coulometry to find out how much Cu(I) was consumed by the dissolving superconductor. The amount of Cu(I) consumed is equal to the moles of Cu^{3+} plus 2 times the moles of O_2^{2-} in the superconductor.

The coulometry is done under Ar to prevent oxidation of Cu(I) by O_2 from the air. If the superconductor contained Cu(I) (but no Cu(III) or peroxide), then the amount of Cu(I) found by coulometry would be greater than the known amount used in the original solution.

15-34. (a) Initial Fe^{2+} in 5.000 mL = (5.000 mL)(0.100 0 M) = 0.500 0 mmol.

$K_2Cr_2O_7$ required for titration of unreacted Fe^{2+}

$\quad$ = (3.22 8 mL)(0.015 93 M $K_2Cr_2O_7$) = 0.0514 2 mmol.

But 1 mmol $K_2Cr_2O_7$ reacts with 6 mmol Fe^{2+} by the reaction

$K_2Cr_2O_7 + 6Fe^{2+} + 14H^+ \rightarrow 2Cr^{3+} + 6 Fe^{3+} + 2K^+ + 14H_2O$.

Therefore, Fe^{2+} left after reaction with $Li_{1+y}CoO_2$

$\quad$ = (0.0514 2 mmol)(6 mmol Fe^{2+}/mmol $K_2Cr_2O_7$) = 0.308 5 mmol.

Fe^{2+} consumed by Co^{3+} = (0.500 0 – 0.308 5) = 0.191 5 mmol.

1 mol Fe^{2+} is consumed by 1 mol Co^{3+}, so Co^{3+} in 25.00 mg solid sample = 0.191 5 mmol.

(b) Co in 25.00 mg solid = (0.564 g Co/g solid)(25.00 g solid) = 14.10 mg

Co in 25.00 mg solid = (14.10 mg)/(58.933 g/mol) = 0.239 3 mmol

From (a), we know that Co^{3+} = 0.191 5 mmol, so

$\quad Co^{2+}$ = 0.239 3 – 0.191 5 = 0.047 8 mmol.

Co oxidation state $= \dfrac{(0.047\ 8\ \text{mmol})(2+) + (0.191\ 5\ \text{mmol})(3+)}{0.239\ 2\ \text{mmol}} = 2.80$

(c) If Co has an average oxidation number of +2.80 and O has an oxidation number of –2, Li must contribute a charge of 4 – 2.80 = 1.20. Therefore, the formula is $Li_{1.20}CoO_2$ and y = 0.20.

(d) Theoretical weight percent for metals in $Li_{1.20}CoO_2$:

Formula mass = 1.20(6.941) + 1(58.933) + 2(15.9994) = 99.261

wt% Li = 100 × 1.20(6.941)/99.261 = 8.39%

wt% Co = 100 × 58.933/99.261 = 59.37%

$\dfrac{\text{wt\% Li}}{\text{wt\% Co}} = 0.141\ 3$

The observed quotient wt% Li/wt% Co is 0.138 8 ±. 0.000 6, which is not exactly equal to the quotient computed from the oxidation number. The difference represents experimental error between the two methods used find the stoichiometry.

15-35. Denote the average oxidation number of Bi as $3 + b$ and the average oxidation number of Cu as $2 + c$.

$$Bi_2^{3+b}Sr_2^{2+}Ca^{2+}Cu_2^{2+c}O_x$$

Positive charge = $6 + 2b + 4 + 2 + 4 + 2c = 16 + 2b + 2c$

The charge must be balanced by $O^{2-} \Rightarrow x = 8 + b + c$.

The formula mass of the superconductor is $760.37 + 15.999\ 4(8 + b + c)$.

One gram contains $1/[760.37 + 15.9994(8 + b + c)]$ moles.

(a)   Experiment A: Initial $Cu^+ = 0.2000$ mmol; final $Cu^+ = 0.1085$ mmol.

Therefore, 102.3 mg of superconductor consumed 0.0915 mmol Cu^+.

$2 \times$ mmol Bi^{5+} + mmol Cu^{3+} in 102.3 mg of superconductor = 0.0915.

Experiment B: Initial $Fe^{2+} = 0.1000$ mmol; final $Fe^{2+} = 0.0577$ mmol. Therefore, 94.6 mg of superconductor consumed 0.0423 mmol Fe^{2+}.

$2 \times$ mmol Bi^{5+} in 94.6 mg of superconductor = 0.0423.

Normalizing to 1 gram of superconductor gives

   Expt A: 2(mmol Bi^{5+}) + mmol Cu^{3+} in 1 g of superconductor = 0.89443

   Expt B: 2(mmol Bi^{5+}) in 1 g of superconductor = 0.44715

It is easier not to get lost in the arithmetic if we suppose that the oxidized bismuth is Bi^{4+} and equate one mole of Bi^{5+} to two moles of Bi^{4+}.

Therefore, we can rewrite the two previous equations as

   mmol Bi^{4+} + mmol Cu^{3+} in 1 g of superconductor = 0.89443   (1)

   mmol Bi^{4+} in 1 g of superconductor = 0.44715   (2)

Subtracting (2) from (1) gives

   mmol Cu^{3+} in 1 g of superconductor = 0.44728   (3)

Equations (2) and (3) tell us that the stoichiometric relationship in the formula of the superconductor is $b/c = 0.44715/0.44728 = 0.9997$.

Since 1 g of superconductor contains 0.44728 mmol Cu^{3+}, we can say

$$\frac{mol\ Cu^{3+}}{mol\ solid} = 2c$$

$$\frac{mol\ Cu^{3+}/mol\ solid}{gram\ solid/mol\ solid} = \frac{2c}{760.37 + 15.9994(8 + b + c)}$$

$$\frac{mol\ Cu^{3+}}{gram\ solid} = \frac{2c}{760.37 + 15.9994(8 + b + c)} = 4.4728 \times 10^{-4} \quad (4)$$

Substituting $b = 0.9997c$ in the denominator of (4) allows us to solve for c:

$$\frac{2c}{760.37 + 15.9994(8 + 1.9997c)} = 4.4728 \times 10^{-4} \Rightarrow c = 0.2001$$

$$\Rightarrow b = 0.9997c = 0.2000$$

The average oxidation numbers are $Bi^{3.2000+}$ and $Cu^{2.2001+}$ and the formula of the compound is $Bi_2Sr_2CaCu_2O_{8.4001}$, since the oxygen stoichiometry derived at the beginning of the solution is $x = 8 + b + c$.

(b)   Propagation of error:

Expt A: 102.3 (±0.2) mg compound consumed 0.0915 (±0.0007) mmol Cu^+

Expt B: 94.6 (±0.2) mg compound consumed 0.0423 (±0.0007) mmol Fe^{2+}

Normalizing to 1 gram of superconductor gives

Expt A: mmol Bi^{4+} + mmol Cu^{3+} in 1 g of superconductor

$$= \frac{0.091\,5\,(\pm 0.000\,7)}{0.102\,3\,(\pm 0.000\,2)} = 0.894\,43\,(\pm 0.007\,06)\,\frac{mmol}{gram}$$

Expt B: mmol Bi^{4+} in 1 g of superconductor

$$= \frac{0.042\,3\,(\pm 0.000\,7)}{0.094\,6\,(\pm 0.000\,2)} = 0.447\,15\,(\pm 0.007\,46)\,\frac{mmol}{gram}$$

$$\frac{mmol\,Cu^{3+}}{g\,superconductor} = 0.894\,43\,(\pm 0.007\,06) - 0.447\,15\,(\pm 0.007\,46)$$

$$= 0.447\,28\,(\pm 0.010\,27)$$

$$\frac{b}{c} = \frac{0.447\,15\,(\pm 0.007\,46)}{0.447\,28\,(\pm 0.010\,27)} = 0.999\,7\,(\pm 0.028\,4)$$

$$\frac{2c}{760.37 + 15.999\,4(8 + [1.999\,7(\pm 0.028\,4)]c)} = 4.472\,8\,(\pm 0.102\,7) \times 10^{-4}$$

$$[4\,471.47\,(\pm 102.7)]\,c = 888.365 + [31.999\,4\,(\pm 0.445)]\,c$$

$$\Rightarrow c = 0.200\,1\,(\pm 0.004\,6)$$

The relative uncertainty in b just given as $0.007\,46/0.447\,15$ is smaller than the relative uncertainty in c, which is $0.010\,27/0.447\,28$.

Uncertainty in $b = \dfrac{0.007\,46/0.447\,15}{0.010\,27/0.447\,28}$ (uncertainty in c)

$$= \frac{0.007\,46/0.447\,15}{0.010\,27/0.447\,28}\,(\pm 0.004\,6) = \pm 0.003\,3$$

$$\Rightarrow b = 0.200\,0\,(\pm 0.003\,3)$$

The average oxidation numbers are $Bi^{+3.200\,0}(\pm 0.003\,3)$ and $Cu^{+2.200\,1}(\pm 0.004\,6)$ and the formula of the compound is $Bi_2Sr_2CaCu_2O_{8.400\,1}(\pm 0.005\,7)$.

CHAPTER 16
ELECTROANALYTICAL TECHNIQUES

16-1. We observe that the silver electrode requires ~0.5 V more negative potential than the platinum electrode for reduction of H_3O^+ to H_2. The extra voltage needed to liberate H_2 at the silver surface is the overpotential required to overcome the activation energy for the reaction. In Table 16-1, we see that the difference in overpotential between Pt and Ag is ~0.5 V for a current density of 100 A/m^2.

16-2. $(0.100 \text{ mol})(96\ 485 \text{ C/mol}) = 9.648 \times 10^3 \text{ C}$

$(9.648 \times 10^3 \text{ C})/(1.00 \text{ C/s}) = 9.648 \times 10^3 \text{ s} = 2.68 \text{ h}$

16-3. $E° = -\Delta G°/2F = -237.13 \times 10^3/[(2)(964\ 85)] = -1.228\ 8 \text{ V}$

"Standard" means that reactants and products are in their standard states (1 bar for gases, pure liquid for water, unit activity for H^+ and OH^-).

16-4. (a) $E = E(\text{cathode}) - E(\text{anode})$

$= \left\{ E°(\text{cathode}) - 0.059\ 16 \log P_{H_2}^{1/2} [OH^-] \right\}$

$- \left\{ E°(\text{anode}) - 0.059\ 16 \log [Br^-] \right\}$

(remember to write both reactions as reductions)

$= \{-0.828 - 0.059\ 16 \log (1.0)^{1/2} [0.10]\}$

$- \{1.078 - 0.059\ 16 \log [0.10]\} = -1.906 \text{ V}$

(b) Ohmic potential $= I \cdot R = (0.100 \text{ A})(2.0\ \Omega) = 0.20 \text{ V}$

(c) $E = E(\text{cathode}) - E(\text{anode}) - I \cdot R - \text{Overpotentials}$

$= -1.906 - 0.20 - (0.20 + 0.40) = -2.71 \text{ V}$

(d) $E(\text{cathode}) = E°(\text{cathode}) - 0.059\ 16 \log P_{H_2}^{1/2} [OH^-]_s$

$= -0.828 - 0.059\ 16 \log (1.0)^{1/2} [1.0] = -0.828 \text{ V}$

$E(\text{anode}) = E°(\text{anode}) - 0.059\ 16 \log [Br^-]_s$

$= 1.078 - 0.059\ 16 \log [0.010] = 1.196 \text{ V}$

$E = E(\text{cathode}) - E(\text{anode}) - I \cdot R - \text{Overpotentials}$

$= -0.828 - 1.196 - 0.20 - (0.20 + 0.40) = -2.82 \text{ V}$

16-5. V_2 is the voltage between the working and reference electrodes, which is held constant. Working: ——o   Reference: ——>   Auxiliary: ——|

16-6. (a) For every mole of Hg produced, one mole of electrons flows.

$1.00 \text{ mL Hg} = 13.53 \text{ g Hg} = 0.067\ 45 \text{ mol Hg} = 0.067\ 45 \text{ mol e}^-$.

(0.067 45 mol) (96 485 C/mol) = 6 508 C.

Work = $q \cdot E$ = (6 508 C) (1.02 V) = 6.64×10^3 J.

(b) The power is 0.209 J/min = 0.003 48 J/s.

$P = I^2R \Rightarrow I = \sqrt{P/R} = \sqrt{(0.003\ 48\ W)/(100\ \Omega)} = 5.902$ mA.

In 1 h the total charge flowing through the circuit is

(5.902 × 10^{-3} C/s)· (3 600 s) = 21.25 C/(96 485 C/mol)

= 2.202×10^{-4} mol of e^-/h = 1.101×10^{-4} mol of Cd/h

= 0.012 4 g Cd/h.

16-7. Hydroxide generated at the cathode and Cl^- in the anode compartment cannot cross the membrane. Na^+ from seawater crosses from the anode to the cathode to preserve charge balance. Therefore, NaOH can be formed free from Cl^-.

16-8. $E = E(\text{cathode}) - E(\text{anode}) - IR - \text{overpotentials}$

Suppose that the open-circuit voltage of each cell is $E(\text{cathode}) - E(\text{anode}) =$ 2.2 V when no current flows. Ohmic loss and overpotentials for the two half-reactions decrease the output of the cell by 0.2 V, giving a net cell voltage of 2.0 V when the cell is delivering current. The cell can be recharged at very low current flow by applying just over 2.2 V in the opposite direction to reverse the cell chemistry. To charge at a significant rate requires additional voltage to overcome ohmic loss and overpotentials. The recharge requires ~0.2 V more than open-circuit voltage, or ~2.4 V. Electrical losses [IR, overpotentials, and concentration polarization in the terms $E(\text{cathode})$ and $E(\text{anode})$] always decrease the magnitude of the voltage that can be delivered by a cell and increase the magnitude of the voltage required to reverse the spontaneous cell reaction.

16-9. Pb(lactate)$_2$ + 2H$_2$O → PbO$_2$(s) + 2 lactate$^-$ + 4H$^+$ + 2e$^-$

$\quad\quad$ Pb^{2+} $\quad\quad\quad\quad\quad\quad\quad$ Pb^{4+}

Lead is oxidized to PbO_2 at the anode.

The mass of lead lactate (FM 385.3) giving 0.111 1 g of PbO_2

(FM = 239.2) is (385.3/239.2)(0.111 1 g) = 0.179 0 g.

% Pb = $\dfrac{0.179\ 0}{0.326\ 8} \times 100$ = 54.77%

16-10. Cathode: $Sn^{2+} + 2e^- \rightleftharpoons Sn(s)$ $E° = -0.141$ V

$E(\text{cathode, vs S.H.E.}) = -0.141 - \dfrac{0.059\,16}{2} \log \dfrac{1}{1.0 \times 10^{-8}} = -0.378$ V

$E(\text{cathode, vs S.C.E.}) = -0.378 - 0.241 = -0.619$ V

The voltage will be more negative if concentration polarization occurs. Concentration polarization means that $[Sn^{2+}]_s < 1.0 \times 10^{-8}$ M.

16-11. When 99.99% of Cd(II) is reduced, the formal concentration will be 1.0×10^{-5} M, and the predominant form is $Cd(NH_3)_4^{2+}$.

$\beta_4 = \dfrac{[Cd(NH_3)_4^{2+}]}{[Cd^{2+}][NH_3]^4} = \dfrac{(1.0 \times 10^{-5})}{[Cd^{2+}](1.0)^4} \Rightarrow [Cd^{2+}] = 2.8 \times 10^{-12}$ M

$Cd^{2+} + 2e^- \rightleftharpoons Cd(s)$ $E° = -0.402$

$E(\text{cathode}) = -0.402 - \dfrac{0.059\,16}{2} \log \dfrac{1}{[Cd^{2+}]} = -0.744$ V

16-12. Ni deposited = $(0.479\,8\text{ g} - 0.477\,5\text{ g}) = 2.3$ mg = 39.19 µmol Ni which would require $(39.19 \times 10^{-6}$ mol Ni$)(2$ e$^-$/Ni$)(96\,485$ C/mol$) = 7.562$ C. Percentage of current going to reduction of $Ni^{2+} = 100 \times \dfrac{7.562\text{ C}}{8.082\text{ C}} = 94\%$. The remainder went into reduction of H^+ to H_2.

16-13. When excess Br_2 appears in the solution, current flows at a low applied potential difference (0.25 V) in the detector circuit by virtue of the reactions

anode: $2Br^- \rightarrow Br_2 + 2e^-$

cathode: $Br_2 + 2e^- \rightarrow 2Br^-$

16-14. A mediator shuttles electrons between analyte and the electrode. After being oxidized or reduced by analyte, the mediator is regenerated at the electrode.

16-15. (a) 0.005 C/s $\times 0.1$ s $= 0.000\,5$ C

$\dfrac{0.000\,5\text{ C}}{96\,485\text{ C/mol}} = 5._2 \times 10^{-9}$ mol e$^-$

(b)   A 0.01 M solution of a two-electron reductant delivers 0.02 moles of electrons/liter.

$\dfrac{5._2 \times 10^{-9}\text{ moles}}{0.02\text{ moles/liter}} = 2._6 \times 10^{-7}$ L $= 0.000\,2_6$ mL $= 0.2_6$ µL

16-16. (a)   mol e$^- = \dfrac{I \cdot t}{F} = \dfrac{(5.32 \times 10^{-3}\text{ C/s})(964\text{ s})}{96\,485\text{ C/mol}} = 5.32 \times 10^{-5}$ mol

(b) One mol e^- reacts with ½ mol Br_2, which reacts with ½ mol cyclohexene
$\Rightarrow$ 2.66×10^{-5} mol cyclohexene.

(c) 2.66×10^{-5} mol/5.00×10^{-3} L = 5.32×10^{-3} M

16-17. $2I^- \rightarrow I_2 + 2e^- \Rightarrow$ one mole of I_2 is created when two moles of electrons flow.
(812 s)(52.6×10^{-3} C/s)/(96 485 C/mol) = 0.442 7 mmol of e^- = 0.221 3 mmol
of I_2. Therefore, there must have been 0.221 3 mmol of H_2S (FM 34.08) = 7.542
mg of H_2S/50.00 mL = 7.542×10^3 μg of H_2S/50.00 mL = 151 μg/mL.

16-18. (a) $C_6H_5N=NC_6H_5 + 4H^+ + 4e^- \rightarrow 2C_6H_5NH_2$

Electron flow =

$$\left(4 \frac{electrons}{C_6H_5N=NC_6H_5}\right)\left(25.9 \frac{nmol}{s}\right)\left(96\ 485 \frac{C}{mol}\right) = 1.00 \times 10^{-2} \text{ C/s}$$

current density = $\dfrac{1.00 \times 10^{-2}\text{ A}}{1.00 \times 10^{-4}\text{ m}^2} = 1.00 \times 10^2$ A/m^2

$\Rightarrow$ overpotential = 0.85 V

(b) E(cathode) = $0.100 - 0.059\ 16 \log \dfrac{[Ti^{3+}]_s}{[TiO^{2+}]_s[H^+]^2}$

$= 0.100 - 0.059\ 16 \log \dfrac{[0.10]}{[0.050][0.10]^2} = -0.036$ V

(c) $O_2 + 4H^+ + 4e^- \rightleftharpoons 2H_2O$ $E° = 1.229$ V

E(anode) = $1.229 - \dfrac{0.059\ 16}{4} \log \dfrac{1}{P_{O_2}[H^+]^4}$

$= 1.229 - \dfrac{0.059\ 16}{4} \log \dfrac{1}{(0.20)[0.10]^4} = 1.160$ V

(d) $E = E$(cathode) $- E$(anode) $- I\cdot R -$ Overpotential
$= -0.036 - 1.160 - (1.00 \times 10^{-2}\text{ A})(52.4\ \Omega) - 0.85 = -2.57$ V

16-19. $F = \dfrac{coulombs}{mol} = \dfrac{I \cdot t}{mol}$

$= \dfrac{[0.203\ 639\ 0(\pm 0.000\ 000\ 4)\text{ A}][18\ 000.075\ (\pm 0.010)\text{ s}]}{[4.097\ 900\ (\pm 0.000\ 003)\text{ g}]/[107.868\ 2\ (\pm 0.000\ 2)\text{g/mol}]}$

$= \dfrac{[0.203\ 639\ 0(\pm 1.96\ 10^{-4}\ \%)][18\ 000.075\ (\pm 5.56\ 10^{-5}\ \%)]}{[4.097\ 900\ (\pm 7.32\ 10^{-5}\ \%)]/[107.868\ 2\ (\pm 1.85\ 10^{-4}\ \%)]}$

$= 9.648\ 667 \times 10^4\ (\pm 2.85 \times 10^{-4}\ \%) = 964\ 86.6_7 \pm 0.2_8$ C/mol

16-20. (a) $H_2SO_3 \underset{\rightleftharpoons}{\overset{pK_1 = 1.86}{}} HSO_3^- \underset{\rightleftharpoons}{\overset{pK_2 = 7.17}{}} SO_3^{2-}$

H_2SO_3 is predominant below pH 1.86. HSO_3^- dominates between pH 1.86 and 7.17. SO_3^{2-} is dominant above pH 7.17.

(b) Cathode: $H_2O + e^- \rightarrow \frac{1}{2}H_2(g) + OH^-$

Anode: $3I^- \rightarrow I_3^- + 2e^-$

(c) $I_3^- + HSO_3^- + H_2O \rightarrow 3I^- + SO_4^{2-} + 3H^+$

$I_3^- + 2S_2O_3^{2-} \rightleftharpoons 3I^- + $ O=$\overset{O}{\underset{O^-}{\overset{\|}{S}}}$—S—S—$\overset{O}{\underset{O^-}{\overset{\|}{S}}}$=O

Thiosulfate   Tetrathionate

(d) In Step 3, I_3^- was generated by a current of 10.0 mA ($=10.0 \times 10^{-3}$ C/s) for 4.00 min (= 240 s).

charge = $I \cdot t = (10.0 \times 10^{-3}$ C/s$)(240$ s$) = 2.40$ C

mol e$^- = I/F = (2.40$ C$)/(96\,485$ C/mol$) = 24.8_7$ μmol e$^-$

The anode reaction generates $\frac{1}{2}$ mol I_3^- for 1 mol e$^-$. Therefore, 24.8_7 μmol e$^-$ will generate $\frac{1}{2}(24.8_7) = 12.4_4$ μmol I_3^-.

In Step 5, 0.500 mL of 0.050 7 M thiosulfate = 25.3_5 μmol $S_2O_3^{2-}$. But 2 mol $S_2O_3^{2-}$ consume 1 mol I_3^-. Therefore, 25.3_5 μmol $S_2O_3^{2-}$ consume $\frac{1}{2}(25.3_5) = 12.6_8$ μmol I_3^-.

We added excess $S_2O_3^{2-}$ in Step 5 and consumed the excess in Step 6. In Step 6, we had to generate I_3^- at 10.0 mA for 131 s to react with excess $S_2O_3^{2-}$.

charge = $I \cdot t = (10.0 \times 10^{-3}$ C/s$)(131$ s$) = 1.31$ C

mol e$^- = I/F = (1.31$ C$)/(96\,485$ C/mol$) = 13.5_8$ μmol e$^-$

mol $I_3^- = \frac{1}{2}(13.5_8) = 6.79$ μmol I_3^-

Here is where we are so far:

Step 3: 12.4_4 μmol I_3^- were generated.

Step 4: x μmol I_3^- were consumed by sulfite in wine.

Step 5: We added enough $S_2O_3^{2-}$ to consume 12.6_8 μmol I_3^-.

Step 6: We had to generate 6.79 μmol I_3^- to consume excess $S_2O_3^{2-}$ from

Step 5. Therefore, I_3^- left after Step 4 = $12.6_8 - 6.79 = 5.8_9$ μmol I_3^-.

We began with 12.4$_4$ μmol I$_3^-$ and 5.8$_9$ μmol I$_3^-$ were left after reaction with sulfite in wine. Therefore, sulfite in wine consumed 12.4$_4$ – 5.8$_9$ = 6.5$_5$ μmol I$_3^-$. But 1 mol I$_3^-$ reacts with 1 mol sulfite. Therefore, the wine contained 6.5$_5$ μmol sulfite in the 2.00 mL injected for analysis.

The wine sample prepared in Step 1 consisted of 9.00 mL wine diluted to 10.00 mL. Therefore, the original wine contained 10.00/9.00 of the amount found in the analysis. That is, 2.000 mL of pure wine contains (10.00/9.00)(6.5$_5$ μmol sulfite) = 7.2$_8$ μmol sulfite.

$$\text{sulfite in wine} = \frac{7.2_8\ \mu\text{mol sulfite}}{2.00\ \text{mL}} = 3.64\ \text{mM}$$

This problem left out a description of the blank titration that should be done in a real analysis. There are components in wine in addition to sulfite that could react with I$_3^-$. For the blank titration, 1 M formaldehyde is added to the wine to bind all sulfite. The sulfite-formaldehyde adduct is not decomposed in 2 M NaOH and does not react with I$_3^-$. The blank titration consists of taking this formaldehyde/wine solution through the entire procedure. We subtract I$_3^-$ consumed by the blank from I$_3^-$ consumed by the wine without formaldehyde.

16-21. (a) Balance carbon: $B = c$; balance halogen: $C = x$; balance nitrogen: $D = n$
Balance oxygen: $o + A = 2B \Rightarrow o + A = 2c \Rightarrow A = 2c - o$
Balance hydrogen: $h + 2A = 3D + E \Rightarrow h + 2(2c - o) = 3n + E$
$$\Rightarrow E = h + 4c - 2o - 3n$$
Charge balance, $F = E - C = h + 4c - 2o - 3n - c = h - c/2 + o - 3n$

(b) To consume Fe^- requires $F/4$ O$_2$, because each O$_2$ consumes 4e$^-$.

(c) $F = (9.43 \times 10^{-3}\ \text{C})/(9.648\ 5 \times 10^4\ \text{C/mol}) = 9.77_4 \times 10^{-8}$ mol e$^-$
mol O$_2$ = $F/4 = 2.22_3 \times 10^{-8}$ mol

(d) The mass of O$_2$ in (c) is $(2.22_3 \times 10^{-8}\ \text{mol})(32.00\ \text{g/mol}) = 7.11_4 \times 10^{-7}$ g. This much O$_2$ was required to react with 13.5 μL of sample. The mass of O$_2$ that would react with 1 L of sample is $(7.11_4 \times 10^{-7}\ \text{g})/(13.5 \times 10^{-6}\ \text{L}) = 0.0527$ g/L = 52.7 mg/L.

(e) The balanced equation oxidation half-reaction is
C$_9$H$_6$NO$_2$ClBr$_2$ + 16H$_2$O → 9CO$_2$ + 3X$^-$ + NH$_3$ + 35H$^+$ + 32e$^-$.

The observed number of electrons in the reaction was $9.77_4 \times 10^{-8}$ mol e⁻, so there must have been $(9.77_4 \times 10^{-8}$ mol e⁻$)/(32$ mol e⁻/mol $C_9H_6NO_2ClBr_2) = 3.05_4 \times 10^{-9}$ mol $C_9H_6NO_2ClBr_2$ in 13.5 μL. The molarity of $C_9H_6NO_2ClBr_2$ is $(3.05_4 \times 10^{-9}$ mo$)/(13.5 \times 10^{-6}$ L$) = 2.26 \times 10^{-4}$ M.

16-22. The Clark electrode measures dissolved oxygen by reducing it to H_2O at a gold tip on a platinum electrode held at –0.75 V with respect to Ag|AgCl. The opening of the body of the electrode is filled with a 10- to 40-μm-long plug of silicone rubber that is permeable to O_2. Current is proportional to the concentration of dissolved O_2 in the external medium. The electrode needs to be calibrated in solutions of known O_2 concentration.

16-23. (a)   The glucose monitor has a test strip with two carbon indicator electrodes and a silver-silver chloride reference electrode. Indicator electrode 1 is coated with glucose oxidase and a mediator. When a drop of blood is placed on the test strip, glucose from the blood is oxidized near indicator electrode 1 by mediator to gluconolactone and the mediator is reduced. With a potential of +0.2 V (with respect to the Ag|AgCl electrode) on the indicator electrode, reduced mediator is re-oxidized at the indicator electrode. The current between indicator electrode 1 and the reference electrode is proportional to the rate of oxidation of the mediator, which is proportional to the concentration of glucose plus any interfering species in the blood. Indicator electrode 2 has mediator, but no glucose oxidase. Current measured between indicator electrode 2 and the reference electrode is proportional to the concentration of interfering species in the blood. The difference between the two currents is proportional to the concentration of glucose in the blood.

(b)   In the absence of a mediator, the rate of oxidation of glucose depends on the concentration of O_2 in the blood. If $[O_2]$ is low, the current will be low and the monitor will give an incorrect, low reading for the glucose concentration. A mediator such as 1,1'-dimethylferrocene can replace O_2 in the glucose oxidation and be subsequently reduced at the indicator electrode. The concentration of mediator is constant and high enough, so variations in electrode current are due mainly to variations in glucose concentration. Also, by lowering the required electrode potential for oxidation of the mediator, there is less possible interference by other species in the blood.

(c) Glucose oxidase is replaced by glucose dehydrogenase, which does not use O_2 as a reactant. The enzyme oxidizes glucose and reduces the PQQ cofactor to $PQQH_2$. $PQQH_2$ is oxidized back to PQQ by a nearby Os^{3+} bound to the polymer chain. A nearby Os^{2+} can exchange electrons with the Os^{3+}. By moving from Os to Os, electrons eventually reach the carbon electrode. The coulometric sensor measures the total number of electrons needed to oxidize all of the glucose in the small blood sample.

(d) Amperometry measures current during the enzyme-catalyzed oxidation of glucose. Current is proportional to the rate of the oxidation reaction. The rates of most chemical reactions increase with increasing temperature. Therefore, the current will increase with increasing temperature of the blood sample. Coulometry measures the total number of electrons released in the oxidation. Glucose releases 2 electrons per molecule, regardless of temperature. The coulometric signal should have no temperature dependence.

(e) 1.00 g glucose/L = 5.55 mM glucose. A volume of 0.300×10^{-6} L contains 1.665 nmol glucose. Each mole of glucose releases $2e^-$ and $2H^+$ during oxidation. Therefore, 2×1.665 nmol = 3.33 nmol e^- are released. The charge is $Q = nF = (3.33 \times 10^{-9} \text{ mol})(96\ 485 \text{ C/mol}) = 321$ μC.

16-24. ω is the rotation rate in radians per second. We need to convert rpm (revolutions per minute) to radians per second.

$$\left(2.00 \times 10^3 \frac{\text{revolutions}}{\text{min}}\right)\left(\frac{1 \text{ min}}{60 \text{ s}}\right)\left(\frac{2\pi \text{ radians}}{\text{revolution}}\right) = 209 \text{ rad/s} = 209 \text{ s}^{-1}$$

(because radian is a dimensionless unit)

$$\delta = 1.61 D^{1/3} \nu^{1/6} \omega^{-1/2}$$

$$= 1.61(2.5 \times 10^{-9} \text{ m}^2/\text{s})^{1/3}(1.1 \times 10^{-6} \text{ m}^2/\text{s})^{1/6}(209 \text{ rad/s})^{-1/2} = 1.5_3 \times 10^{-5} \text{ m}$$

To calculate current density, we need to express the concentration of the species reacting at the electrode in mol/m³ instead of mol/L. Since 1 L is the volume of a 10-cm cube, there are 1 000 L in 1 m³. The concentration of $K_4Fe(CN)_6$ is 50.0 mM $= \left(0.050\ 0 \frac{\text{mol}}{\text{L}}\right)\left(1\ 000 \frac{\text{L}}{\text{m}^3}\right) = 50.0$ mol/m³.

Current density $= 0.62 nFD^{2/3} \nu^{-1/6} \omega^{1/2} C_o$

$$= 0.62(1)\left(96\ 485 \frac{\text{C}}{\text{mol}}\right)\left(2.5 \times 10^{-9} \frac{\text{m}^2}{\text{s}}\right)^{2/3}\left(1.1 \times 10^{-6} \frac{\text{m}^2}{\text{s}}\right)^{-1/6}\left(209 \frac{1}{\text{s}}\right)^{1/2}\left(50.0 \frac{\text{mol}}{\text{m}^3}\right)$$

$$= 7.8_4 \times 10^2 \frac{\text{C}}{\text{m}^2 \cdot \text{s}} = 7.8_4 \times 10^2 \frac{\text{A}}{\text{m}^2}$$

16-25. (a)   (b)

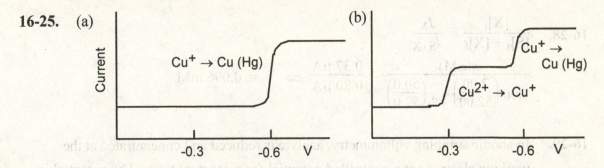

(c) The potential for the reaction Cu(I) → Cu(Hg) will change if Pt is used, since the product obviously cannot be copper amalgam.

16-26. (a) Charging current arises from charging or discharging of the electric double layer at the electrode-solution interface. Faradaic current arises from oxidation or reduction reactions.

(b) Charging current decays more rapidly than Faradaic current. If we wait 1 s after a potential step, the charging current decays to near zero and the Faradaic current is still significant. The ratio of the desired signal (Faradaic current) to the undesired background (charging current) is larger at 1 s than it was at earlier times. If we wait too long, both signals become too small to measure.

(c) In square wave voltammetry, an anodic pulse follows each cathodic pulse and the signal is the difference between the two. The anodic pulse oxidizes the product of each cathodic pulse, thereby replenishing the electroactive species at the electrode surface for the next pulse. The concentration of analyte available at the electrode surface is therefore greater in square wave voltammetry.

16-27. Electrons flowing in 3.4 min $=$

$$\frac{(14 \times 10^{-6} \text{ C/s})(60 \text{ s/min})(3.4 \text{ min})}{96\,485 \text{ C/mol}} = 2.9_6 \times 10^{-8} \text{ mol e}^-$$

For the reaction $Cd^{2+} + 2e^- \rightarrow Cd(\text{in Hg})$,

moles of $Cd^{2+} = \frac{1}{2}$ moles of $e^- = 1.4_8 \times 10^{-8}$ mol

moles of Cd^{2+} in 25 mL of 0.50 mM solution $= 1.25 \times 10^{-5}$ mol

percentage of Cd^{2+} reduced $= \dfrac{1.4_8 \times 10^{-8}}{1.25 \times 10^{-5}} \times 100 = 0.11_8\%$

16-28. $\dfrac{[X]_i}{[S]_f + [X]_f} = \dfrac{I_X}{I_{S+X}}$

$$\dfrac{x(mM)}{3.00\left(\dfrac{2.00}{52.00}\right) + x\left(\dfrac{50.0}{52.0}\right)} = \dfrac{0.37\ \mu A}{0.80\ \mu A} \Rightarrow x = 0.096\ mM$$

16-29. In anodic stripping voltammetry, analyte is reduced and concentrated at the working electrode at a controlled potential for a constant time. The potential is then ramped in a positive direction to reoxidize the analyte, during which time current is measured. The height of the oxidation wave is proportional to the original concentration of analyte. Stripping is the most sensitive voltammetric technique because analyte is concentrated from a dilute solution. The longer the period of concentration, the more sensitive is the analysis.

16-30. (a) Concentration (deposition) stage: $Cu^{2+} + 2e^- \rightarrow Cu(s)$

   (b) Stripping stage: $Cu(s) \rightarrow Cu^{2+} + 2e^-$

   (c) All solutions were made up to the same volume, so $[X]_i = [X]_f \equiv x$. Prepare a graph of I vs. $[S]_f$ using data measured from the figure in the problem. The intercept is at -313 ppb, so the original concentration of Cu^{2+} is 313 ppb.

Added standard (ppb)	Current (μA)
0	0.59_9
100	0.77_4
200	0.94_3
300	1.12_8
400	1.31_4
500	1.54_4

$y = 0.001\,866\,x + 0.583\,9$

Intercept = -313 ppb

16-31.

	A	B	C	D	E
1	Standard Addition Constant Volume Least-Squares Spreadsheet				
2	x	y			
3	Added Fe(III)	Relative			
4	(pM)	peak height			
5	0	1.00			
6	50	1.56			
7	100	1.98			
8	B10:C12 = LINEST(B5:B7,A5:A7,TRUE,TRUE)				
9		LINEST output:			
10	m	0.0098	1.0233	b	
11	s_m	0.0008	0.0522	s_b	
12	R^2	0.9932	0.0572	s_y	
13	x-intercept = -b/m =	-104.422			
14	n =	3	B14 = COUNT(A6:A7)		
15	Mean y =	1.513	B15 = AVERAGE(B5:B7)		
16	$\div(x_i - \text{mean } x)^2 =$	5000	B16 = DEVSQ(A5:A7)		
17	Std deviation of				
18	x-intercept =	13.174			
19	B18 = (C12/ABS(B10))*SQRT((1/B14) + B15^2/(B10^2*B16))				

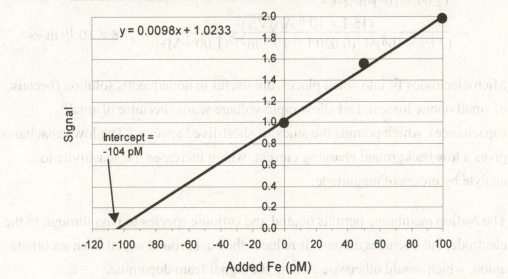

$y = 0.0098x + 1.0233$

Intercept = -104 pM

Signal

Added Fe (pM)

Cells B13 and B18 of the spreadsheet tell us that [Fe(III)] = 104 ± 13 pM.

16-32.   Peak B: $RNHOH \rightarrow RNO + 2H^+ + 2e^-$

Peak C: $RNO + 2H^+ + 2e^- \rightarrow RNHOH$

There was no RNO present before the initial scan.

16-33.

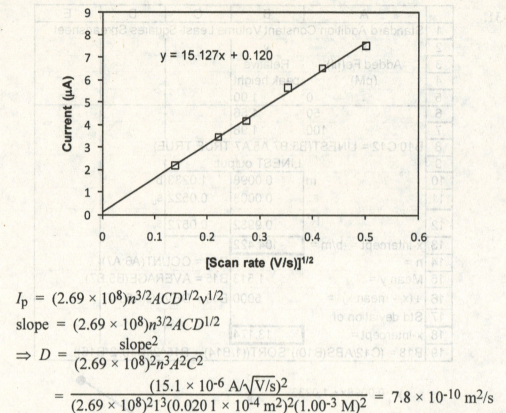

$$I_p = (2.69 \times 10^8)n^{3/2}ACD^{1/2}v^{1/2}$$

$$\text{slope} = (2.69 \times 10^8)n^{3/2}ACD^{1/2}$$

$$\Rightarrow D = \frac{\text{slope}^2}{(2.69 \times 10^8)^2 n^3 A^2 C^2}$$

$$= \frac{(15.1 \times 10^{-6} \text{ A}/\sqrt{\text{V/s}})^2}{(2.69 \times 10^8)^2 1^3 (0.020\ 1 \times 10^{-4} \text{ m}^2)^2 (1.00^{-3} \text{ M})^2} = 7.8 \times 10^{-10} \text{ m}^2/\text{s}$$

16-34. Microelectrodes fit into small places, are useful in nonaqueous solution (because of small ohmic losses), and allow rapid voltage scans (because of small capacitance), which permits the study of short-lived species. The low capacitance gives a low background charging current, which increases the sensitivity to analyte by orders of magnitude.

16-35. The Nafion membrane permits neutral and cationic species to pass through to the electrode, but excludes anions. It reduces the background signal from ascorbate anion, which would otherwise swamp the signal from dopamine.

16-36. $ROH + SO_2 + B \rightarrow BH^+ + ROSO_2^-$

$H_2O + I_2 + ROSO_2^- + 2B \rightarrow ROSO_3^- + 2BH^+I^-$

One mole of H_2O plus one mole of I_2 are required for the oxidation of the alkyl sulfite to an alkyl sulfate in the second reaction.

16-37. The bipotentiometric detector maintains a constant current (~10 µA) between two detector electrodes, while measuring the voltage needed to sustain that current. Before the equivalence point, the solution contains I⁻, but little I_2. To maintain a current of 10 µA, the cathode potential must be negative enough to reduce some component of the solvent system (perhaps $CH_3OH + e^- \rightleftharpoons CH_3O^- + \frac{1}{2}H_2(g)$). At the equivalence point, excess I_2 suddenly appears and current can be carried at low voltage by the reactions below. The abrupt voltage drop marks the end point.

Cathode: $I_3^- + 2e^- \rightarrow 3I^-$

Anode: $3I^- \rightarrow I_3^- + 2e^-$

CHAPTER 17
FUNDAMENTALS OF SPECTROPHOTOMETRY

17-1. (a) double   (b) halve   (c) double

17-2. (a) $E = h\nu = hc/\lambda = (6.626\,2 \times 10^{-34}\text{ J s})(2.997\,9 \times 10^8\text{ m s}^{-1})/(650 \times 10^{-9}\text{ m})$

$= 3.06 \times 10^{-19}$ J/photon $= 184$ kJ/mol

(b) For $\lambda = 400$ nm, $E = 299$ kJ/mol.

17-3. $\nu = c/\lambda = 2.997\,9 \times 10^8\text{ m s}^{-1}/562 \times 10^{-9}\text{ m} = 5.33 \times 10^{14}$ Hz

$\tilde{\nu} = 1/\lambda = 1.78\ 10^6\text{ m}^{-1}$ (1 m/100 cm) $= 1.78\ 10^4\text{ cm}^{-1}$

$E = h\nu = (6.626\,2 \times 10^{-34}\text{ J s})(5.33 \times 10^{14}\text{ s}^{-1}) = 3.53 \times 10^{-19}$ J/photon

$= 213$ kJ/mol (after multiplication by Avogadro's number).

17-4. Microwave energies correspond to molecular rotation energies. Infrared energies correspond to vibrational energies. Visible light can promote electrons to excited states (in colored compounds). Ultraviolet light also promotes electrons and can even break chemical bonds.

17-5. From the definition of index of refraction, we can write

$c_{\text{vacuum}} = n \cdot c_{\text{air}}$

$\lambda_{\text{vacuum}} \cdot \nu = n \cdot \lambda_{\text{air}} \cdot \nu$

$\lambda_{\text{air}} = \lambda_{\text{vacuum}}/n$

$\nu = c/\lambda_{\text{vacuum}} = 5.088\,491\,0$ and $5.083\,335\,8 \times 10^{14}$ Hz

$\lambda_{\text{air}} = \lambda_{\text{vacuum}}/n = 588.985\,54$ and $589.582\,86$ nm

$\tilde{\nu}_{\text{air}} = 1/\lambda_{\text{air}} = 1.697\,834\,5$ and $1.696\,114\,4\ \ 10^4\text{ cm}^{-1}$

17-6. Transmittance (T) is the fraction of incident light that is transmitted by a substance: $T = P/P_0$, where P_0 is incident irradiance and P is transmitted irradiance. Absorbance is logarithmically related to transmittance: $A = -\log T$. When all light is transmitted, absorbance is zero. When no light is transmitted, absorbance is infinite. Absorbance is proportional to concentration. Molar absorptivity is the constant of proportionality between absorbance at a particular wavelength and the product cb, where c is concentration and b is pathlength.

17-7. An absorption spectrum is a graph of absorbance vs. wavelength.

17-8. The color of transmitted light is the complement of the color that is absorbed. If blue-green light is absorbed, red light is transmitted.

17-9.

Curve	Absorption peak (nm)	Predicted color (Table 17-1)	Observed color
A	760	green	green
B	700	green	blue-green
C	600	blue	blue
D	530	violet	violet
E	500	red or purple red	red
F	410	green-yellow	yellow

17-10. If absorbance is too high, too little light reaches the detector for accurate measurement. If absorbance is too low, there is too little difference between sample and reference for accurate measurement.

17-11. $\varepsilon = A/bc = 0.822/[(1.00 \text{ cm})(2.31 \times 10^{-5} \text{ M})] = 3.56 \times 10^4 \text{ M}^{-1} \text{ cm}^{-1}$

17-12. Violet blue, according to Table 17-1.

17-13. [Fe] in reference cell $= \left(\dfrac{10.0}{50.0}\right)(6.80 \; 10^{-4}) = 1.36 \times 10^{-4}$ M. Setting the absorbances of sample and reference equal to each other gives $\varepsilon_s b_s c_s = \varepsilon_r b_r c_r$. But $\varepsilon_s = \varepsilon_r$, so $(2.48 \text{ cm})c_s = (1.00 \text{ cm})(1.36 \times 10^{-4} \text{ M}) \Rightarrow c_s = 5.48 \times 10^{-5}$ M. This is a 1/4 dilution of runoff, so [Fe] in runoff $= 2.19 \times 10^{-4}$ M.

17-14. (a) Measured from graph:

$\sigma \approx 1.3 \times 10^{-20} \text{ cm}^2$ at 325 nm $\qquad \sigma \approx 3.5 \times 10^{-19} \text{ cm}^2$ at 300 nm

at 325 nm: $T = e^{-(8 \times 10^{18} \text{ cm}^{-3})(1.3 \times 10^{-20} \text{ cm}^2)(1 \text{ cm})} = 0.90$

$A = -\log T = 0.045$

at 300 nm: $T = e^{-(8 \times 10^{18} \text{ cm}^{-3})(3.5 \times 10^{-19} \text{ cm}^2)(1 \text{ cm})} = 0.061$

$A = -\log T = 1.22$

(b) $T = e^{-n\sigma b} \qquad 0.14 = e^{-(8 \times 10^{18} \text{ cm}^{-3})\,\sigma\,(1 \text{ cm})}$

$\Rightarrow \sigma = 2.4_{576} \times 10^{-19} \text{ cm}^2$

If n is decreased by 1%, $T = e^{-(7.92 \times 10^{18} \text{ cm}^{-3})(2.457_6 \times 10^{-19} \text{ cm}^2)(1 \text{ cm})}$

$= 0.142_8$

Increase in transmittance is $\dfrac{0.142\,8 - 0.14}{0.14} = 2.0\%$.

Note that the fractional increase in transmittance is greater than the fractional decrease in ozone concentration.

(c) $T_{winter} = e^{-(290 \text{ D.U.})(2.69 \times 10^{16} \text{ molecules/cm}^3/\text{D.U.})(2.5 \times 10^{-19} \text{ cm}^2)(1 \text{ cm})}$

$= 0.142$

$T_{summer} = e^{-(350)(2.69 \times 10^{16})(2.5 \times 10^{-19})(1)} = 0.095$

Fractional increase in transmittance is $(0.142 - 0.095) / (0.095) = 49\%$.

17-15. Neocuproine reacts with Cu(I) and prevents it from forming a complex with ferrozine that would give a false positive result in the analysis of iron.

17-16. (a) $c = A/\varepsilon b = 0.427/[(6\,130 \text{ M}^{-1} \text{ cm}^{-1})(1.000 \text{ cm})] = 6.97 \times 10^{-5} \text{ M}$

(b) The sample had been diluted $\times 10 \Rightarrow 6.97 \times 10^{-4} \text{ M}$.

(c) $\dfrac{x \text{ g}}{(292.16 \text{ g/mol})(5.00 \times 10^{-3} \text{ L})} = 6.97 \times 10^{-4} \text{ M} \Rightarrow x = 1.02 \text{ mg}$

17-17. Yes

17-18. (a) $\varepsilon = \dfrac{A}{cb} = \dfrac{0.267 - 0.019}{(3.15 \times 10^{-6} \text{ M})(1.000 \text{ cm})} = 7.87 \times 10^{4} \text{ M}^{-1} \text{ cm}^{-1}$

(b) $c = \dfrac{A}{\varepsilon b} = \dfrac{0.175 - 0.019}{(7.87 \times 10^{4} \text{ M}^{-1} \text{ cm}^{-1})(1.000 \text{ cm})} = 1.98 \times 10^{-6} \text{ M}$

17-19. (a) The absorbance due to the colored product from nitrite added to sample C is $0.967 - 0.622 = 0.345$. The concentration of colored product due to added nitrite in sample C is $\dfrac{(7.50 \times 10^{-3} \text{ M})(10.0 \times 10^{-6} \text{ L})}{0.054 \text{ L}} = 1.389 \times 10^{-6} \text{ M}$.

$\varepsilon = A/bc = 0.345/[(1.389 \times 10^{-6})(5.00)] = 4.97 \times 10^{4} \text{ M}^{-1} \text{ cm}^{-1}$

(b) 7.50×10^{-8} mol of nitrite (from 10.0 µL added to sample C) gives $A = 0.345$. In sample B, x mole of nitrite in food extract gives $A = 0.622 - 0.153 = 0.469$.

$\dfrac{x \text{ mol}}{7.50 \times 10^{-8} \text{ mol}} = \dfrac{0.469}{0.345} \Rightarrow x = 1.020 \times 10^{-7} \text{ mol NO}_2^- = 4.69 \text{ µg}$

17-20. (a) Prior to the equivalence point, all added Fe(III) binds to the protein to form a red complex whose absorbance is measured in the figure. After the equivalence point, there are no more binding sites available on the protein. The slight increase in absorbance arises from the color of the iron titrant.

(b) 163×10^{-6} L of 1.43×10^{-3} M Fe(III) $= 2.33 \times 10^{-7}$ mol Fe(III)

(c) 1.17×10^{-7} mol transferrin in 2.00×10^{-3} L $\Rightarrow 5.83 \times 10^{-5}$ M transferrin

17-21. Theoretical equivalence point =

$$\frac{\left(2 \dfrac{\text{mol Ga}}{\text{mol transferrin}}\right)\left(\dfrac{0.003\,57 \text{ g transferrin}}{81\,000 \text{ g transferrin/mol transferrin}}\right)}{0.006\,64 \dfrac{\text{mol Ga}}{\text{L}}} = 13.3 \text{ }\mu\text{L}$$

Observed end point ≈ intersection of lines taken from first 6 points and last 4 points in the following graph = 12.2 μL, corresponding to $\dfrac{12.2}{13.3}$ = 91.7% of 2 Ga/transferrin = 1.83 Ga/transferrin. In the absence of oxalate, there is no evidence for specific binding of Ga to the protein, since the slope of the curve is small and does not change near 1 or 2 Ga/transferrin.

17-22. (a) In the graph below, least-squares lines were put through the first 6 points and the last 3 points. Their intersection is the estimated end point of 21.4 μL. The moles of TCNQ in the titration are (0.700 mL)(1.00 × 10⁻⁴ M TCNQ) = 70.0 nmol TCNQ. This many moles of Au(0) are present in 21.4 μL of nanoparticle solution. The mass of nanoparticles in 21.4 μL is (21.4 μL)(1.43 g/L) = 30.6 μg nanoparticles. To find the moles of Au(0) in 1.00g of nanoparticles, set up a proportion:

$$\frac{70.0 \times 10^{-9} \text{ mol Au(0)}}{30.6 \times 10^{-6} \text{ g nanoparticles}} = \frac{x \text{ mol Au(0)}}{1.00 \text{ g nanoparticles}} \Rightarrow x = 2.29 \text{ mmol Au(0)}$$

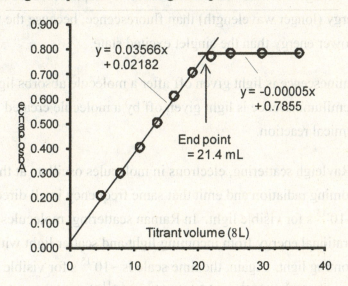

(b) 1.00 g of nanoparticles is estimated to contain 0.25 g $C_{12}H_{25}S$ (FM 201.40), which is 1.24 mmol $C_{12}H_{25}S$.

(c) From (a), the mass of Au(0) in 1.00 g is (2.29 mmol Au(0))(196.97 g/mol) = 0.451 g. The mass of Au(I) is estimated as the difference 1.00 – 0.451 – 0.25 = 0.299 g = 1.52 mmol Au(I). The calculated mole ratio Au(I):$C_{12}H_{25}S$ is 1.52/1.24 = 1.23. Ideally, this mole ratio should be 1.00.

17-23. $n \rightarrow \pi^*(T_1)$:

$$E = h\nu = h\frac{c}{\lambda} = (6.626\,1 \times 10^{-34}\ \text{J·s}) \frac{2.997\,9 \times 10^8\ \text{s}^{-1}}{397 \times 10^{-9}\ \text{m}} = 5.00 \times 10^{-19}\ \text{J}$$

To convert to J/mol, multiply by Avogadro's number:

5.00×10^{-19} J/molecule $\times\ 6.022 \times 10^{23}$ molecules/mol = 301 kJ/mol.

$n \rightarrow \pi^*(S_1)$:

$$E = (6.626\,1 \times 10^{-34}\ \text{J·s}) \frac{2.997\,9 \times 10^8\ \text{s}^{-1}}{355 \times 10^{-9}\ \text{m}} = 5.60 \times 10^{-19}\ \text{J} = 337\ \text{kJ/mol}.$$

The difference between the T_1 and S_1 states is 337 – 301 = 36 kJ/mol.

17-24. Fluorescence is emission of light with no change in the electronic spin state of the molecule (for example, singlet → singlet). In phosphorescence, the electronic spin does change during emission (for example, triplet → singlet). Phosphorescence is less probable, so molecules spend more time in the excited state prior to phosphorescence than to fluorescence. That is, phosphorescence has a longer lifetime than fluorescence. Phosphorescence also comes at lower energy (longer wavelength) than fluorescence, because the triplet excited state is at lower energy than the singlet excited state.

17-25. Luminescence is light given off after a molecule absorbs light. Chemiluminscence is light given off by a molecule created in an excited state in a chemical reaction.

17-26. In Rayleigh scattering, electrons in molecules oscillate at the frequency of incoming radiation and emit that same frequency in all directions. The time scale is $\sim 10^{-15}$ s for visible light. In Raman scattering, molecules extract a quantum of vibrational energy from incoming light and scatter light with less energy than the incoming light. Again, the time scale is $\sim 10^{-15}$ s for visible light. Fluorescence occurs in 10^{-8} to 10^{-4} s, which is 10^7 to 10^{11} times longer than scattering.

17-27. Phosphorescence is emitted at longer wavelength than fluorescence. Absorption is at shortest wavelength.

17-28. In an excitation spectrum, the exciting wavelength (λ_{ex}) is varied while the detector wavelength (λ_{em}) is fixed. In an emission spectrum, λ_{ex} is held constant and λ_{em} is varied. The excitation spectrum resembles an absorption spectrum because emission intensity is proportional to absorption of the exciting radiation.

17-29. The spreadsheet computes relative fluorescence given by

$$\text{relative intensity} = \frac{I}{k'P_0} = 10^{-\varepsilon_{ex}b_1c}(1 - 10^{-\varepsilon_{ex}b_2c})10^{-\varepsilon_{em}b_3c}.$$

	A	B	C
1	Anthracene fluorescence response		
2			
3	ε_{ex} =	9000	$M^{-1}\,cm^{-1}$
4	ε_{em} =	50	$M^{-1}\,cm^{-1}$
5			
6	b_1 =	0.3	cm
7	b_2 =	0.4	cm
8	b_3 =	0.5	cm
9			
10		Relative	
11		fluorescence	
12	c (M)	$I/k'P_o$	Intensity/c
13	1.E-08	8.288E-05	8288
14	2.E-08	0.0001658	8288
15	4.E-08	0.0003314	8286
16	6.E-08	0.000497	8284
17	8.E-08	0.0006626	8282
18	1.E-07	0.0008281	8281
19	2.E-07	0.0016544	8272
20	4.E-07	0.0033019	8255
21	6.E-07	0.0049426	8238
22	8.E-07	0.0065764	8221
23	1.E-06	0.0082034	8203
24	2.E-06	0.0162369	8118
25	4.E-06	0.0318052	7951
26	6.E-06	0.0467266	7788
27	8.E-06	0.0610222	7628
28	1.E-05	0.0747124	7471
29	2.E-05	0.1347552	6738
30	4.E-05	0.2195664	5489
31	6.E-05	0.2689283	4482
32	8.E-05	0.2934519	3668
33	1.E-04	0.300872	3009
34	2.E-04	0.2307768	1154
35	4.E-04	0.0783318	196
36	6.E-04	0.0230136	38
37	8.E-04	0.0065982	8
38	1.E-03	0.0018832	2

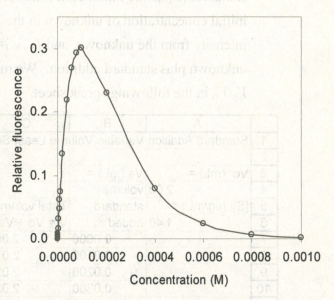

Fluorescence increases with concentration and then decreases because of self-absorption. Most of the absorption occurs at the excitation wavelength, and a little comes at the emission wavelength. Column C of the spreadsheet gives the relative fluorescence intensity divided by concentration. If intensity were proportional to concentration, then column C would be constant. We see that it is constant at low concentration, and falls by ~5% at ~5 μM. The calibration curve in the text goes up to 0.6 μM, which is in the linear range.

17-30.   The equation for standard addition is

$$\underbrace{I_{S+X}\left(\frac{V}{V_0}\right)}_{\substack{\text{Function to plot}\\\text{on } y\text{-axis}}} = I_X + \frac{I_X}{[X]_i}\underbrace{[S]_i\left(\frac{V_s}{V_0}\right)}_{\substack{\text{Function to plot}\\\text{on } x\text{-axis}}}$$

where V_0 is the volume of unknown in the cuvet (2.00 mL), V_s is the volume of standard added (0 to 40 μL), V is the total volume of unknown plus added standard, $[S]_i$ is the initial concentration of standard (1.40 μg Se/mL), $[X]_i$ is the initial concentration of unknown in the 2.00-mL solution, I_X is the fluorescence intensity from the unknown, and I_{S+X} is the fluorescence intensity from the unknown plus standard addition. We make a graph of $I_{S+X} V/V_0$ versus $[S]_i V_s/V_0$ in the following spreadsheet.

	A	B	C	D	E	F
1	Standard Addition Variable Volume Least-Squares Spreadsheet					
2						
3	Vo (mL) =	Vs (mL) =		x		y
4	2.00	volume				
5	[S]i (μg/mL) =	standard	Total volume	x-axis function	I(s+x) =	y-axis function
6	1.40	added	V = Vo + Vs	Si*Vs/Vo	signal	I(s+x)*V/Vo
7		0.0000	2.000	0.00000	41.4	41.4
8		0.0100	2.010	0.00700	49.2	49.4
9		0.0200	2.020	0.01400	56.4	57.0
10		0.0300	2.030	0.02100	63.8	64.8
11		0.0400	2.040	0.02800	70.3	71.7
12						
13	B16:D18 = LINEST(F7:F11,D7:D11,TRUE,TRUE)					
14						
15		LINEST output:				
16	m	1084.6	41.670	b		
17	sm	14.9	0.255	sb		
18	R²	0.9994	0.330	sy		
19	x-intercept					
20	= -b/m =	-0.03842				
21						
22	n =	5	B22 = COUNT(B7:B11)			
23	Mean y =	56.8546	B23 = AVERAGE(F7:F11)			
24	Σ(xi - mean x)² =	0.00049	B24 = DEVSQ(D7:D11)			
25						
26	Std deviation of					
27	x-intercept =	0.000732				
28	B27 =(C18/ABS(B16))*SQRT((1/B22) + B23^2/(B16^2*B24))					

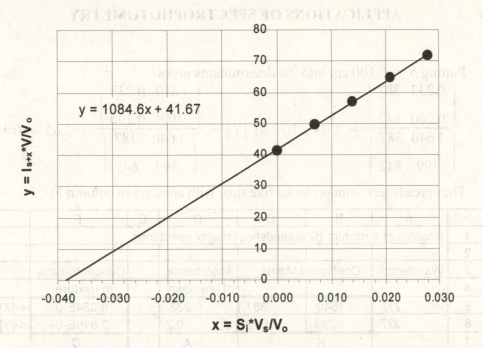

The x-intercept is found from the equation of the straight line by setting $y = 0$:

$0 = 1084.6x + 41.67 \Rightarrow x = -0.038\ 42$. The concentration of Se in the unknown is 0.038 42 μg/mL. All Se from 0.108 g of Brazil nuts was dissolved in 10.0 mL of solvent, which contained (10.0 mL)(0.038 42 μg/mL) = 0.384 2 μg Se. The wt% Se in the nuts is $100 \times (0.384\ 2 \times 10^{-6}\ \text{g}/0.108\ \text{g}) = 3.56 \times 10^{-4}$ wt%.

The standard deviation of the x-intercept is

$$\begin{array}{c}\text{Standard deviation}\\ \text{of } x\text{-intercept}\end{array} = \frac{s_y}{|m|}\sqrt{\frac{1}{n}+\frac{\bar{y}^2}{m^2\ \Sigma(x_i-\bar{x})^2}}$$

where s_y is the standard deviation of y, m is the slope, n is the number of data points $(= 5)$, $\bar{y}$ is the mean value of y for the 5 points, x_i are individual values of x, and $\bar{x}$ is the mean value of x for the 5 points. Cell B27 of the spreadsheet gives a standard deviation of 0.000 732 for the x-intercept.

The relative uncertainty in the intercept is 0.000 732/0.038 42 = 1.91%. This is the relative uncertainty in wt% Se if other sources of error are insignificant. Uncertainty in wt% = $(0.0191)(3.56 \times 10^{-4}$ wt%$) = 6.8 \times 10^{-6}$ wt%. Answer = $3.56\ (\pm 0.06_8) \times 10^{-4}$ wt%.

The confidence interval is $\pm t \times$ (standard deviation) where t is Student's t (Table 4-2) for $n - 2 = 3$ degrees of freedom. The 95% confidence interval is $\pm(3.182)(0.06_8 \times 10^{-4}$ wt%$) = \pm 0.22 \times 10^{-4}$ wt%. The value $t = 3.182$ was taken from Table 4-2 for 3 degrees of freedom. Answer = $3.56\ (\pm 0.22) \times 10^{-4}$ wt%.

CHAPTER 18
APPLICATIONS OF SPECTROPHOTOMETRY

18-1. Putting $b = 0.100$ cm into the determinants gives

$$[X] = \frac{\begin{vmatrix} 0.233 & 387 \\ 0.200 & 642 \end{vmatrix}}{\begin{vmatrix} 1640 & 387 \\ 399 & 642 \end{vmatrix}} = 8.03 \times 10^{-5} \text{ M} \quad [Y] = \frac{\begin{vmatrix} 1640 & 0.233 \\ 399 & 0.200 \end{vmatrix}}{\begin{vmatrix} 1640 & 387 \\ 399 & 642 \end{vmatrix}} = 2.62 \times 10^{-4} \text{ M}$$

The spreadsheet solution looks like this, with answers in column F:

	A	B	C	D	E	F	G
1	Analysis of a mixture by spreadsheet matrix operations						
2							
3	Wavelength	Coefficient Matrix		Absorbance		Concentrations	
4				of unknown		in mixture	
5	272	1640	387	0.233		8.034E-05	<-[X]
6	327	399	642	0.2		2.616E-04	<-[Y]
7		K		A		C	

18-2.

	A	B	C	D	E	F	G	H
1	Analysis of a Mixture When You Have More Data Points than Components of the Mixture							
2				Measured			Calculated	
3				Absorbance			Absorb-	
4	Wave-	Absorbance of Standard		of Mixture	Molar Absorptivity		ance	
5	length	MnO4	Cr2O7	Am	MnO4	Cr2O7	Acalc	[Acalc-Am]^2
6	266	0.042	0.410	0.766	420.0	4100.0	0.7650	1.017E-06
7	288	0.082	0.283	0.571	820.0	2830.0	0.5723	1.763E-06
8	320	0.168	0.158	0.422	1680.0	1580.0	0.4217	1.132E-07
9	350	0.125	0.318	0.672	1250.0	3180.0	0.6706	2.050E-06
10	360	0.056	0.181	0.366	560.0	1810.0	0.3690	9.106E-06
11							sum =	1.405E-05
12	Standards		Concentrations in the mixture					
13	[Mn](M)=		(to be found by Solver)					
14	1.00E-04		[MnO4] =	8.356E-05				
15	[Cr](M)=		[Cr2O7] =	1.780E-04				
16	1.00E-04							
17	Pathlength		E6 = B6/($A$19*$A$14)					
18	(cm) =		F6 = C6/($A$19*$A$16)					
19	1.000		G6 = E6*$A$19*$D$14+F6*$A$19*$D$15					
20			H6 =(G6-D6)^2					

18-3. If the spectra of two compounds with a constant total concentration cross at any wavelength, all mixtures with the same total concentration will go through that same point, called an isosbestic point. The appearance of isosbestic points in a

238

chemical reaction is good evidence that we are observing one main species being converted to one other major species.

18-4. As VO^{2+} is added (traces 1-9), the peak at 439 decreases and a new one near 485 nm develops, with an isosbestic point at 457 nm. When VO^{2+}/xylenol orange > 1, the peak near 485 nm decreases and a new one at 566 nm grows in, with an isosbestic point at 528 nm. This sequence is logically interpreted by the sequence

$$M \quad + \quad \underset{439 \text{ nm}}{L} \quad \rightarrow \quad \underset{485 \text{ nm}}{ML}$$

$$\underset{485 \text{ nm}}{ML} \quad + \quad M \quad \rightarrow \quad \underset{566 \text{ nm}}{M_2L}$$

where M is vanadyl ion and L is xylenol orange. The structure of xylenol orange in Table 11-3 shows that it has metal-binding groups on both ends of the molecule, and could form an M_2L complex.

18-5. Convert T to A ($= -\log T$) and then convert A to ε ($= A/bc = A/[(0.005)(0.01)]$)

	Absorbance			ε (M^{-1} cm^{-1})	
	2 022	1 993 cm^{-1}		2 022	1 993 cm^{-1}
A	0.508 6	0.098 54	A	10 170	1 971
B	0.011 44	0.699 0	B	228.8	13 980

For the mixture, $A_{2022} = -\log(0.340) = 0.468\,5$ and $A_{1993} = -\log(0.383) = 0.416\,8$. Equation 18-6 gives $[A] = 9.11 \times 10^{-3}$ M and $[B] = 4.68 \times 10^{-3}$ M.

18-6.

	A	B	C	D	E	F	G	H
1	Solving Simultaneous Linear Equations with Excel Matrix Operations							
2								
3	Wavelength	Coefficient Matrix			Absorbance		Concentrations	
4		TB	STB	MTB	of unknown		in mixture	
5	455	4800	11100	18900	0.412		1.2194E-05	<-[TB]
6	485	7350	11200	11800	0.350		9.2953E-06	<-[STB]
7	545	36400	13900	4450	0.632		1.3243E-05	<-[MTB]
8			K		A		C	
9								
10	1. Highlight block of blank cells required for solution (G5:G7)							
11	2. Type the formula "= MMULT(MINVERSE(B5:D7),E5:E7)"							
12	3. Press CONTROL+SHIFT+ENTER on a PC or COMMAND+RETURN on a Mac							
13	4. The answer appears in cells G5:G7							

18-7.

	A	B	C	D	E	F	G	H	I
1	Solving 4 Simultaneous Linear Equations with Excel Matrix Operations								
2	Wave-								
3	length	Coefficient Matrix			Ethyl-	Absorbance		Conc. in	
4	(μm)	p-xylene	m-xylene	o-xylene	benzene	of unknown		mixture	
5	12.5	1.5020	0.0514	0	0.0408	0.1013		0.0627	p-xylene
6	13.0	0.0261	1.1516	0	0.0820	0.09943		0.0795	m-xylene
7	13.4	0.0342	0.0355	2.532	0.2933	0.2194		0.0759	o-xylene
8	14.3	0.0340	0.0684	0	0.3470	0.03396		0.0761	Ethylbz
9			K				A		C
10									
11	1. Highlight block of blank cells required for solution (H5:H8)								
12	2. Type the formula "= MMULT(MINVERSE(B5:E8),F5:F8)"								
13	3. Press CONTROL+SHIFT+ENTER on a PC or COMMAND+RETURN on a Mac								
14	4. The answer appears in cells H5:H8								

18-8. The quantity of HIn is small compared to aniline and sulfanilic acid. Calling aniline B and sulfanilic acid HA, we can write

$$B \quad + \quad HA \quad \overset{K}{\rightleftharpoons} \quad BH^+ \quad + \quad A^-$$

Initial mmol: $\quad$ 2.00 $\qquad$ 1.500 $\qquad$ — $\qquad$ —

Final mmol: $\qquad$ $2.00 - x$ $\quad$ $1.500 - x$ $\qquad$ x $\qquad$ x

$$K = \frac{K_a K_b}{K_w} = \frac{(10^{-3.232})(K_w/10^{-4.601})}{K_w} = 23.39$$

$$\frac{x^2}{(2.00 - x)(1.500 - x)} = 23.39 \Rightarrow x = 1.372 \text{ mmol}$$

$$pH = pK_{BH^+} + \log\frac{[B]}{[BH^+]} = 4.601 + \log\frac{2.00 - 1.372}{1.372} = 4.26$$

For HIn we can write:

absorbance $= 0.110 = (2.26 \times 10^4)(5.00)\,[\text{HIn}] + (1.53 \times 10^4)\,(5.00)[\text{In}^-]$.

Substituting $[\text{HIn}] = 1.23 \times 10^{-6} - [\text{In}^-]$ gives $[\text{In}^-] = 7.94 \times 10^{-7}$ and

$[\text{HIn}] = 4.36 \times 10^{-7}$. The Henderson-Hasselbalch equation for HIn is therefore

$$pH = pK_{HIn} + \log\frac{[\text{In}^-]}{[\text{HIn}]} \Rightarrow 4.26 = pK_{HIn} + \log\frac{7.94 \times 10^{-7}}{4.36 \times 10^{-7}} \Rightarrow pK_{HIn} = 4.00.$$

18-9. (a) $A_{620} = \varepsilon_{620}^{HIn^-}\, b[\text{HIn}^-] + \varepsilon_{620}^{In^{2-}}\, b[\text{In}^{2-}]$

$\qquad\qquad A_{434} = \varepsilon_{434}^{HIn^-}\, b[\text{HIn}^-] + \varepsilon_{434}^{In^{2-}}\, b[\text{In}^{2-}]$

The solution of these two equations is given by Equation 18-6 in the text:

$$[\text{HIn}^-] = \frac{1}{D}\left(A_{620}\varepsilon_{434}^{In^{2-}}b - A_{434}\varepsilon_{620}^{In^{2-}}b\right)$$

$$[In^{2-}] = \frac{1}{D}\left(A_{434}\varepsilon_{620}^{HIn^-}b - A_{620}\varepsilon_{434}^{HIn^-}b\right)$$

where $D = b^2\left(\varepsilon_{620}^{HIn^-}\varepsilon_{434}^{In^{2-}} - \varepsilon_{620}^{In^{2-}}\varepsilon_{434}^{HIn^-}\right)$

Dividing the expression for $[In^{2-}]$ by the expression for $[HIn^-]$ gives

$$\frac{[In^{2-}]}{[HIn^-]} = \frac{A_{434}\varepsilon_{620}^{HIn^-} - A_{620}\varepsilon_{434}^{HIn^-}}{A_{620}\varepsilon_{434}^{In^{2-}} - A_{434}\varepsilon_{620}^{In^{2-}}}$$

Dividing numerator and denominator on the right side by A_{434} gives

$$\frac{[In^{2-}]}{[HIn^-]} = \frac{\varepsilon_{620}^{HIn^-} - R_A\varepsilon_{434}^{HIn^-}}{R_A\varepsilon_{434}^{In^{2-}} - \varepsilon_{620}^{In^{2-}}} = \frac{R_A\varepsilon_{434}^{HIn^-} - \varepsilon_{620}^{HIn^-}}{\varepsilon_{620}^{In^{2-}} - R_A\varepsilon_{434}^{In^{2-}}}$$

(b)   Mass balance for indicator: $[HIn^-] + [In^{2-}] = F_{In}$

Dividing both sides by $[HIn^-]$ gives

$$\frac{[HIn^-]}{[HIn^-]} + \frac{[In^{2-}]}{[HIn^-]} = \frac{F_{In}}{[HIn^-]} \Rightarrow 1 + R_{In} = \frac{F_{In}}{[HIn^-]} \Rightarrow [HIn^-] = \frac{F_{In}}{R_{In} + 1}$$

Acid dissociation constant of indicator:

$$K_{In} = \frac{[In^{2-}][H^+]}{[HIn^-]}$$

Substituting $F_{In}/(R_{In} + 1)$ for $[HIn^-]$ gives

$$K_{In} = \frac{[In^{2-}][H^+](R_{In} + 1)}{F_{In}} \Rightarrow [In^{2-}] = \frac{K_{In}F_{In}}{[H^+](R_{In} + 1)}$$

(c)   Equation A in the problem defines R_{In} as $[In^{2-}]/[HIn^-]$. So,

$$K_{In} = \frac{[In^{2-}][H^+]}{[HIn^-]} = R_{In}[H^+] \Rightarrow [H^+] = K_{In}/R_{In}$$

(d)   From the acid dissociation reaction of carbonic acid, we can write

$$K_1 = \frac{[HCO_3^-][H^+]}{[CO_2(aq)]} \Rightarrow [HCO_3^-] = \frac{K_1[CO_2(aq)]}{[H^+]}$$

From the acid dissociation reaction of bicarbonate, we can write

$$K_2 = \frac{[CO_3^{2-}][H^+]}{[HCO_3^-]} \Rightarrow [CO_3^{2-}] = \frac{K_2[HCO_3^-]}{[H^+]}$$

Substituting in the expression for $[HCO_3^-]$ gives

$$[CO_3^{2-}] = \frac{K_1K_2[CO_2(aq)]}{[H^+]^2}$$

(e)   Charge balance:

$$[Na^+] + [H^+] = [OH^-] + [HIn^-] + 2[In^{2-}] + [HCO_3^-] + 2[CO_3^{2-}]$$

$$F_{Na} + [H^+] =$$
$$\frac{K_w}{[H^+]} + \frac{F_{In}}{R_{In} + 1} + 2\frac{K_{In}F_{In}}{[H^+](R_{In} + 1)} + \frac{K_1[CO_2(aq)]}{[H^+]} + 2\frac{K_1K_2[CO_2(aq)]}{[H^+]^2}$$

(f) From part (c) we know that $[H^+] = K_{In}/R_{In}$. We calculate R_{In} from part (a):

$$R_{In} = \frac{R_A\varepsilon_{434}^{HIn^-} - \varepsilon_{620}^{HIn^-}}{\varepsilon_{620}^{In^{2-}} - R_A\varepsilon_{434}^{In^{2-}}} = \frac{(2.84)(8.00 \times 10^3) - (0)}{(1.70 \times 10^4) - (2.84)(1.90 \times 10^3)} = 1.95_8$$

$$[H^+] = K_{In}/R_{In} = (2.0 \times 10^{-7})/1.95_8 = 1.0_2 \times 10^{-7}\ M$$

Substituting this value of $[H^+]$ into the mass balance in part (e) produces an equation in which the only unknown is $[CO_2(aq)]$:

$$F_{Na} + [H^+] =$$
$$\frac{K_w}{[H^+]} + \frac{F_{In}}{R_{In} + 1} + 2\frac{K_{In}F_{In}}{[H^+](R_{In} + 1)} + \frac{K_1[CO_2(aq)]}{[H^+]} + 2\frac{K_1K_2[CO_2(aq)]}{[H^+]^2}$$

$$92.0 \times 10^{-6} + 1.0_2 \times 10^{-7} =$$
$$\frac{(6.7 \times 10^{-15})}{(1.0_2 \times 10^{-7})} + \frac{(50.0 \times 10^{-6})}{1.95_8 + 1} + 2\frac{(2.0 \times 10^{-7})(50.0 \times 10^{-6})}{(1.0_2 \times 10^{-7})(1.95_8 + 1)}$$
$$+ \frac{(3.0 \times 10^{-7})[CO_2(aq)]}{(1.0_2 \times 10^{-7})} + 2\frac{(3.0 \times 10^{-7})(3.3 \times 10^{-11})[CO_2(aq)]}{(1.0_2 \times 10^{-7})^2}$$

$$9.21 \times 10^{-5} = 6.56 \times 10^{-8} + 1.69 \times 10^{-5} + 6.62 \times 10^{-5} +$$
$$+ 2.94\ [CO_2(aq)] + 0.001\ 9\ [CO_2(aq)]$$
$$\Rightarrow [CO_2(aq)] = 3.0_4 \times 10^{-6}\ M$$

(g) The ions in solution are Na^+, HIn^-, In^{2-}, HCO_3^-, CO_3^{2-}, H^+, and OH^-. We know that $[Na^+] = 92.0\ \mu M$ and $[H^+] = 0.10\ \mu M$. If the total cation charge is 92.1 μM, the total anion charge must be 92.1 μM, and the ionic strength must be ~92 μM $\approx 10^{-4}$ M. (The ionic strength is not exactly 92.1 μM because some anions have a charge of -2, which will increase the ionic strength from 92.1 μM.) An ionic strength of 10^{-4} M is low enough that the activity coefficients are close to 1.00.

We can calculate the exact ionic strength from the following expressions derived above:

$$[OH^-] = \frac{K_w}{[H^+]} = 0.07\ \mu M$$

$$[HIn^-] = \frac{F_{In}}{R_{In} + 1} = 16.9\ \mu M; \quad [In^{2-}] = \frac{K_{In}F_{In}}{[H^+](R_{In} + 1)} = 33.1\ \mu M$$

$$[HCO_3^-] = \frac{K_1[CO_2(aq)]}{[H^+]} = 2.94\ [CO_2(aq)] = 8.9\ \mu M$$

$$[CO_3^{2-}] = \frac{K_1 K_2 [CO_2(aq)]}{[H^+]^2} = 0.001\ 9\ [CO_2(aq)] = 0.003\ \mu M$$

$$\text{Ionic strength} = \frac{1}{2} \sum_i c_i z_i^2 =$$

$$\frac{1}{2} \left\{ [Na^+] \cdot 1^2 + [H^+] \cdot 1^2 + [OH^-] \cdot 1^2 + [HIn^-] \cdot 1^2 + [In^{2-}] \cdot 2^2 + [HCO_3^-] \cdot 1^2 + [CO_3^{2-}] \cdot 2^2 \right\}$$

$$= 125\ \mu M$$

18-10. The Scatchard plot is a graph of $[PX]/[X]$ versus $[PX]$. Data in the following spreadsheet are plotted in the figure in the textbook. The slope is -4.0×10^9 M^{-1}, giving a binding constant $K = 4.0 \times 10^9$ M^{-1}. The fraction of saturation in column E is $S = [PX]/P_0$, where $P_0 = 10.0$ nM. S ranges from 0.29 to 0.84.

	A	B	C	D	E	F
1	Data for Scatchard plot			P_o =	10	nM
2				Fraction of saturation		
3	[PX] (nM)	[X] (nM)	[PX]/[X]		= [PX]/P_o	
4	2.87	0.120	24.0		0.29	
5	3.80	0.192	19.8		0.38	
6	4.66	0.296	15.7		0.47	
7	5.54	0.450	12.3		0.55	
8	6.29	0.731	8.61		0.63	
9	6.77	1.22	5.54		0.68	
10	7.52	1.50	5.02		0.75	
11	8.45	3.61	2.34		0.84	

18-11. (a)

	A	B	C	D	E	F	G
1	Estradiol - Albumin Scatchard Plot						
2	abscissa	ordinate					
3	P ($\div$M)	Xo/X		Highlight cells B15:C17			
4	6.3	1.26		Type			
5	10.0	1.62		"=LINEST(B4:B12,A4:A12,TRUE,TRUE)"			
6	20.0	2.16		Press CTRL+SHIFT+ENTER (on PC)			
7	30.0	2.51		Press COMMAND+RETURN (on Mac)			
8	40.0	3.34					
9	50.0	3.33		Student's t (95% confidence,			
10	60.0	4.19		7 degrees of freedom) =			
11	70.0	4.13		2.364624	=TINV(0.05,7)		
12	80.0	4.36		95% confidence interval =			
13				0.008248	= t*s_m = D11*B16		
14		LINEST output					
15	m	0.042885	1.243476	b			
16	s_m	0.003488	0.166224	s_b			
17	R^2	0.955742	0.259413	s_y			

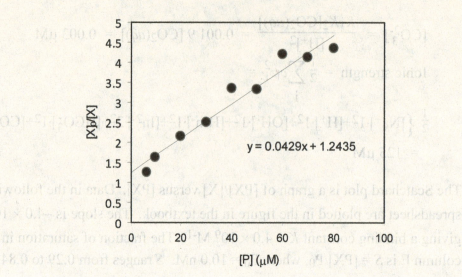

The slope is 0.042 88 with a standard deviation of 0.003 49 in cells B15 and B16. However, the units on the abscissa are μM, so the slope is really $(0.042\ 88 \pm 0.003\ 49)/10^{-6}\ M^{-1} = (4.288 \pm 0.349) \times 10^4\ M^{-1}$. To find the 95% confidence interval, we need Student's t for $9 - 2 = 7$ degrees of freedom, which is $t = 2.365$ in cell D11. The 95% confidence interval is $(0.349)(2.365) = 0.825$. The final result is $K = (4.3 \pm 0.8) \times 10^4\ M^{-1}$.

(b) Estradiol is X. The quotient $[X]_0/[X]$ is 1.26 at the first point and 4.36 at the last point. The fraction of free estradiol is $[X]/[X]_0 = 1/1.26 = 0.79$ at the first point and $1/4.36 = 0.23$ at the last point. The fraction of bound estradiol is $1 - 0.79 = 0.21$ at the first point and $1 - 0.23 = 0.77$ at the last point.

18-12. (a) We will make the substitutions $[\text{complex}] = A/\varepsilon$ and $[I_2] = [I_2]_{\text{tot}} - [\text{complex}]$ in the equilibrium expression:

$$K = \frac{[\text{complex}]}{[I_2][\text{mesitylene}]} = \frac{A/\varepsilon}{([I_2]_{\text{tot}} - [\text{complex}])\,[\text{mesitylene}]}$$

$$K[I_2]_{\text{tot}} - K[\text{complex}] = \frac{A}{\varepsilon[\text{mesitylene}]}$$

Making the substitution $[\text{complex}] = A/\varepsilon$ once more on the left-hand side gives

$$K[I_2]_{\text{tot}} - \frac{KA}{\varepsilon} = \frac{A}{\varepsilon[\text{mesitylene}]}$$

Multiplying both sides by ε and dividing by $[I_2]_{\text{tot}}$ gives the desired result:

$$\varepsilon K - \frac{KA}{[I_2]_{\text{tot}}} = \frac{A}{[I_2]_{\text{tot}}\,[\text{mesitylene}]}$$

(b) The graph of $A/([\text{mesitylene}][I_2]_{\text{tot}})$ versus $A/[I_2]_{\text{tot}}$ is an excellent straight line with a slope of -0.464 and an intercept of 4.984×10^3. Since slope = $-K$, the equilibrium constant is 0.464. The molar absorptivity is ε = intercept/$K \Rightarrow \varepsilon = 1.074 \times 10^4$ M^{-1} cm^{-1}.

18-13. After running Solver, the average value of K in cell E10 is 0.464 and ε = 1.073×10^4 M^{-1} cm^{-1}. The Scatchard plot in the previous problem gave $K = 0.464$ and $\varepsilon = 1.074 \times 10^4$ M^{-1} cm^{-1}.

	A	B	C	D	E
1	Equilibrium constant for reaction of I_2 with mesitylene				
2					
3	[Mesitylene] (M)	[I_2]tot (M)	A	[Complex] = A/ε	K_{eq}
4	1.6900	7.82E-05	0.369	3.437E-05	0.46443
5	0.9218	2.56E-04	0.822	7.657E-05	0.46349
6	0.6338	3.22E-04	0.787	7.331E-05	0.46439
7	0.4829	3.57E-04	0.703	6.549E-05	0.46473
8	0.3900	3.79E-04	0.624	5.813E-05	0.46480
9	0.3271	3.93E-04	0.556	5.179E-05	0.46353
10				Average =	0.46423
11	Guess for ε:			Standard Dev =	0.00058
12	1.073E+04			Stdev/Average =	0.00125
13					
14	D4 = C4/$A$12				
15	E4 = D4/(A4*(B4-D4)) = [complex]/([Mesitylene][Free I_2])				
16	E10 = AVERAGE(E4:E9)				
17	E11 = STDEV(E4:E9)				
18	E12 = E11/E10				
19	Use Solver to vary ε (cell A12) until cell E12 is minimized				

18-14. (a) Maximum absorbance occurs at $X_{SCN^-} = 0.500$

$\Rightarrow$ stoichiometry $= 1 : 1$ ($n = 1$)

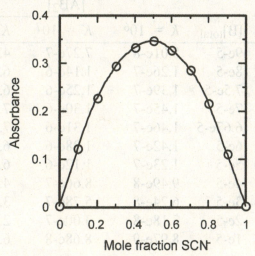

(b) The curved maximum indicates that the equilibrium constant is not very large.

(c) The different acid concentrations give both solutions the same ionic strength (= 16.0 mM).

18-15. The Job plot peak is at a xylenol orange mole fraction of 0.40, suggesting the stoichiometry (xylenol orange)$_2$Zr$_3$ which has a mole fraction of

$$\frac{\text{xylenol orange}}{\text{xylenol orange} + \text{Zr(IV)}} = \frac{2}{2+3} = 0.4$$

The plot could have been improved by obtaining data points at mole fractions of 0.35 and 0.45 to verify the location of the maximum.

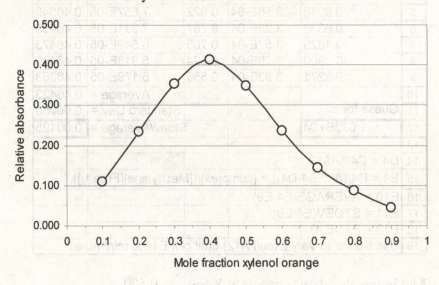

18-16. (a) Here are the results:

		[AB$_2$]			
[A]$_{total}$	[B]$_{total}$	$K = 10^6$	$K = 10^7$	$K = 10^8$	Mole fraction A
1e-5	9e-5	8.01e-8	7.27e-7	4.02e-6	0.1
2e-5	8e-5	1.26e-7	1.14e-6	6.26e-6	0.2
2.5e-5	7.5e-5	1.39e-7	1.25e-6	6.83e-6	0.25
3e-5	7e-5	1.45e-7	1.30e-6	7.12e-6	0.3
3.33e-5	6.67e-5	1.46e-7	1.31e-6	7.17e-6	0.333
4e-5	6e-5	1.42e-7	1.28e-6	6.99e-6	0.4
5e-5	5e-5	1.23e-7	1.12e-6	6.20e-6	0.5
6e-5	4e-5	9.49e-8	8.66e-7	4.97e-6	0.6
7e-5	3e-5	6.24e-8	5.78e-7	3.51e-6	0.7
8e-5	2e-5	3.18e-8	3.00e-7	2.00e-6	0.8
9e-5	1e-5	8.97e-9	8.68e-8	6.70e-7	0.9

(b) The maximum occurs at a mole fraction of $A = 1/3$, since the stoichiometry is 1:2. The greater the equilibrium constant, the greater the extent of reaction and the steeper the curve. When the equilibrium constant is too small, the curve is so shallow that it does not at all resemble two intersecting lines.

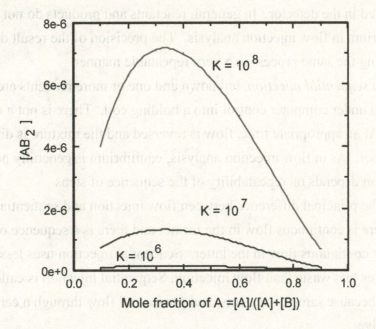

18-17. The mole fraction of thymine varies from 0.10 to 0.90 in increments of 0.10 as we go down the table. Job's plot reaches a broad peak at a mole fraction of 0.50, which is consistent with 1:1 complex formation. Job's plot gives us no information on the structure of the product except for its stoichiometry.

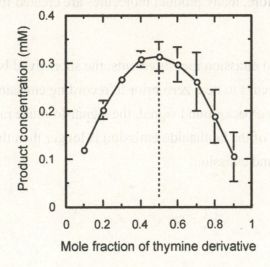

18-18.   In *flow injection analysis*, unknown is injected into a continuously flowing stream that could contain a reagent which reacts with analyte. Alternatively, one or more reagents can be added to the flow downstream of sample injection. The sample and reagents travel through a coil in which mixing and reaction occur. Absorbance or fluorescence or some other property of the flowing stream is measured in the detector. In general, reactants and products do not reach equilibrium in flow injection analysis. The precision of the result depends on executing the same process in a very repeatable manner.

In *sequential injection*, unknown and one or more reagents are sequentially injected under computer control into a holding coil. There is not a continuous flow. At an appropriate time, flow is reversed and the mixture is directed through a detector. As in flow injection analysis, equilibrium is generally not reached and precision depends on repeatability of the sequence of steps.

The principal difference between flow injection and sequential injection is that there is continuous flow in the former and there is a sequence of steps without continuous flow in the latter. Sequential injection uses less reagent and generates less waste than flow injection. Sequential injection is called "lab-on-a-valve" because sample, reagents, and product all flow through a central multi-port valve.

18-19.   Each molecule of analyte bound to antibody 1 also binds one molecule of antibody 2 that is linked to one enzyme molecule. Each enzyme molecule catalyzes many cycles of reaction in which a colored or fluorescent product is created. Therefore, many product molecules are created for every analyte molecule.

18-20.   In time-resolved emission measurements, the short-lived background fluorescence decays to near zero prior to recording emission from the lanthanide ion. By reducing background signal, the signal-to-noise ratio is increased. Also, the wavelength of the lanthanide emission is longer than the wavelength of much of the background emission.

18-21.

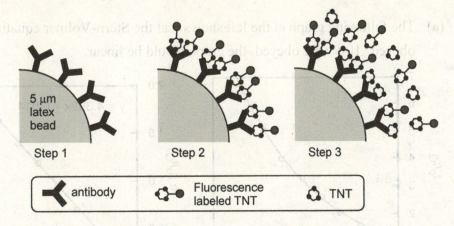

In Step 1, antibodies for TNT are attached to latex beads. In Step 2, the antibody is saturated with a fluorescent derivative of TNT. Excess fluorescent derivative is removed. In Step 3, the beads are incubated with TNT, which displaces some fluorescent derivative from binding sites on the antibodies. The suspension of beads is then injected into the flow cytometer. As each bead passes in front of the detector, it is excited by a laser and its fluorescence is measured. The graph in the textbook shows median bead fluorescence versus TNT concentration in a series of standards. The more TNT in the standard, the less fluorescence remains associated with the beads.

18-22. The graph of K_{sv} versus pH has a plateau at low pH near $K_{sv} \approx 100$ and a plateau at high pH near $K_{sv} \approx 1350$. The quencher, 2,6-dimethylphenol, is a weak acid whose pK_a is expected to be near 10. A logical interpretation is that the basic form, A^-, is a strong quencher with $K_{sv} \approx 1350$, and the acidic form, HA, is a weak quencher with $K_{sv} \approx 100$. We estimate pK_a as the midpoint in the curve at $K_{sv} \approx (1350 - 100)/2 = 625$. At this point, pH ≈ 10.8, which is our estimate for pK_a. The literature value is 10.63.

The smooth curve in the graph is a least-squares fit to the equation

$$K_{sv} = K_{sv}^{HA} \underbrace{\left(\frac{[H^+]}{[H^+] + K_a} \right)}_{\substack{\text{Fraction in form HA}}} + K_{sv}^{A^-} \underbrace{\left(\frac{K_a}{[H^+] + K_a} \right)}_{\substack{\text{Fraction in form A}^-}}$$

$$\underbrace{\phantom{K_{sv} = K_{sv}^{HA} \left(\frac{[H^+]}{[H^+] + K_a} \right)}}_{\text{Quenching by HA}} \underbrace{\phantom{K_{sv}^{A^-} \left(\frac{K_a}{[H^+] + K_a} \right)}}_{\text{Quenching by A}^-}$$

The least-squares fit gave $K_{sv}^{HA} = 69.0 \text{ M}^{-1}$, $K_{sv}^{A^-} = 1\,370 \text{ M}^{-1}$, and $pK_a = 10.78$.

18-23. (a) The following graph at the left shows that the Stern-Volmer equation is not obeyed. If it were obeyed, the graph would be linear.

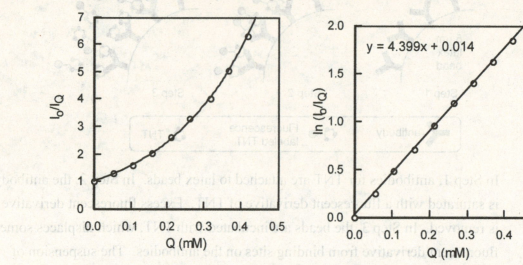

(b) The graph at the right above shows that Equation 4 is obeyed. Ideally, the intercept should be zero. The slope of the graph is $N_{av}/([S] - [CMC])$.

Given that [Q] was expressed in mM, we will express [S] and [CMC] in mM: $4.399 = N_{av} / ([20.8] - [8.1]) \rightarrow N_{av} = 55.9$

(c) $[M] = ([S] - [CMC]) / N_{av} = ([20.8] - [8.1]) / 55.9 = 0.227$ mM

$\overline{Q} = [Q]/[M] = 0.200$ mM $/ 0.227$ mM $= 0.881$ molecules per micelle

(d) $P_n = \dfrac{\overline{Q}^n}{n!} e^{-\overline{Q}}$. For $n = 0$, $P_0 = e^{-0.881} = 0.414$

$P_1 = \dfrac{(0.881)^1}{1!} e^{-0.881} = 0.365 \qquad P_2 = \dfrac{(0.881)^2}{2!} e^{-0.881} = 0.161$

CHAPTER 19
SPECTROPHOTOMETERS

19-1. The light source provides ultraviolet, visible, or infrared radiation. The monochromator selects a narrow band of wavelengths to pass on to the sample. As the experiment progresses, the monochromator scans through a desired range of wavelengths. The beam chopper is a rotating mirror that alternately directs light to the sample or reference. The sample cuvet holds the sample of interest. The reference cuvet is an identical cell containing pure solvent or a reagent blank. The mirror after the reference cuvet and the semitransparent mirror after the sample cuvet pass both beams of light through to the detector, which could be a photomultiplier tube that generates an electric current proportional to the photon irradiance. The amplifier increases the detector signal for display.

19-2. An excited state of the lasing material is pumped to a high population by light, an electric discharge, or other means. Photons emitted when the excited state decays to a less populated lower state stimulate emission from other excited molecules. The stimulated emission has the same energy and phase as the incident photon. In the laser cavity, most light is retained by reflective end mirrors. Some light is allowed to escape from one end. Properties of laser light: monochromatic, bright, collimated, polarized, coherent

19-3. Deuterium. Silicon carbide globar.

19-4. Resolution increases in proportion to the number of grooves that are illuminated and to the diffraction order. The number of grooves can be increased with a more finely ruled grating (closer grooves) and with a longer grating. The diffraction order is optimized by appropriate choice of the blaze angle of the grating.

Dispersion is proportional to the diffraction order and inversely proportional to the spacing between lines in the grating. The closer the lines, the greater the dispersion.

The optimum wavelength selected by a grating is the one for which the diffraction condition $n\lambda = d(\sin\theta + \sin\phi)$ is satisfied by specular reflection when the angle of incidence is equal to the angle of reflection ($\alpha = \beta$).

19-5. A filter removes higher order diffraction (different wavelengths) at the same angle as the desired diffraction.

251

19-6.   Advantage - increased ability to resolve closely spaced spectral peaks. Disadvantage - more noise because less light reaches detector.

19-7.   (a) A photomultiplier tube has a photosensitive cathode that emits an electron when struck by a photon. Electrons from the cathode are accelerated by a positive electric potential toward the first dynode. When an accelerated electron strikes the dynode, more electrons are emitted from the dynode. This multiplication process continues through several successive dynodes until ~10^6 electrons are finally collected at the anode for each photon striking the photocathode. The signal is the current measured at the anode.

(b) Each photodiode in a linear array has p-type silicon on a substrate of n-type silicon. Reverse bias draws electrons and holes away from the junction, which is a depletion region with few electrons and holes. The junction acts as a capacitor, with charge on either side of the depletion region. At the beginning of a measurement cycle, each diode is fully charged. Free electrons and holes created when radiation is absorbed in the semiconductor migrate to regions of opposite charge, partially discharging the capacitor. Charge left in each capacitor is measured at the end of a collection cycle by measuring the current needed to recharge each capacitor.

(c) A charge coupled device is made from p-doped Si on an n-doped substrate. The structure is capped with an insulating layer of SiO_2, on top of which is a two-dimensional pattern of conducting Si electrodes. When light is absorbed in the p-doped region, an electron is introduced into the conduction band and a hole is left in the valence band. The electron is attracted to the region beneath the positive electrode, where it is stored. The hole migrates to the n-doped substrate, where it combines with an electron. Each electrode can store ~10^5 electrons. After the desired observation time, electrons stored in each pixel of the top row of the array are moved into a serial register at the top and then moved, one pixel at a time, to the top right position, where the charge is read out. Then the next row is moved up and read out, and the sequence is repeated until the entire array has been read.

19-8.   DTGS has a permanent electric polarization. That is, one face of the crystal has a positive charge and the opposite face has a negative charge. When the temperature of the crystal changes by absorption of infrared light, the polarization (the voltage difference between the two faces) changes. The change in voltage is the detector signal.

19-9. (a) $n\lambda = d(\sin\theta + \sin\phi)$

$1 \cdot 600 \times 10^{-9}$ m $= d(\sin 40° + \sin(-30°)) \Rightarrow d = 4.20 \times 10^{-6}$ m

Lines/cm $= 1/(4.20 \times 10^{-4}$ cm$) = 2.38 \times 10^3$ lines/cm

(b) $\lambda = 1/(1\,000$ cm$^{-1}) = 10^{-3}$ cm $\Rightarrow d = 7.00 \times 10^{-3}$ cm $\Rightarrow 143$ lines/cm

19-10. 10^3 grooves/cm means $d = 10^{-5}$ m $= 10$ μm

$$\text{Dispersion} = \frac{n}{d\cos\phi} = \frac{1}{(10\ \mu m)\cos 10°} = 0.102\ \frac{\text{radians}}{\mu m}$$

$$0.102\ \frac{\text{radians}}{\mu m} \times \frac{180°}{\pi\ \text{radians}} = 5.8°/\mu m$$

19-11. (a) $\text{Resolution} = \dfrac{\lambda}{\Delta\lambda} = \dfrac{512.245}{0.03} = 1.7 \times 10^4$

(b) $\Delta\lambda = \dfrac{\lambda}{10^4} = \dfrac{512.23}{10^4} = 0.05$ nm

(c) $\text{Resolution} = nN = (4)(8.00$ cm $\times 1\,850$ cm$^{-1}) = 5.9 \times 10^4$

(d) 250 lines/mm $= 4$ μm/line $= d$

$$\frac{\Delta\phi}{\Delta\lambda} = \frac{n}{d\cos\phi} = \frac{1}{(4\ \mu m)\cos 3°} = 0.250\ \frac{\text{radians}}{\mu m} = 14.3°/\mu m$$

For $\Delta\lambda = 0.03$ nm, $\Delta\phi = (14.3°/\mu m)(3 \times 10^{-5}\ \mu m) = 4.3 \times 10^{-4}$ degrees.

For 30th order diffraction, the dispersion will be 30 times greater, or $0.013°$.

19-12. (a) True transmittance $= 10^{-1.500} = 0.031\,6$. With 0.50% stray light, the apparent transmittance is

$$\text{Apparent transmittance} = \frac{P+S}{P_0+S} = \frac{0.031\,6 + 0.005\,0}{1 + 0.005\,0} = 0.036\,4$$

The apparent absorbance is $-\log 0.036\,4 = 1.439$.

(b) Apparent absorbance $= 1.999$

Apparent transmittance $= 10^{-1.999} = 0.010\,023\,05$

$$\text{Apparent transmittance} = \frac{P+S}{P_0+S} = \frac{0.010+S}{1+S} = 0.010\,023\,05$$

$\Rightarrow S = 2.328 \times 10^{-5}$ or 0.002 328%

(c) For true absorbance $= 2$,

$$\text{Apparent transmittance} = \frac{P+S}{P_0+S} = \frac{0.010 + 0.000\,000\,5}{1 + 0.000\,000\,5} = 0.010\,000\,49$$

Apparent absorbance is $-\log T = -\log(0.010\ 000\ 49) = 1.999\ 978$

Absorbance error $= 2 - 1.999\ 978 = 0.000\ 022$

For true absorbance $= 3$,

Apparent transmittance $= \dfrac{P + S}{P_0 + S} = \dfrac{0.001 + 0.000\ 000\ 5}{1 + 0.000\ 000\ 5} = 0.001\ 000\ 495$

Apparent absorbance is $-\log T = -\log(0.001\ 000\ 495) = 2.999\ 785$

Absorbance error $= 3 - 2.999\ 785 = 0.000\ 215$

19-13. $b = \dfrac{30}{2 \cdot 1}\left(\dfrac{1}{1\ 906 - 698\ \text{cm}^{-1}}\right) = 0.124\ 2\ \text{mm}$

(Air between the plates has refractive index of 1.)

19-14. $M = \sigma T^4 = [5.669\ 8 \times 10^{-8}\ \text{W}/(m^2 K^4)]T^4$

T(K)	M(W/m^2)
77	1.99
298	447

19-15. (a) $M_\lambda = \dfrac{2\pi hc^2}{\lambda^5}\left(\dfrac{1}{e^{hc/\lambda kT} - 1}\right)$

at $T = 1\ 000$ K:

λ (μm)	M_λ (W/m^3)
2.00	8.79×10^9
10.00	1.164×10^9

(b) $M_\lambda \Delta\lambda = (8.79 \times 10^9\ \text{W/m}^3)(0.02 \times 10^{-6}\ \text{m}) = 1.8 \times 10^2\ \text{W/m}^2$ at 2.00 μm

(c) $M_\lambda \Delta\lambda = (1.164 \times 10^9\ \text{W/m}^3)(0.02 \times 10^{-6}\ \text{m}) = 2.3 \times 10^1\ \text{W/m}^2$ at 10.00 μm

(d) at $T = 100$ K:

λ (μm)	M_λ (W/m^3)
2.00	6.69×10^{-19}
10.00	2.111×10^3

$\dfrac{M_{2.00\ \mu\text{m}}}{M_{10.00\ \mu\text{m}}} = \dfrac{8.79 \times 10^9\ \text{W/m}^3}{1.164 \times 10^9\ \text{W/m}^3} = 7.55$ at 1 000 K

$\dfrac{M_{2.00\ \mu\text{m}}}{M_{10.00\ \mu\text{m}}} = \dfrac{6.69 \times 10^{-19}\ \text{W/m}^3}{2.111 \times 10^3\ \text{W/m}^3} = 3.17 \times 10^{-22}$ at 100 K

At 100 K, there is virtually no emission at 2.00 μm compared to 10.00 μm, whereas at 1 000 K, there is a great deal of emission at both wavelengths.

19-16. $A = \dfrac{L}{c \ln 10}\left(\dfrac{1}{\tau} - \dfrac{1}{\tau_0}\right)$

$= \dfrac{0.210 \text{ m}}{(3.00 \times 10^8 \text{ m/s}) \ln 10}\left(\dfrac{1}{16.06 \times 10^{-6} \text{ s}} - \dfrac{1}{18.52 \times 10^{-6} \text{ s}}\right) = 2.51 \times 10^{-6}$

19-17. $n_1 \sin \theta_1 = n_2 \sin \theta_2$, where $n_1 = 1.50$ and $n_2 = 1.33$

(a) If $\theta_1 = 30°$, $\theta_2 = 34°$

(b) If $\theta_1 = 0°$, $\theta_2 = 0°$ (no refraction)

19-18.   Light inside the fiber strikes the wall at an angle greater than the critical angle for total reflection. Therefore, all light is reflected back into the core and continues to be reflected from wall-to-wall as it moves along the fiber. If the bending angle is not too great, the angle of incidence will still exceed the critical angle and light will not leave the core.

19-19.   When traveling from medium 1 into medium 2, the critical angle for total internal reflection is $\sin \theta_{\text{critical}} = n_2/n_1$, where n_i is the refractive index of medium i. For the solvent/silica interface, $n_1 = 1.50$ and $n_2 = 1.46$, so $\sin \theta_{\text{critical}} = 1.46/1.5$ $= 0.973\ 3$. $\theta_{\text{critical}} = \sin^{-1}(0.973\ 3) = 1.339$ radians from the Excel function ASIN(0.9733). Degrees $= 180 \times \dfrac{\text{radians}}{\pi} = 180 \times \dfrac{1.339}{\pi} = 76.7°$.

For the silica/air interface, $n_1 = 1.45$ and $n_2 = 1.00$, so $\sin \theta_{\text{critical}} = 1.00/1.46 =$ $0.684\ 9$. $\theta_{\text{critical}} = \sin^{-1}(0.684\ 9) = 0.754\ 5$ radians. Degrees $= 180 \times \dfrac{0.754\ 5}{\pi} =$ $43.2°$.

The angle in the photograph is ~55°, which exceeds the critical angle at the silica/air interface but does not exceed the critical angle at the solvent/silica interface. Total internal reflection in the photo must be from the silica/air interface.

19-20.   Light is transmitted through the diamond crystal waveguide by total internal reflection at the upper and lower surfaces. The upper surface is in contact with a fluid channel containing sorbent beads that retain caffeine from a soft drink flowing through the channel. Caffeine in the beads absorbs some of the evanescent wave from the totally internally reflected radiation, decreasing the radiant power transmitted through the waveguide. The integrated area of the absorption spectrum of transmitted power is proportional to the concentration of caffeine in the soft drink.

19-21. Sensitivity increases as the number of reflections inside the waveguide increases, because there is some attenuation at each reflection. For a constant angle of incidence, the number of reflections increases as the thickness of the waveguide decreases.

Three reflections

Six reflections in the same
length when waveguide
thickness is cut in half

19-22. (a) The value of θ_i, called the critical angle (θ_c), is such that $(n_1/n_2)\sin \theta_c = 1$. For $n_1 = 1.52$ and $n_2 = 1.50$, $\theta_c = 80.7°$. That is, θ must be $\geq 80.7°$ for total internal reflection.

(b) $\dfrac{\text{power out}}{\text{power in}} = 10^{-\ell(\text{dB/m})/10} = 10^{-(20.0\ \text{m})(0.010\ 0\ \text{dB/m})/10} = 0.955$

19-23. (a) $n_{\text{core}} \sin \theta_i = n_{\text{cladding}} \sin \theta_r$

For total reflection, $\sin \theta_r \geq 1 \Rightarrow \sin \theta_i \geq \dfrac{n_{\text{cladding}}}{n_{\text{core}}}$

For $n_{\text{cladding}} = 1.400$ and $n_{\text{core}} = 1.600$, $\sin \theta_i \geq \dfrac{1.400}{1.600} \Rightarrow \theta_i \geq 61.04°$

(b) For $n_{\text{cladding}} = 1.400$ and $n_{\text{core}} = 1.800$, $\theta_i \geq 51.06°$

19-24. Angle of incidence = angle of reflection = 45°. Angle of refraction ≡ θ. $n_{\text{prism}} \sin 45° = n_{\text{air}} \sin \theta$. If total reflection occurs, there is no refracted light. This happens if $\sin \theta > 1$, or $\dfrac{n_{\text{prism}} \sin 45°}{n_{\text{air}}} > 1$. Using $n_{\text{air}} = 1$ gives $n_{\text{prism}} > \sqrt{2}$. As long as $n_{\text{prism}} > \sqrt{2}$, no light will be transmitted through the prism and all light will be reflected.

19-25. (a) The Teflon tube acts as an optical fiber because the internal solution has a higher refractive index (1.33) than the walls (1.29). The tube is a 4.5-m-long sample cell that can be conveniently coiled to fit in a reasonable volume and guide the incident radiation all the way through the tube. The long pathlength allows us to obtain a measurable absorbance for a very low concentration of analyte.

(b) $n_{\text{core}} \sin \theta_i = n_{\text{cladding}} \sin \theta_r$

For total reflection, $\sin \theta_r \geq 1 \Rightarrow \sin \theta_i \geq \dfrac{n_{\text{cladding}}}{n_{\text{core}}}$

For $n_{\text{cladding}} = 1.29$ and $n_{\text{core}} = 1.33$, $\sin \theta_i \geq \dfrac{1.29}{1.33} \Rightarrow \theta_i \geq 76°$

(c) $A = \varepsilon bc = (4.5 \times 10^4 \text{ M}^{-1} \text{ cm}^{-1})(450 \text{ cm})(1.0 \times 10^{-9} \text{ M}) = 0.020$

19-26. (a) The following diagram shows the path of a light wave through the waveguide. The length of one bounce, ℓ, satisfies the equation $(0.60 \text{ μm})/\ell$ = $\tan 20°$, giving $\ell = 1.648$ μm. The hypotenuse of the triangle, h, satisfies the equation $(0.60 \text{ μm})/h = \sin 20°$, giving $h = 1.754$ μm. The number of intervals of length ℓ in 3.0 cm is $(3.0 \text{ cm})/(1.648 \text{ μm}) = 1.820 \times 10^4$. Therefore, the pathlength covered by the light is $h \times (1.820 \times 10^4) = 3.19$ cm.

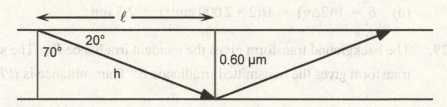

$\dfrac{\text{power out}}{\text{power in}} = 10^{-\ell(\text{dB/m})/10} = 10^{-(3.19 \text{ cm})(0.050 \text{ dB/cm})/10} = 0.964$

(b) Wavelength $= \lambda_0/n$, where λ_0 is the wavelength in vacuum and n is the refractive index. Wavelength $= (514 \text{ nm})/1.5 = 343 \text{ nm}$. The frequency is unchanged from that in vacuum:

$\nu = c/\lambda = (2.997\,9 \times 10^8 \text{ m/s})/(514 \times 10^{-9} \text{ m}) = 5.83 \times 10^{14} \text{ Hz}$

19-27. (a)

λ (μm)	n	λ (μm)	n
0.2	1.550 5	2	1.438 1
0.4	1.470 1	3	1.419 2
0.6	1.458 0	4	1.389 0
0.8	1.453 3	5	1.340 5
1	1.450 4	6	1.258 0

(b) $dn/d\lambda$ is greater for blue light (~ 400 nm) than for red light (~ 600 nm)

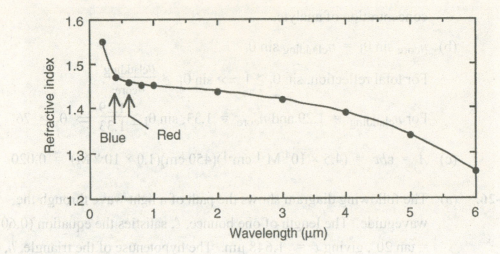

19-28. (a) $\Delta = \pm 2$ cm

(b) Resolution refers to the ability to distinguish closely spaced peaks.

(c) Resolution $\approx 1/\Delta = 0.5$ cm^{-1}

(d) $\delta = 1/(2\Delta\nu) = 1/(2 \times 2\,000$ cm$^{-1}) = 2.5$ µm

19-29. The background transform gives the incident irradiance P_o. The sample transform gives the transmitted irradiance P. Transmittance is P/P_o, not $P-P_o$.

19-30. White noise is independent of frequency. One source is random motion of charge carriers in electronic circuits. $1/f$ noise decreases with increasing frequency. Drift and flicker of the lamp of a spectrophotometer are sources of $1/f$ noise. Line noise results from disturbances at discrete frequencies. The 60 Hz wall frequency is the most common electromagnetic disturbance seen in electronic instruments.

19-31. In beam chopping, the beam is alternately directed through the sample and reference cells. Slow drift of source intensity should be cancelled because the intensity seen by the sample and reference are almost the same. More rapid flicker of the lamp might not be cancelled.

19-32. The difference voltage is ideally very close to zero if the sample and reference are the same because the same lamp intensity goes through each compartment. If the sample absorbs some radiation, the difference voltage should respond to sample absorption with very little noise from source flicker.

19-33. To increase the ratio from 8 to 20 (a factor of $20/8 = 2.5$) requires $2.5^2 = 6.25$
≈ 7 scans.

19-34. (a) $(100 \pm 1) + (100 \pm 1) = 200 \pm \sqrt{2}$, since $e = \sqrt{e_1^2 + e_2^2} = \sqrt{1^2 + 1^2} = \sqrt{2}$

(b) $(100 \pm 1) + (100 \pm 1) + (100 \pm 1) + (100 \pm 1) = 400 \pm 2$, since
$e = \sqrt{1^2 + 1^2 + 1^2 + 1^2} = 2$. The signal-to-noise ratio is $400:2 = 200:1$.

(c) The initial measurement has signal/noise $= 100/1$.
Averaging n measurements gives

$$\text{average signal} = \frac{n \cdot 100}{n} = 100$$

$$\text{average noise} = \frac{\sqrt{n}}{n} = 1/\sqrt{n}$$

$$\frac{\text{average signal}}{\text{average noise}} = \frac{100}{1/\sqrt{n}} = 100\sqrt{n},$$

which is $\sqrt{n}$ times greater than the original value of signal/noise.

19-35. The theoretical signal-to-noise (S/N) ratio should increase in proportion to the
square root of the number of cycles that are averaged.

Number of cycles $= n$	$\sqrt{n}$	Predicted relative S/N ratio
1	1	$\equiv 1$
100	10.00	10.00
300	17.32	17.32
1 000	31.62	31.62

If the observed S/N $= 60.0$ for the average of 1 000 cycles, then the predicted
S/N for the other experiments are shown in the following table:

Number of cycles	Predicted S/N ratio	Observed S/N ratio
1 000	60.0 (observed)	60.0
300	$60.0\left(\frac{17.32}{31.62}\right) = 32.9$	35.9
100	$60.0\left(\frac{10.00}{31.62}\right) = 19.0$	20.9
1	$60.0\left(\frac{1}{31.62}\right) = 1.90$	1.95

20-1. Temperature is more critical in emission spectroscopy, because the small population of the excited state varies substantially as the temperature is changed. The population of the ground state does not vary much.

20-2. Furnaces give increased sensitivity and require smaller sample volumes, but give poorer reproducibility with manual sample introduction. Automated sample introduction gives good precision.

20-3. Drying (~20–100°C) removes water from the sample. Ashing (~100–500°C) is intended to remove as much matrix as possible without evaporating analyte. Atomization (~500–2 000°C) vaporizes analyte (and most of the rest of the sample) for the atomic absorption measurement. For some samples, ashing temperatures might be much higher than 500°C to remove more of the matrix, but it must be demonstrated that the ashing does not remove analyte.

20-4. The plasma operates at higher temperature than a flame and the environment is Ar, not combustion gases. The plasma decreases chemical interference (such as oxide formation) and allows emission instead of absorption to be used. Lamps are not required and simultaneous multi-element analysis is possible. Self-absorption is reduced in the plasma because the temperature is more uniform. Disadvantages of the plasma are increased cost of equipment and operation.

20-5. Doppler broadening occurs because an atom moving toward the radiation source sees a higher frequency than one moving away from the source. Increasing temperature gives increased speeds (more broadening) and increased mass gives decreased speeds (less broadening).

20-6. (a) A beam chopper alternately blocks or exposes the lamp to the flame and detector. When the lamp is blocked, signal is due to background. When the lamp is exposed, signal is due to analyte plus background. The difference is the desired analytical signal.

(b) The flame or furnace is alternately exposed to a D_2 lamp and the hollow-cathode lamp. Absorbance from the D_2 lamp is due to background. Absorbance from the hollow-cathode lamp is due to analyte plus background. The difference is the desired signal.

(c) When a magnetic field parallel to the viewing direction is applied to the furnace, the analytical signal is split into two components that are separated from the analytical wavelength, and one component at the analytical wavelength. The component at the analytical wavelength is not observed because of its polarization. The other two components have the wrong wavelength to be observed. Analyte is essentially "invisible" to the detector when the magnetic field is applied, and only background is seen. Corrected signal is that observed without a field minus that observed with the field.

20-7. Spectral interference refers to the overlap of analyte signal with signals due to other elements or molecules in the sample or with signals due to the flame or furnace. Chemical interference occurs when a component of the sample decreases the extent of atomization of analyte through some chemical reaction. Isobaric interference is the overlap of different species with nearly the same mass-to-charge ratio in a mass spectrum. Ionization interference refers to a loss of analyte atoms through ionization.

20-8. La^{3+} acts as releasing agent by binding tightly to PO_4^{3-} and freeing Pb^{2+}.

20-9. (a) A collision cell guides ions to the entrance of the mass separator and reduces the spread of ion kinetic energies by a factor of 10.

(b) A dynamic reaction cell contains a reactive gas such as NH_3, CH_4, N_2O, CO, or O_2 and its electric field is configured to select lower and upper masses of ions to pass through the cell. Plasma species which interfere with some elements can be reduced by as many as 9 orders of magnitude by reactions such as electron transfer ($^{40}Ar^{16}O^+ + NH_3 \rightarrow NH_3^+ + Ar + O$) and proton transfer ($^{40}ArH^+ + NH_3 \rightarrow NH_4^+ + Ar$). The reactive gas can also be used to shift the analyte signal from a position at which interference occurs (for example, $^{40}Ar^{16}O^+$ interferes with $^{56}Fe^+$) to one where there is no interference ($^{56}Fe^+ + N_2O \rightarrow ^{56}Fe^{16}O^+ + N_2$).

(c) When $^{87}Sr^+$ is converted to $^{87}Sr^{19}F^+$, which has a mass of 106, it no longer overlaps with $^{87}Rb^+$ in the mass spectrum.

20-10. The extent to which an element is ablated, transported to the plasma, and atomized depends on the matrix in which it is found. Different elements in a given matrix might not behave in the same manner. The most reliable calibration

is for the analyte of interest to be measured in the same matrix as the unknown—
if that is possible.

20-11. In the excitation spectrum, we are looking at emission over a band of
wavelengths 1.6 nm wide, while exciting the sample with different narrow bands
(0.03 nm) of laser light. The sample absorbs light only when the laser frequency
coincides with the atomic frequency. Therefore, emission is observed only when
the narrow laser line is in resonance with the atomic levels. In the emission
spectrum, the sample is excited by a fixed laser frequency and then emits
radiation. The monochromator bandwidth is not narrow enough to discriminate
between emission at different wavelengths, so a broad envelope is observed.

20-12. For Pb:

$$\left(104 \pm 17 \frac{\text{pg Pb}}{\text{g snow}}\right)\left(11.5 \frac{\text{g snow}}{\text{cm}^2}\right) = 1\,196 \pm 196 \frac{\text{pg Pb}}{\text{cm}^2}$$

$$\left(1\,196 \pm 196 \frac{\text{pg Pb}}{\text{cm}^2}\right)\left(\frac{1\,\text{ng}}{1\,000\,\text{pg}}\right) = 1.2 \pm 0.2 \frac{\text{ng Pb}}{\text{cm}^2}$$

Similarly, we multiply each of the other concentrations by 11.5 g snow/cm^2 to
find Tl: 0.005 ± 0.001; Cd: 0.04 ± 0.01; Zn: 2.0 ± 0.3;
Al: $7 (\pm 2) \times 10^1$ ng/cm^2.

20-13. $\lambda = \dfrac{hc}{\Delta E} = \dfrac{(6.626 \times 10^{-34}\,\text{J·s})(2.998 \times 10^8\,\text{m/s})}{3.371 \times 10^{-19}\,\text{J}} = 5.893 \times 10^{-7}\,\text{m} = 589.3\,\text{nm}$

20-14. We derive the value for 6 000 K as follows:

$\Delta E = h\nu = \dfrac{hc}{\lambda} = \dfrac{(6.626\,1 \times 10^{-34}\,\text{J · s})(2.997\,9 \times 10^8\,\text{m/s})}{500 \times 10^{-9}\,\text{m}} = 3.97 \times 10^{-19}\,\text{J}$

$\dfrac{N^*}{N_0} = \dfrac{g^*}{g_0}e^{-\Delta E/kT} = \dfrac{g^*}{g_0}e^{-(3.97 \times 10^{-19}\,\text{J})/(1.381 \times 10^{-23}\,\text{J/K})(6\,000\text{K})} = \dfrac{g^*}{g_0}(8.3 \times 10^{-3})$

If $g^*/g_0 = 3$, then $N^*/N_0 = 3(8.3 \times 10^{-3}) = 0.025$.

20-15. Doppler linewidth: $\Delta\lambda = \lambda(7 \times 10^{-7})\sqrt{T/M}$
For $\lambda = 589$ nm, $M = 23$ (sodium) at $T = 2\,000$ K,
  $\Delta\lambda = (589\,\text{nm})(7 \times 10^{-7})\sqrt{(2\,000)/23} = 0.003_8\,\text{nm}$
For $\lambda = 254$ nm, $M = 201$ (mercury) at $T = 2\,000$ K,
  $\Delta\lambda = (254\,\text{nm})(7 \times 10^{-7})\sqrt{(2\,000)/201} = 0.000\,5_6\,\text{nm}$

20-16. (a) $\Delta E = h\nu = \dfrac{hc}{\lambda} = \dfrac{(6.626\ 1 \times 10^{-34}\ J \cdot s)(2.997\ 9 \times 10^8\ m/s)}{422.7 \times 10^{-9}\ m}$

$= 4.699 \times 10^{-19}\ J/molecule = 283.0\ kJ/mol$

(b) $\dfrac{N^*}{N_0} = \dfrac{g^*}{g_0} e^{-\Delta E/kT} = 3 e^{-(4.699 \times 10^{-19}\ J)/(1.381 \times 10^{-23}\ J/K)(2\ 500K)} = 3.67 \times 10^{-6}$

(c) At $2\ 515$ K, $N^*/N_0 = 3.98 \times 10^{-6} \Rightarrow 8.4\%$ increase from $2\ 500$ to $2\ 515$ K

(d) At $6\ 000$ K, $N^*/N_0 = 1.03 \times 10^{-2}$

20-17.

Element:	Na	Cu	Br
Excited state energy (eV):	2.10	3.78	8.04
Wavelength (nm):	591	328	154
Degeneracy ratio (g^*/g_0):	3	3	2/3
N^*/N_0 at $2\ 600$ K in flame:	2.6×10^{-4}	1.4×10^{-7}	1.8×10^{-16}
N^*/N_0 at $6\ 000$ K in plasma:	5.2×10^{-2}	2.0×10^{-3}	1.2×10^{-7}

Calculations: wavelength $= hc/\Delta E$ $\qquad N^*/N_0 = (g^*/g_0)\ e^{-\Delta E/kT}$

Br is not readily observed in atomic absorption, because its lowest excited state requires far-ultraviolet radiation for excitation. Nitrogen and oxygen in the air absorb far-ultraviolet energy and would have to be excluded from the optical path. The excited state lies at such high energy that it is not sufficiently populated to provide adequate intensity for optical emission.

20-18. The dissociation energy of YC is greater than that of BaC, so the equilibrium $BaC + Y \rightleftharpoons Ba + YC$ is driven to the right, increasing the concentration of free Ba atoms in the gas phase.

20-19. Area of pit $= \pi(20 \times 10^{-4}\ cm)^2 = 1.26 \times 10^{-5}\ cm^2$

Power $= 2.4 \times 10^{-3}\ J/10 \times 10^{-9}\ s = 2.4 \times 10^5\ W$

Power density $= 2.4 \times 10^5\ W/1.26 \times 10^{-5}\ cm^2 = 1.9 \times 10^{10}\ W/cm^2 = 20\ GW/cm^2$

Ablated mass $=$ volume $\times$ density $=$ depth $\times$ area $\times$ density $=$

$(1 \times 10^{-4}\ cm)(1.26 \times 10^{-5}\ cm^2)(4\ g/cm^3) = 5\ ng$

20-20. Analyte and standard are lost in equal proportions, so their ratio remains constant.

20-21.

	A	B	C	D
1	Standard Addition Constant Volume Least-Squares			
2	x	y		
3	Volume added (mL)	Absorbance		
4	0.00	0.151		
5	1.00	0.185		
6	3.00	0.247		
7	5.00	0.300		
8	8.00	0.388		
9	10.00	0.445		
10	15.00	0.572		
11	20.00	0.723		
12	B14:C16 = =LINEST(B4:B11,A4:A11,TRUE,TRUE)			
13		LINEST output:		
14	m	0.0282	0.1579	b
15	s_m	0.0003	0.0031	s_b
16	R^2	0.9993	0.0057	s_y
17	x-intercept = -b/m =	-5.599		
18	n =	8	= COUNT(A4:A11)	
19	Mean y =	0.376	= AVERAGE(B4:B11)	
20	$\Sigma(x_i - \text{mean } x)^2 =$	343.5	= DEVSQ(A4:A8)	
21	Std deviation of			
22	x-intercept =	0.1630		
23	B22=(C16/ABS(B14))*SQRT((1/B18) + B19^2/(B14^2*B20))			

The *x*-intercept is -5.60 ± 0.16 mL. Standard [Ca] = 20.0 µg Ca/mL. The intercept corresponds to Ca = $(5.60 \pm 0.16 \text{ mL})(20.0 \text{ µg Ca/mL}) = 112.0 \pm 3.2$ µg. This is the mass of Ca in 5.00 mL of unknown. The total volume of unknown was 100.0 mL, so mass of Ca in total unknown = $(100.0 \text{ mL/5.00 mL})(112.0 \pm 3.2 \text{ µg Ca}) = 2240 \pm 64$ µg Ca.

$$\text{wt\% Ca in cereal} = \frac{100 \times (2240 \pm 64 \text{ µg Ca})}{0.521\ 6 \text{ g cereal}} = 0.429 \pm 0.012 \text{ wt\%}.$$

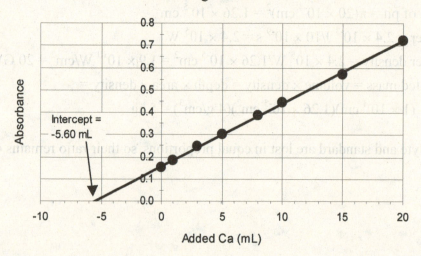

20-22. (a) [S] in unknown mixture = $(8.24 \ \mu g/mL)\left(\dfrac{5.00}{50.0}\right) = 0.824 \ \mu g/mL$

Standard mixture has equal concentrations of X and S:

$$\frac{A_X}{[X]} = F\left(\frac{A_S}{[S]}\right) \Rightarrow \frac{0.930}{[X]} = F\left(\frac{1.000}{[S]}\right) \Rightarrow F = 0.930$$

Unknown mixture:

$$\frac{A_X}{[X]} = F\left(\frac{A_S}{[S]}\right) \Rightarrow \frac{1.690}{[X]} = 0.930\left(\frac{1.000}{[0.824 \ mg/mL]}\right) \Rightarrow [X] = 1.497 \ \mu g/mL$$

But X was diluted by a factor of 10.00/50.0, so the original concentration in the unknown was $(1.49_7 \ \mu g/mL)\left(\dfrac{50.0}{10.00}\right) = 7.49 \ \mu g/mL$.

(b) Standard mixture has equal concentrations of X and S:

$$\frac{A_X}{[X]} = F\left(\frac{A_S}{[S]}\right) \Rightarrow \frac{0.930}{[3.42]} = F\left(\frac{1.000}{[1.00]}\right) \Rightarrow F = 0.271_9$$

Unknown mixture:

$$\frac{A_X}{[X]} = F\left(\frac{A_S}{[S]}\right) \Rightarrow \frac{1.690}{[X]} = 0.271_9\left(\frac{1.000}{[0.824 \ \mu g/mL]}\right) \Rightarrow [X] = 5.12_2$$

$\mu g/mL$

But X was diluted by a factor of 10.00/50.0, so the original concentration in the unknown was $(5.12_2 \ \mu g/mL)\left(\dfrac{50.0}{10.00}\right) = 25.6 \ \mu g/mL$.

20-23.

	A	B	C	D	E	F	G	H	I
1	Least-Squares Spreadsheet								
2		x	y						
3	Highlight cells B10:C12	µg K/mL	signal						
4	Type "= LINEST(C4:C8,	0	0						
5	B4:B8,TRUE,TRUE)	5	124						
6	For PC, press	10	243						
7	CTRL+SHIFT+ENTER	20	486						
8	For Mac, press	30	712						
9	COMMAND+RETURN	LINEST output:							
10	m	23.7672	4.0259	b					
11	s_m	0.2256	3.8090	s_b					
12	R²	0.9997	5.4338	s_y					
13									
14	n =	5	B14 = COUNT(B4:B8)						
15	Mean y =	313	= AVERAGE(C4:C8)						
16	Σ(x_i - mean x)² =	580	= DEVSQ(B4:B8)						
17									
18	Measured y =	417	Input						
19	k = Number of replicate measurements of y =	1	Input						
20	Derived x =	17.3758	= (B18-C10)/B10						
21	s_x =	0.2539	= (C12/B10)*SQRT((1/B19)+(1/B14)+((B18-B15)^2)/(B10^2*B16))						

Cells B20 and B21 give us [unknown] = 17.4 ± 0.3 µg/mL for an emission intensity of 417.

20-24.   (a) CsCl provides Cs atoms which ionize to $Cs^+ + e^-$ in the plasma. Electrons in the plasma inhibit ionization of Sn. Therefore, emission from atomic Sn is not lost to emission from Sn^+.

(b)

	A	B	C	D	E	F
1	Tin in canned food - *Anal. Bioanal. Chem.* **2002**, *374*, 235					
2	Calibration data for 189.927 nm					
3						
4	Conc (µg/L)	Signal intensity			Output from LINEST	
5	0	4.0			slope	intercept
6	10	8.5		Parameter	0.781651	0.863321
7	20	19.6		Std Dev	0.018508	1.556732
8	30	23.6		R^2	0.996648	3.213618
9	40	31.1				
10	60	41.7				
11	100	78.8				
12	200	159.1				
13						
14	Select cells E6:F8					
15	Enter the formula = LINEST(B5:B12,A5:A12,TRUE,TRUE)					
16	CONTROL+SHIFT+ENTER on PC or COMMAND+RETURN on Mac					

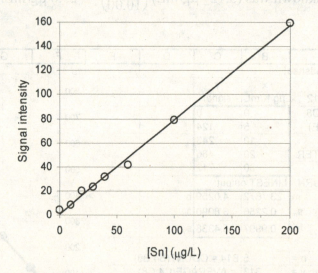

(c) For the 189.927 nm Sn emission line, spike recoveries are all near 100 µg/L, which is near 100%. None of the elements in the table appears to interfere significantly at 189.927 nm. For the 235.485 nm emission line, interference from an emission line of Fe is so serious that the Sn signal cannot be measured. Several other elements interfere enough to reduce the accuracy of the Sn measurement. These elements include Cu, Mn, Zn, Cr, and, perhaps,

Mg. The 189.927 nm line is clearly the better of the two wavelengths for minimizing interference.

(d) Limit of detection = minimum detectable concentration = $3s/m$
where s is the standard deviation of the replicate samples and m is the slope of the calibration curve. Putting in the values $s = 2.4$ units and $m = 0.782$ units per (μg/L) gives

$$\text{limit of detection} = \frac{3s}{m} = \frac{3(2.4 \text{ units})}{0.782 \text{ units}/(\mu g/L)} = 9.2 \ \mu g/L$$

$$\text{limit of quantitation} = \frac{10s}{m} = \frac{10(2.4 \text{ units})}{0.782 \text{ units}/(\mu g/L)} = 30.7 \ \mu g/L$$

It would be reasonable to quote a limit of detection as 9 μg/L and a limit of quantitation as 31 μg/L.

(e) A 2-g food sample ends up in a volume of 50 mL. The limit of quantitation is 30.7 μg Sn/L for the solution. A 50-mL volume with Sn at the limit of quantitation contains (0.050 L)(30.7 μg Sn/L) = 1.54 μg Sn. The quantity of Sn per unit mass of food is

$$\frac{(1.54 \ \mu g \ Sn)(1 \ mg/1 \ 000 \ \mu g)}{(2.0 \ g \ food)(1 \ kg/1 \ 000 \ g)} = 0.77 \ \frac{mg \ Sn}{kg \ food} = 0.77 \ ppm$$

20-25. Standard addition graph: plot signal versus Ti or S concentration.

Ti (ppm)	Signal		S (ppm)	Signal
0.00	0.86		0.0	0.0174
3.00	1.10		37.0	0.0221
6.00	1.34		74.0	0.0268
12.00	1.82		148.0	0.0362

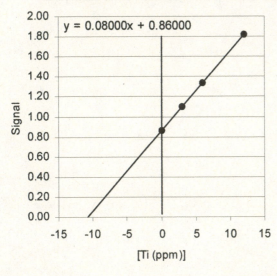

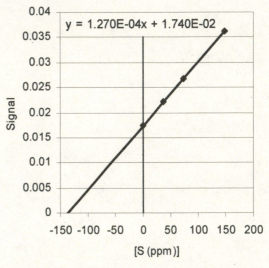

Ti standard addition graph: negative intercept = 0.860/0.0800 = 10.75 mg/L

S standard addition graph: negative intercept = 0.017 4/0.000 127 = 137.0 mg/L

Ti atomic mass = 47.867   S atomic mass = 32.065

[Ti] = (10.75 mg/L)/(47.867 g/mol) = 2.246×10^{-4} M

[S] = (137.0 mg/L)/(32.065 g/mol) = 4.273×10^{-3} M

[Transferrin] = [S]/39 = 1.096×10^{-4} M

Ti/transferrin = $(2.246 \times 10^{-4}$ M$)/(1.096 \times 10^{-4}$ M$) = 2.05$

21-1. Gaseous molecules are ionized by collisions with 70-eV electrons in the ion source. The ions are accelerated out of the source by a voltage, V. All ions have nearly the same kinetic energy ($\frac{1}{2}mv^2$, where m is mass and v is velocity), so the heavier ions have lower velocity. Ions then enter a magnetic field (B) and are deflected so they travel through the arc of a circle whose radius is $(\sqrt{2V(m/z)/e})/B$, where z is the number of charges on the ion and e is the elementary charge. By varying the magnetic field, ions of different m/z are deflected through the slit leading to the detector. At the detector, ion impacts liberate electrons from a cathode. The electrons are amplified by a series of dynodes (as in a photomultiplier tube). The mass spectrum is a graph of detector signal versus m/z.

21-2. For the electron impact spectrum, pentobarbital is bombarded by electrons with an energy of 70 electron volts. The molecular ion ($m/z = 226$) produced by the impact has enough energy to break into fragments and little $M^{+\bullet}$ is observed. Large peaks correspond to the most stable cation fragments. For chemical ionization, pentobarbital reacts with CH_5^+, which is a potent proton donor, but does not have excess kinetic energy. The dominant peak is usually MH^+ ($m/z = 227$). In the case of pentobarbital, some fragmentation is observed even in the chemical ionization spectrum.

21-3. 1 dalton (Da) $\equiv$ 1/12 of the mass of $^{12}C = \left(\dfrac{(1/12) \times 12 \text{ g/mol (exactly)}}{6.022\ 14 \times 10^{23} \text{ mol}^{-1}} \right)$

$= 1.660\ 54 \times 10^{-24}$ g

$[5.03\ (\pm 0.14) \times 10^{10}$ Da$][1.660\ 54 \times 10^{-24}$ g/Da$]$

$= 8.35\ (\pm 0.23) \times 10^{-14}$ g $= 83.5\ (\pm 2.3)$ fg

21-4. The atomic mass in the periodic table is a weighted average of the masses of all the isotopes of that element. We can estimate the relative abundance of the two major isotopes of Ni from the heights of their mass spectral peaks. The heights of the peaks that I measured from an earlier version of this illustration are 42.6 mm for ^{58}Ni and 17.1 mm for ^{60}Ni. The weighted average is atomic mass

$$= (^{58}\text{Ni mass})(\% \text{ abundance of } ^{58}\text{Ni}) + (^{60}\text{Ni mass})(\% \text{ abundance of } ^{60}\text{Ni})$$

$$= (57.935\ 3)\left(\frac{42.6}{42.6 + 17.1}\right) + (59.933\ 2)\left(\frac{17.1}{42.6 + 17.1}\right) = 58.51$$

The atomic mass in the periodic table is 58.69. This main reason for disagreement is that we neglected the existence of ^{61}Ni (1.13% natural abundance), ^{62}Ni (3.59%), and ^{64}Ni (0.90%).

21-5. Resolving power $= \dfrac{m}{m_{1/2}} = \dfrac{2\ 846.3}{0.19} = 1.5 \times 10^4$.

We should be able to barely distinguish two peaks differing by 1 Da at a mass of 1.5×10^4 Da. Therefore, we should be able to distinguish two peaks at 10 000 and 10 001 Da.

21-6. The overlap at the base of the peaks is approximately 10% in the mass spectrum. The resolving power is approximately $m/\Delta m \approx 31/0.010 \approx 3\ 100$.

21-7. Resolving power by 10% valley formula: $m/\Delta m = 906.49/0.000\ 45 = 2.0 \times 10^6$
Resolving power by half-width formula: $m/m_{1/2} = 906.49/0.000\ 27 = 3.4 \times 10^6$
The mass of an electron, 0.000 55 Da, is greater than the mass difference between the two compounds. The mass difference of the compounds is 82% of the mass of one electron.

21-8. $C_5H_7O^+$: $\quad 5 \times 12.000\ 00$

$\qquad\qquad\qquad +7 \times\ \ 1.007\ 825$

$\qquad\qquad\qquad +1 \times 15.994\ 91$

$\quad -e^-\text{ mass} \quad \underline{-1 \times\ \ 0.000\ 55}$

$\qquad\qquad\qquad\qquad 83.049\ 14$

$C_6H_{11}^+$: $\quad 6 \times 12.000\ 00$

$\qquad\qquad\qquad +11 \times\ 1.007\ 825$

$\quad -e^-\text{ mass} \quad \underline{-1 \times\ \ 0.000\ 55}$

$\qquad\qquad\qquad\qquad 83.085\ 52$

$C_6H_{11}^+$ is a closer match than $C_5H_7O^+$ to the observed mass of 83.086 5 Da.

21-9. $^{31}P^+ = ^{31}P - e^- = 30.973\ 76 - 0.000\ 55 = 30.973\ 21$ (observed: 30.973_5)

To measure m/z, I enlarged the figure and sketched a Gaussian curve over each signal by eye. I then measured the position of the center of the peak with a millimeter scale ruler.

$^{15}N^{16}O^+ = ^{15}N + ^{16}O - e^- = 15.000\ 11 + 15.994\ 91 - 0.000\ 55$
$= 30.994\ 47$ (observed: 30.994_6)

$^{14}N^{16}OH^+ = ^{14}N + ^{16}O + ^1H - e^- =$
$14.003\ 07 + 15.994\ 91 + 1.007\ 82 - 0.000\ 55 = 31.005\ 25$

(observed: 31.005_6)

21-10. ^{79}Br abundance $\equiv a = 0.506\ 9$ $\qquad$ ^{81}Br abundance $\equiv b = 0.493\ 1$

Abundance of $C_2H_2^{79}Br_2 = a^2 = 0.256\ 9_5$

Abundance of $C_2H_2^{79}Br^{81}Br = 2ab = 0.499\ 9_0$

Abundance of $C_2H_2^{81}Br_2 = b^2 = 0.243\ 1_5$

Relative abundances: $M^+ : M+1 : M+2 = 1 : 1.946 : 0.946\ 3$

Figure 21-7 shows the stick diagram.

21-11. ^{10}B abundance $\equiv a = 0.199$ $\qquad$ ^{11}B abundance $\equiv b = 0.801$

Abundance of $^{10}B_2H_6 = a^2 = 0.039\ 6_0$

Abundance of $^{10}B^{11}BH_6 = 2ab = 0.318_8$

Abundance of $^{11}B_2H_6 = b^2 = 0.641_6$

Relative abundances: $M^+ : M+1 : M+2 = 1 : 8.05 : 16.20$

21-12. (a)

phenobarbital, $C_{11}H_{18}N_2O_3$

$R + DB = c - h/2 + n/2 + 1$

$R + DB = 11 - 18/2 + 2/2 + 1 = 4$

The molecule has one ring + three double bonds.

(b)

$C_{12}H_{15}BrNPOS$

$R + DB = c - h/2 + n/2 + 1$

$R + DB = 12 - \dfrac{15 + 1}{2} + \dfrac{1 + 1}{2} + 1 = 6$

The molecule has two rings + four double bonds. Note that h includes H + Br, and n includes N + P. S, like O, does not contribute to the count.

(c)

A fragment in a mass spectrum

$C_3H_5^+$

$$R + DB = c - h/2 + n/2 + 1$$

$$R + DB = 3 - 5/2 + 1 = 1\tfrac{1}{2} \quad \text{Huh?}$$

We come out with a fraction instead of an integer because the species is an ion in which one C makes three bonds instead of four.

21-13. (a)

— Cl C_6H_5Cl: $M^{+\bullet} = 112$

The pair of peaks at $m/z = 112$ and 114 strongly suggest that the molecule contains 1 Cl.

$$\text{rings + double bonds} = c - h/2 + n/2 + 1 = 6 - 6/2 + 1 = 4$$
$$\uparrow$$
$$h \text{ includes H + Cl}$$

Expected intensity of M+1 is $1.08(6) + 0.012(5) = 6.54\%$
                              carbon   hydrogen

Observed intensity of M+1 = $69/999 = 6.9\%$

Expected intensity of M+2 = $0.005\ 8(6)(5) + 32.0(1) = 32.2\%$
                              carbon   chlorine

Observed intensity of M+2 = $329/999 = 32.9\%$

The M+3 peak is the isotopic partner of the M+2 peak. M+3 contains ^{37}Cl plus either 1 ^{13}C or 1 ^{2}H. Therefore, the expected intensity of M+3 (relative to M+2) is $1.08(6) + 0.012(5) = 6.54\%$ of predicted intensity of M+2 = $(0.065\ 4)(32.2) = 2.11\%$ of $M^{+\bullet}$.

Observed intensity of M+3 is $21/999 = 2.1\%$.

(b) Cl— — Cl $C_6H_4Cl_2$: $M^{+\bullet} = 146$

The peaks at $m/z = 146$, 148, and 150 look like the isotope pattern from 2 Cl in Figure 21-7.

$$\text{rings + double bonds} = c - h/2 + n/2 + 1 = 6 - 6/2 + 1 = 4$$

Expected intensity of M+1 is $1.08(6) + 0.012(4) = 6.53\%$
                              carbon   hydrogen

Observed intensity of M+1 = $56/999 = 5.6\%$

Expected intensity of M+2 = $0.005\ 8(6)(5) + 32.0(2) = 64.2\%$
                              carbon   chlorine

Observed intensity of M+2 = 624/999 = 62.5%

The M+3 peak is the isotopic partner of the M+2 peak. M+3 contains 1 ^{35}Cl + 1 ^{37}Cl plus either 1 ^{13}C or 1 ^{2}H. Therefore, the expected intensity of M+3 (relative to M+2) is 1.08(6) + 0.012(4) = 6.53% of predicted intensity of M+2 = (0.065 3)(64.2) = 4.19% of $M^{+\bullet}$.

Observed intensity of M+3 is 33/999 = 3.3%.

Expected intensity of M+4 from $C_6H_4{}^{37}Cl_2$ is 5.11(2)(1) = 10.22% of $M^{+\bullet}$. The small contribution from $^{12}C_4{}^{13}C_2H_4{}^{35}Cl^{37}Cl$ is based on the predicted intensity of M+2. It is 0.005 8(6)(5) = 0.174% of 64.2% = 0.11%.

Total expected intensity of M+4 is 10.22% + 0.11% = 10.33% of $M^{+\bullet}$

Observed intensity = 99/999 = 9.9%.

Expected intensity of M+5 from $^{12}C_5{}^{13}CH_4{}^{37}Cl_2$ and $^{12}C_6H_3{}^{2}H^{37}Cl_2$ is based on the predicted intensity of M+4. M+5 should have 1.08(6) + 0.012(4) = 6.53% of M+4 = 6.53% of 10.33% = 0.67%.

Observed intensity = 5/999 = 0.5%.

(c) C_6H_7N: $M^{+\bullet}$ = 93

The peak at m/z = 93 was chosen as the molecular ion, because it is the tallest peak in the cluster and it has plausible isotope peaks at M+1 and M+2. The significant peak at M–1 could be from loss of 1 H. The tiny stuff at M–2 and M–3 could be noise or, possibly, loss of more than 1 H.

With an odd mass, the nitrogen rule tells us that there are an odd number of N atoms in the molecule.

rings + double bonds = $c - h/2 + n/2 + 1$ = 6 – 7/2 + 1/2 + 1 = 4

Expected intensity of M+1 is $\underset{\text{carbon}}{1.08(6)} + \underset{\text{hydrogen}}{0.012(7)} + \underset{\text{nitrogen}}{0.369(1)}$ = 6.93%

Observed intensity of M+1 = 71/999 = 7.1%

Expected intensity of M+2 = 0.005 8(6)(5) = 0.17%

Observed intensity of M+2 = 2/999 = 0.2%

(d) $(CH_3)_2Hg$ C_2H_6Hg: $M^{+\bullet}$ = 228

There are six strong peaks in an unfamiliar pattern. Given that only elements from Table 21-1 are admissible, we notice that Hg has six significant isotopes. By convention, we take the lightest isotope, ^{198}Hg, for the molecular ion at m/z = 228. This leaves just 30 Da for the rest of the molecule, which could be composed of two methyl groups.

In computing rings + double bonds, we include Hg as a Group 6 atom (like O or S) because it makes 2 bonds.

rings + double bonds $= c - h/2 + n/2 + 1 = 2 - 6/2 + 1 = 0$.

The peak at $m/z = 228$ is $M^{+\bullet} = (CH_3)_2{}^{198}Hg$.

Small peaks at $m/z = 227$ and 226 could arise from loss of one or two H atoms. If $(CH_3)_2{}^{198}Hg$ loses H atoms, then all the species at higher mass, such as $(CH_3)_2{}^{199}Hg$, will also lose H atoms. That is, each isotopic molecule is going to contribute some intensity to peaks of lower mass. It makes no sense for us to get too carried away with the analysis of the isotopic pattern, because each peak derives intensity from peaks at lower and higher mass.

The peak at $m/z = 229$ is M+1, composed mainly of $(CH_3)_2{}^{199}Hg$, with some $({}^{12}CH_3)({}^{13}CH_3){}^{198}Hg + {}^{12}C_2H_5{}^{2}H^{198}Hg$. Just considering Hg, the predicted intensity, based on M^+, is $\frac{16.87}{9.97} \times 100 = 169.2\%$ of $M^{+\bullet}$. The observed intensity is $215/130 = 165\%$ of $M^{+\bullet}$. In this calculation, the fraction $\frac{16.87}{9.97}$ is the ratio of the abundances of ^{199}Hg to ^{198}Hg. The peak at $m/z = 230$ is M+2, composed mainly of $(CH_3)_2{}^{200}Hg$. The predicted ^{200}Hg isotopic intensity, based on M^+, is $\frac{23.10}{9.97} \times 100 = 231.7\%$ of $M^{+\bullet}$.

Observed intensity of M+2 $= 291/130 = 224\%$ of $M^{+\bullet}$.

Just considering Hg isotopes, we expect the peaks at M, M+1, M+2, M+3, M+4, and M+6 to have the ratios $9.97 : 16.87 : 23.10 : 13.18 : 29.86 : 6.87 = 1 : 1.69 : 2.32 : 1.32 : 2.99 : 0.69$.

Observed intensity ratio $= 1 : 1.65 : 2.24 : 1.29 : 2.81 : 0.64$.

(e) CH_2Br_2: $M^{+\bullet} = 172$

The three peaks at $m/z = 172$, 174 and 176, with approximate ratios $1 : 2 : 1$ looks like the pattern from 2 Br atoms in Figure 21-7.

rings + double bonds $= c - h/2 + n/2 + 1 = 1 - 4/2 + 1 = 0$
$$\uparrow$$
$$h \text{ includes H + Br}$$

Expected intensity of M+1 is $1.08(1) + 0.012(2) = 1.10\%$
$$\text{carbon} \quad \text{hydrogen}$$

Observed intensity of M+1 $= 12/531 = 2.3\%$. It is possible that this peak at $m/z = 173$ also has contributions from $CH^{79}Br^{81}Br$. We have no way to compute the intensity at $m/z = 173$ if some of this peak comes from

CH^{79}Br^{81}Br. Given this ambiguity, we will just compare the theoretical pattern for 2 Br atoms to the observed pattern:

Theoretical intensity of M+2 = 97.3(2) = 194.6%

Observed intensity of M+2 = 999/531 = 188%

Theoretical intensity of M+4 = 47.3(2)(1) = 94.6%

Observed intensity of M+4 = 497/531 = 93.6%

(f) 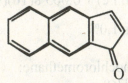 1,10-Phenanthroline, C$_{12}$H$_8$N$_2$: M$^{+\cdot}$ = 180

The strongest peak in the high-mass cluster is at m/z = 180, which could be the molecular ion. It has plausible isotopic peaks at 181 and 182. The significant peak at m/z = 179 could be from loss of 1 H.

The intensity ratio M+1/M$^{+\cdot}$ = 138/999 = 13.8%. We estimate that the number of C atoms is 13.8/1.08 = 12.8.

If the molecule contains 13 C atoms, the formula might be C$_{13}$H$_8$O, which would have 13 − 8/2 + 1 = 10 rings plus double bonds. The expected intensity of M+1 would be 1.08(13) + 0.012 (8) + 0.038(1) = 14.2%. The expected intensity of M+2 would be 0.005 8(13)(12) + 0.205(1) = 1.1%. Observed intensity of M+2 = 9/999 = 0.9%. The formula C$_{13}$H$_8$O fits the data and a conceivable structure is

If the molecule contains 12 C atoms, the formula might be C$_{12}$H$_4$O$_2$, which would have 12 − 4/2 + 1 = 11 rings plus double bonds. A molecule with this many rings + double bonds would be pretty implausible.

If the molecule contains nitrogen, it must contain an even number of N atoms because the molecule has an even mass. A possible formula is C$_{12}$H$_8$N$_2$, which would have 12 − 8/2 + 2/2 + 1 = 10 rings plus double bonds. This turns out to be the correct formula, and the structure is shown at the beginning of this answer. The predicted intensity of M+1 is 1.08(12) + 0.012(8) + 0.369(2) = 13.8%, which is exactly equal to the observed intensity. The expected intensity of M+2 is 0.005 8(12)(11) = 0.8%. Observed intensity = 0.9%.

(g) —Fe— Ferrocene, $C_{10}H_{10}Fe$: $M^{+\cdot} = 186$

The strongest peak at high mass is at $m/z = 186$, which could be the molecular ion. It has plausible isotopic peaks at 187 and 188. Significant peaks at $m/z = 184$ and 185 could be from loss of H. Calling $M^{+\cdot} = 186$, we find the following ratios of peak intensities:

M–2	M–1	$M^{+\cdot}$	M+1	M+2
8.3	1.6	100	13.2	1.0

From the intensity ratio $M+1/M^{+\cdot} = 13.2\%$, we could estimate that the number of C atoms 1s $13.8/1.08 = 12.8$. From this we could propose formulas like $C_{13}H_{14}O$ or $C_{12}H_{10}O_2$.

Alternatively, noting the significant intensity of M–2, we could propose that the molecule has Fe in it, which, in fact, it does. For the formula $C_{10}H_{10}Fe$, we predict that M–2 will have an intensity of $\frac{5.845}{91.754} \times 100 = 6.37\%$ of $M^{+\cdot}$, which is not terribly far from the observed value of 8.3%. The intensity at M+1 will have a contribution from ^{57}Fe and from ^{13}C and ^{2}H. The ^{57}Fe contribution is $2.119/91.754 = 2.31\%$ of $M^{+\cdot}$. The other contributions are $1.08(10) + 0.012(10) = 10.92\%$. The total intensity predicted at M+1 is 13.23% and the observed intensity is 13.2%. The predicted intensity at M+2 is $\frac{0.282}{91.754} \times 100$ (from Fe) $+ 0.005\ 8(10)(9)$ (from C) $= 0.83\%$, and the observed intensity is 1.0%.

21-14. The compound is dibromochloromethane:

212	$CH^{81}Br_2{}^{37}Cl$		94	$CH^{81}Br$
210	$CH^{81}Br_2{}^{35}Cl + CH^{79}Br^{81}Br^{37}Cl$		93	$C^{81}Br$
208	$CH^{79}Br^{81}Br^{35}Cl + CH^{79}Br_2{}^{37}Cl$		92	$CH^{79}Br$
206	$CH^{79}Br_2{}^{35}Cl$		91	$C^{79}Br$
175	$CH^{81}Br_2$		81	^{81}Br
173	$CH^{79}Br^{81}Br$		79	^{79}Br
171	$CH^{79}Br_2$		50	$CH^{37}Cl$
162	$^{81}Br_2$		49	$C^{37}Cl$
160	$^{79}Br^{81}Br$		48	$CH^{35}Cl$
158	$^{79}Br_2$		47	$C^{35}Cl$
131	$CH^{81}Br^{37}Cl$		37	^{37}Cl
129	$CH^{81}Br^{35}Cl + CH^{79}Br^{37}Cl$		35	^{35}Cl
127	$CH^{79}Br^{35}Cl$			

21-15.   The CO_2 that we exhale is derived from oxidation of the food we eat. The chart shows that the group of plants called C_3 plants has less ^{13}C than the groups called C_4 and CAM plants. If the diet in the United States contains more C_4 and CAM plants and the diet in Europe contains more C_3 plants, then the difference in ^{13}C content of exhaled CO_2 might be explained.

21-16.   (a)   Mass of proton + electron = 1.007 276 467 + 0.000 548 580

$\qquad$ =1.007 825 047 Da. To the number of significant digits in Table 1, the masses of the proton and electron are equal to the mass of ^1H.

(b)   mass of proton + neutron + electron

$\qquad$ = 1.007 276 467 + 1.008 664 916 + 0.000 548 580 = 2.016 489 963 Da

mass of ^2H in table = 2.014 10 Da.

The ^2H atom is 0.002 39 Da lighter than the sum of its elementary particles.

(c)   Mass difference = (0.002 39 Da) (1.660 5 × 10^{-27} kg/Da)

$\qquad$ = 3.97 × 10^{-30} kg

$E = mc^2 = $ (3.97 × 10^{-30} kg)(2.997 9 × 10^8 m/s)^2 = 3.57 × 10^{-13} J

mc^2 is the binding energy of a single nucleus. For a mole, the energy is

(3.57 × 10^{-13} J)(6.022 × 10^{23} mol^{-1}) = 2.15 × 10^{11} J/mol = 2.15 × 10^8 kJ/mol.

(d)   Binding energy for atom = (13.6 eV)(1.602 18 × 10^{-19} J/eV) = 2.18 × 10^{-18} J

To convert to a mole: (2.18 × 10^{-18} J)(6.022 × 10^{23} mol^{-1}) = 1.31 × 10^6

J/mol = 1.31 × 10^3 kJ/mol. The ratio of the nuclear binding energy to the electron binding energy is (2.15 × 10^8 kJ/mol)/(1.31 × 10^3 kJ/mol)

$\qquad$ = 1.64 × 10^5.

(e) $\dfrac{\text{nuclear binding energy}}{\text{bond energy}}$ ≈ (2.15 × 10^8 kJ/mol)/(400 kJ/mol) = 5 × 10^5

21-17. ^{79}Br abundance ≡ a = 0.506 9 $\qquad$ ^{81}Br abundance + b = 0.493 1

Abundance of $CH^{79}Br_3 = a^3$ = 0.130 2$_5$

Abundance of $CH^{79}Br_2{}^{81}Br = 3a^2b$ = 0.380 1$_0$

Abundance of $CH^{79}Br^{81}Br_2 = 3ab^2$ = 0.369 7$_5$

Abundance of $CH^{81}Br_3 = b^3$ = 0.119 9$_0$

Relative abundances: M$^+$: M+1 : M+2 : M+3 = 0.342 7 : 1 : 0.972 8 : 0.315 4

21-18. ^{28}Si abundance ≡ a = 0.922 30 $\quad$ ^{29}Si ≡ b = 0.046 83 $\quad$ ^{30}Si ≡ c = 0.030 87

$(a + b + c)^3 = a^3 + 3a^2b + 3a^2c + 3ab^2 + 6abc + 3ac^2 + b^3 + 3b^2c + 3bc^2 + c^3$

	A	B	C	D
1				
2				
3	Silicon			
4	a =	a^3 =	Relative abundance	Composition
5	0.92230	0.784543	1.000000	28Si 28Si 28 Si
6	b =	3a^2b =		(mass = 84)
7	0.04683	0.119506	0.152326	28Si 28Si 29 Si
8	c =	3a^2c =		(mass = 85)
9	0.03087	0.078778	0.100412	28Si 28Si 30 Si
10		3ab^2 =		(mass = 86)
11		0.006068	0.007734	28Si 29Si 29 Si
12		6abc =		(mass = 86)
13		0.008000	0.010197	28Si 29Si 30 Si
14		3ac^2 =		(mass = 87)
15		0.002637	0.003361	28Si 30Si 30 Si
16		b^3 =		(mass = 88)
17		0.000103	0.000131	29Si 29Si 29 Si
18		3b^2c =		(mass = 87)
19		0.000203	0.000259	29Si 29Si 30 Si
20		3bc^2=		(mass = 88)
21		0.000134	0.000171	29Si 30Si 30 Si
22		c^3 =		(mass = 89)
23		2.9418E-05	0.000037	30Si 30Si 30 Si
24				(mass = 90)
25	Check: sum of terms in column B =			
26		1		

mass: 84   85   86   87   88   89   90
intensity: 1   0.152 3   0.108 1   0.010 33   0.003 62   0.000 171   0.000 037

21-19. In a double-focusing mass spectrometer, ions ejected from the source pass through an electrostatic sector that selects ions with a narrow band of kinetic energies to continue into the magnetic sector. The electric sector acts as an energy filter and the magnetic sector acts as a momentum filter.

21-20. From Box 21-2, we know that an ion of $m/z = 500$ accelerated through a potential difference of V volts attains a velocity of $\sqrt{2zeV/m}$. We need to express the mass in kg. The footnote of Table 21-1 gives the conversion factor.

$$500 \text{ Da} \times 1.661 \times 10^{-27} \text{ kg/Da} = 8.30 \times 10^{-25} \text{ kg}$$

$$\text{velocity} = \sqrt{\frac{2zeV}{m}} = \sqrt{\frac{2(1)(1.602 \times 10^{-19} \text{ C})(5.00 \times 10^3 \text{ V})}{8.30 \times 10^{-25} \text{ kg}}}$$

$$= 4.39 \times 10^4 \text{ m/s}$$

To figure out the units, remember that work (joules) = $E \cdot q$ = volts·coulombs. So the product $C \times V = J = m^2kg/s^2$. Putting these units into the square root gives velocity in m/s.

The time needed to travel 2.00 m is (2.00 m)/(4.39 × 10⁴ m/s) = 45.6 μs. If we repeated a cycle each time this heaviest ion reaches the detector, we could collect 1/(45.6 μs) = 2.20 × 10⁴ spectra per second.

If we double the mass in the square root to get up to 1 000 Da, the velocity decreases by $1\sqrt{2}$ and the frequency goes down by $1\sqrt{2}$ to 1.56 × 10⁴ spectra per second.

21-21. The reflectron improves resolving power by ensuring that all ions of the same mass reach the detector grid at the same time. Ions from the ion source have some spread of kinetic energy. Faster ions penetrate deeper into the reflectron and therefore spend more time there before being turned around. The reflectron essentially allows slower ions to catch up to faster ions.

21-22. (a) $\lambda = \dfrac{kT}{(\sqrt{2}\sigma P)} = \dfrac{(1.38 \times 10^{-23} \text{ J/K})(300 \text{ K})}{(\sqrt{2}(\pi(10^{-9} \text{ m})^2)(10^{-5} \text{ Pa}))} = 93 \text{ m}$

(The answer is in meters if you substitute m²·kg·s⁻² for J and kg·m⁻¹·s⁻² for Pa from Table 1-2.)

(b) $\lambda = \dfrac{kT}{(\sqrt{2}\sigma P)} = \dfrac{(1.38 \times 10^{-23} \text{ J/K})(300 \text{ K})}{(\sqrt{2}(\pi(10^{-9} \text{ m})^2)(10^{-8} \text{ Pa}))} = 93 \text{ km}$

21-23. Ions seen in electrospray usually existed in solution prior to electrospray. Atmospheric pressure chemical ionization creates ions in the corona discharge around the high voltage needle.

21-24. In collisionally activated dissociation, ions are accelerated through an electric field and directed into a region with a significant pressure of gas molecules. Collisions transfer enough energy to break molecules into fragments.

Collisionally activated dissociation can be conducted "up front" at the entrance to the mass separator, or in the middle section (the collision cell) in tandem mass spectrometry.

21-25. A reconstructed total ion chromatogram shows the current from all ions above a selected mass displayed as a function of time. The chromatogram is "reconstructed" by summing the intensities for all observed values of m/z. The total ion chromatogram shows everything coming off the column. An extracted ion chromatogram displays detector current for just one or a few values of m/z as a function of time. The intensity displayed is extracted from the full mass spectrum recorded at each time interval. A selected ion chromatogram also displays detector current for just one or a small number of m/z values. However, for a selected ion chromatogram, the detector is not measuring the signal for all values of m/z in each time interval. The detector is set at just the desired values of m/z and collects that information for the whole time. The extracted ion chromatogram and the selected ion chromatogram are selective for an analyte of interest (plus anything else that gives a signal at the same m/z). The selected ion chromatogram has improved signal-to-noise ratio because the most time is spent detecting signal at the selected mass.

21-26.   In selected reaction monitoring, an ion of one m/z value is selected by the first mass separator. This ion is directed to a collision cell in which it undergoes collisionally activated dissociation to produce fragment ions. One of those fragment ions is then selected by a second mass separator and passed through to the detector. The detector is just responding to one product ion from the selected precursor ion. This technique is called MS/MS because it involves two consecutive mass separation steps. The signal/noise ratio is improved because the noise level is very low. There are few sources of the precursor ion other than the desired analyte, and it is very unlikely that other precursor ions of the selected m/z can decompose to give the same product ion selected by the second mass separator.

21-27.   (a) Ibuprofen can readily dissociate to form a carboxylate anion, so I would choose the negative ion mode. It would be harder to form a cation.

   The carboxylate anion should exist in neutral solution, since pK_a is probably around 4. In sufficiently acidic solution, the carboxylate will be protonated. I would use a neutral chromatography solvent to ensure a good supply of analyte anions.

(b) The formula of the molecular ion, M^-, is $C_{13}H_{17}O_2^-$. The intensity expected at M+1 is $1.08(13) + 0.012(17) + 0.038(2) = 14.32$.

        carbon   hydrogen   oxygen

21-28. The analysis follows the same steps as Table 21-3. The work is set out in the following table. Peaks A and B give $n_A = 12$ and peaks H and I give $n_H = 19$. The combination of peaks G and H give $n_G \approx 21$, which makes no sense and will be ignored. Assigning peaks A, B, C... as $n = 12, 13, 14...$ gives the sensible, constant molecular masses in the last column of the table. The mean value, disregarding peak G, is 15 126.

Analysis of electrospray mass spectrum of α-chain of hemoglobin

Peak	Observed m/z $\equiv m_n$	$m_{n+1} - 1.008$	$m_n - m_{n+1}$	Charge $= n =$ $\dfrac{m_{n+1} - 1.008}{m_n - m_{n+1}}$	Molecular mass $= n \times (m_n - 1.008)$
A	1 261.5	1 163.6	96.9	$12.0_1 \approx 12$	15 126
B	1 164.6	—	—	[13]	15 127
C	—	—	—	[14]	—
D	—	—	—	[15]	—
E	—	—	—	[16]	—
F	—	—	—	[17]	—
G	834.3	796.1	37.2	21.4 [18]	14 999
H	797.1	756.2	39.9	$18.9_5 \approx 19$	15 126
I	757.2	—	—	[20]	15 124
				mean =	15 100
				mean without peak G =	15 126

21-29. The separation between adjacent peaks is 0.27, 0.28, 0.25, 0.24, 0.24, 0.24, 0.27, 0.23, 0.24, 0.25, 0.26, and 0.24 m/z units, giving a mean value of 0.25_1. If species differing by 1 Da are separated by 0.25_1 m/z unit, the species must carry 4 charges ($z = 4$). The mass of the tallest peak must be $4(1\ 962.12) = 7\ 848.48$ Da.

21-30. Expected intensities for 37:3, whose formula is $[MNH_4]^+ = C_{37}H_{72}ON$

$X + 1 = 0.012n_H + 1.08n_C + 0.369n_N + 0.038n_O$

    $= 0.012(72) + 1.08(37) + 0.369(1) + 0.038(1) = 41.2\%$ (observed $= 35.8\%$)

$X + 2 = 0.005\ 8n_C(n_C-1) + 0.205n_O$

    $= 0.005\ 8(37)(36) + 0.205(1) = 7.9\%$ (observed $= 7.0\%$)

Expected intensities for 37:3, whose formula is $[MH]^+ = C_{37}H_{69}O$

$X + 1 = 0.012n_H + 1.08n_C + 0.038n_O$

$= 0.012(69) + 1.08(37) + 0.038(1) = 40.8\%$ (observed $= 23.0\%$)

$X + 2 = 0.005\ 8n_C(n_C-1) + 0.205n_O$

$= 0.005\ 8(37)(36) + 0.205(1) = 7.9\%$ (observed $= 8.0\%$)

Expected intensities for 37:2, whose formula is $[MNH_4]^+ = C_{37}H_{74}ON$

$X + 1 = 0.012n_H + 1.08n_C + 0.369n_N + 0.038n_O$

$= 0.012(74) + 1.08(37) + 0.369(1) + 0.038(1) = 41.3\%$ (observed $= 40.8\%$)

$X + 2 = 0.005\ 8n_C(n_C-1) + 0.205n_O$

$= 0.005\ 8(37)(36) + 0.205(1) = 7.9\%$ (observed $= 3.7\%$)

Expected intensities for 37:2, whose formula is $[MH]^+ = C_{37}H_{71}O$

$X + 1 = 0.012n_H + 1.08n_C + 0.369n_N + 0.038n_O$

$= 0.012(71) + 1.08(37) + 0.038(1) = 40.8\%$ (observed $= 33.4\%$)

$X + 2 = 0.005\ 8n_C(n_C - 1) + 0.205n_O$

$= 0.005\ 8(37)(36) + 0.205(1) = 7.9\%$ (observed $= 8.4\%$)

21-31. Selected reaction monitoring chooses the molecular ion ClO_3^- ($m/z = 83$) with the mass separator Q1. In collision cell Q2, this species could possibly undergo the following decomposition:

$$^{35}ClO_3^- \xrightarrow[\text{collisions}]{\text{high energy}} {}^{35}ClO_2^- + {}^{35}ClO^- + {}^{35}Cl^-$$

$m/z = 83 \qquad\qquad m/z = 67 \qquad m/z = 51 \qquad m/z = 35$

Quadrupole Q3 selects only $m/z = 67$. The measurement is specific for ClO_3^- because there are probably few compounds in water producing ions at $m/z = 83$, and *very few* of them are likely to decompose into $m/z = 67$. None of the species ClO_2^-, BrO_3^-, or IO_3^- can produce $m/z = 83$ to be selected by Q1.

21-32. (a) Consider the term $A_xC_xm_x$, which applies to the unknown:

$A_xC_xm_x$

$= \left(\dfrac{\mu\text{mol isotope A}}{\mu\text{mol isotope A} + \mu\text{mol isotope B}}\right)\left(\dfrac{\mu\text{mol V}}{\text{g unknown}}\right)(\text{g unknown})$

$= \left(\dfrac{\mu\text{mol isotope A}}{\mu\text{mol isotope A} + \mu\text{mol isotope B}}\right)(\mu\text{mol V})$

$= \left(\dfrac{\mu\text{mol isotope A}}{\mu\text{mol isotope A} + \mu\text{mol isotope B}}\right)(\mu\text{mol isotope A} + \mu\text{mol isotope B})$

$= \mu\text{mol isotope A in the unknown.}$

Similarly, $B_xC_xm_x = \mu$mol isotope B in the unknown, $A_sC_sms_x = \mu$mol isotope A in the spike, and $B_sC_sm_s = \mu$mol isotope B in the unknown.

When we mix the unknown and the spike, the isotope ratio is

$$R = \frac{\mu mol\ A}{\mu mol\ B} = \frac{\mu mol\ A\ in\ unknown + \mu mol\ A\ in\ spike}{\mu mol\ B\ in\ unknown + \mu mol\ B\ in\ spike}$$

$$= \frac{A_X C_X m_X + A_S C_S m_S}{B_X C_X m_X + B_S C_S m_S}.$$

(b) Cross-multiplying Equation A gives

$$R(B_X C_X m_X + B_S C_S m_S) = A_X C_X m_X + A_S C_S m_S$$

$$RB_X C_X m_X + RB_S C_S m_S = A_X C_X m_X + A_S C_S m_S$$

$$RB_X C_X m_X - A_X C_X m_X = A_S C_S m_S - RB_S C_S m_S$$

$$C_X = \frac{A_S C_S m_S - RB_S C_S m_S}{RB_X m_X - A_X m_X}$$

$$C_X = \left(\frac{C_S m_S}{m_X}\right)\left(\frac{A_S - RB_S}{RB_X - A_X}\right)$$

(c) $A = {}^{51}V$ and $B = {}^{50}V$

Atom fractions in unknown: $A_X = 0.9975$ and $B_X = 0.0025$

Atom fractions in spike: $A_S = 0.6391$ and $B_S = 0.3609$

$$C_X = \left(\frac{C_S m_S}{m_X}\right)\left(\frac{A_S - RB_S}{RB_X - A_X}\right)$$

$$= \left(\frac{(2.2435\ \mu mol\ V/g)(0.41946\ g)}{0.40167\ g}\right)\left(\frac{0.6391 - (10.545)(0.3609)}{(10.545)(0.0025) - 0.9975}\right)$$

$$= 7.6394\ \mu mol\ V/g$$

(d) $C_X = \left(\frac{(2.2435\ \mu mol\ V/g)(0.41946\ g)}{0.40167\ g}\right)\left(\frac{0.6391 - (10.545)(0.3609)}{(10.545)(0.0025 - 0.9975)}\right)$

$$= \left(\frac{(2.2435\ \mu mol\ V/g)(0.41946\ g)}{0.40167\ g}\right)\left(\frac{0.6391 - 3.805_7}{0.026_{36} - 0.9975}\right)$$

$$= (2.3429)\left(\frac{-3.16_6}{-0.971_1}\right)$$

$$= 7.63_9\ \mu mol\ V/g$$

22-1. Three extractions with 100 mL are more effective than one extraction with 300 mL.

22-2. Adjust the pH to 3 so the acid is in its neutral form (CH_3CO_2H), rather than its anionic form ($CH_3CO_2^-$).

22-3. (a) The EDTA complex is anionic (AlY^-), whereas the 8-hydroxyquinoline complex is neutral (AlL_3).

(b) The EDTA complex is anionic (AlY^-), so we need a hydrophobic cation such as $(C_8H_{17})_3NH^+$ to try to bring hydrophobic AlY^- into the organic solvent.

22-4. The complexation reaction $mHL + M^{m+} \rightleftharpoons ML_m + mH^+$ is driven to the right at high pH by consumption of H^+. This consumption increases the fraction of metal in the form ML_m, which is extracted into organic solvent.

22-5. The form that is extracted into organic solvent is ML_n. The formation of ML_n is favored by increasing the formation constant (β). ML_n is also favored by increasing K_a, which increases the fraction of ligand in the form L^-. Increasing K_L decreases the fraction of ligand in the aqueous phase, thereby decreasing the formation of ML_n. Increasing $[H^+]$ decreases the concentration of L^- available for complexation.

22-6. When pH > pK_{BH^+}, the predominant form is B, which is extracted into the organic phase. When pH > pK_a for HA, the predominant form is A^-, which is extracted into the aqueous phase.

22-7. (a) $S_{H_2O} \rightleftharpoons S_{CHCl_3}$ $\qquad K = [S]_{CHCl_3}/[S]_{H_2O} = 4.0$

$[S]_{CHCl_3} = K[S]_{H_2O} = (4.0)(0.020 \text{ M}) = 0.080 \text{ M}$

(b) $\dfrac{\text{mol S in CHCl}_3}{\text{mol S in H}_2\text{O}} = \dfrac{(0.080 \text{ M})(10.0 \text{ mL})}{(0.020 \text{ M})(80.0 \text{ mL})} = 0.50$

22-8. Fraction remaining $= \left(\dfrac{V_1}{V_1+KV_2}\right)^n = \left(\dfrac{80.0}{80.0+(4.0)(10.0)}\right)^6 = 0.088$

22-9. (a) $D = \dfrac{[B]_{C_6H_6}}{[B]_{H_2O}+[BH^+]_{H_2O}}$

(b) D is the quotient of total concentrations in the phases.

   K is the quotient of concentrations of neutral species (B) in the phases.

(c) $D = \dfrac{K \cdot K_a}{K_a + [H^+]} = \dfrac{(50.0)(1.0 \times 10^{-9})}{(1.0 \times 10^{-9}) + (1.0 \times 10^{-8})} = 4.5$

(d) D will be greater at pH 10 because a greater fraction of B is neutral.

22-10.   From Equation 22-12, $D \approx \dfrac{[ML_n]_{org}}{[M^{n+}]_{aq}} = K_{extraction} \dfrac{[HL]_{org}^n}{[H^+]_{aq}^n}$

Comparing this result to Equation 22-13 gives $K_{extraction} = \dfrac{K_M \beta K_a^n}{K_L^n}$

Constant	Effect on $K_{extraction}$	Reason
K_M	increase	ML_n is more soluble in organic phase.
β	increase	Ligand binds metal more tightly and ML_n is the organic-soluble form.
K_a	increase	Ligand dissociates to L^- more easily, increasing ML_n formation.
K_L	decrease	HL is more soluble in organic phase, where it is not available to react with $M^{n+}(aq)$.

22-11.   (a) $D = K[H^+]/([H^+] + K_a) = 3 \cdot 10^{-4.00}/(10^{-4.00} + 1.52 \times 10^{-5}) = 2.60$ at pH 4.00.

   Fraction remaining in water $= q = V_1/(V_1 + DV_2) = 100/[100 + 2.60(25)] = 0.606$. Therefore, the molarity in water is $0.606\,(0.10\text{ M}) = 0.060\,6$ M. The total moles of solute in the system is $(0.100\text{ L})(0.10\text{ M}) = 0.010$ mol. The fraction of solute in benzene is 0.394, so the molarity in benzene is $0.394\,(0.010\text{ mol})/0.025\text{ L} = 0.16$ M.

   (b)   At pH 10.0: $D = 1.97 \times 10^{-5}$, $q = 0.999\,995\,1$, molarity in water $= 0.10$ M, and molarity in benzene $= 2 \times 10^{-6}$ M.

22-12. $D = C/[H^+]^n$, where $C = K_M \beta K_a^n [HL]_{org}^n / K_L^n$

   $D_1 = 0.01 = C/[H^+]_1^2$ and $D_2 = 100 = C/[H^+]_2^2$

   $D_2/D_1 = 10^4 = [H^+]_1^2/[H^+]_2^2 \Rightarrow [H^+]_1/[H^+]_2 = 10^2 \Rightarrow \Delta pH = 2$ pH units

22-13.   (a)   Since there is so much more dithizone than Cu, it is safe to say that $[HL]_{org} = 0.1$ mM.

$$D = \frac{K_M \beta K_a^n}{K_L^n} \frac{[HL]_{org}^n}{[H^+]_{aq}^n} = \frac{(7 \times 10^4)(5 \times 10^{22})(3 \times 10^{-5})^2}{(1.1 \times 10^4)^2} \frac{(1 \times 10^{-4})^2}{[H^+]^2}$$

$$= 2.6 \times 10^4 \text{ at pH 1 and } 2.6 \times 10^{10} \text{ at pH 4}$$

(b) $q = V_1/(V_1+DV_2) = 100/[100 + 2.6 \times 10^4 (10)] = 3.8 \times 10^{-4}$

22-14. (a) $D = \dfrac{[ML_2]_{org}}{[ML_2]_{aq}} = \dfrac{C_{org}V_{org}}{C_{aq}V_{aq}} \Rightarrow C_{org} = D\,C_{aq}\dfrac{V_{aq}}{V_{org}}$

$$\% \text{ extracted} = \frac{100\,C_{org}}{C_{aq} + C_{org}} = \frac{100\,D\,C_{aq}\dfrac{V_{aq}}{V_{org}}}{C_{aq} + D\,C_{aq}\dfrac{V_{aq}}{V_{org}}} = \frac{100\,D\dfrac{V_{aq}}{V_{org}}}{1 + D\dfrac{V_{aq}}{V_{org}}}$$

(b) Spreadsheet for pH dependence of dithizone extraction

	A	B	C	D	E
1	K(M) =	pH	H	D = Dist.coeff	% extracted
2	70000	1	1.00E-01	2.60E-02	0.05
3	Beta =	2	1.00E-02	2.60E+00	4.95
4	5E+18	2.2	6.31E-03	6.54E+00	11.57
5	Ka =	2.4	3.98E-03	1.64E+01	24.73
6	0.00003	2.6	2.51E-03	4.13E+01	45.21
7	K(L) =	2.8	1.58E-03	1.04E+02	67.46
8	11000	3	1.00E-03	2.60E+02	83.89
9	[HL]org =	3.2	6.31E-04	6.54E+02	92.90
10	0.00001	3.4	3.98E-04	1.64E+03	97.05
11	V(org) =	3.6	2.51E-04	4.13E+03	98.80
12	2	3.8	1.58E-04	1.04E+04	99.52
13	V(aq) =	4	1.00E-04	2.60E+04	99.81
14	100	5	1.00E-05	2.60E+06	100.00
15					
16	C2 = 10^-B2				
17	D2 = ($A$2*$A$4*$A$6^2*$A$10^2)/($A$8^2*C2^2)				
18	E2 = (D2*$A$12/$A$14)/(1+(D2*$A$12/$A$14))*100				

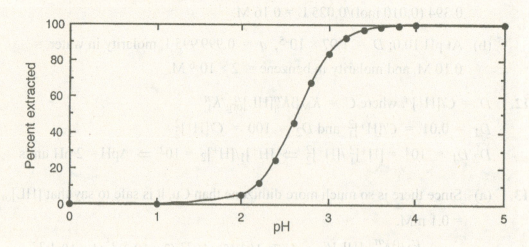

22-15.

	A	B	C	D	E	F
1	Liquid-liquid extraction efficiency					
2						
3	$V_2 =$	50	mL (volume of extraction solvent)			
4	$V_1 =$	50	mL (volume to be extracted)			
5	$K =$	2	(partition coefficient = $[S]_2/[S]_1$)			
6	Divide V_2 into n equal portions for n extractions					
7	Theoretical maximum fraction extracted = $1-q_{limit}$ = $1-exp([V_2/V_1)K]$					
8		$1-q_{limit} =$	0.864665			
9						
10		V_2/n				
11		individual	q =	1-q =	% of limiting	
12		extraction	fraction	fraction	fraction	
13	n	volume	remaining	extracted	extracted	
14	1	50.0	0.333	0.667	77.1	
15	2	25.0	0.250	0.750	86.7	
16	3	16.7	0.216	0.784	90.7	
17	4	12.5	0.198	0.802	92.8	
18	5	10.0	0.186	0.814	94.1	
19	6	8.3	0.178	0.822	95.1	
20	7	7.1	0.172	0.828	95.7	
21	8	6.3	0.168	0.832	96.2	
22	9	5.6	0.164	0.836	96.6	
23	10	5.0	0.162	0.838	97.0	
24	C14 = (\$B\$4/(\$B\$4+B14*\$B\$5))^A14					
25	q = [$V_1/(V_1 + (V_2/n)K$)^n					

The theoretical limit for fraction extracted is in cell C8. 95% of the theoretical fraction extracted is $(0.95)(0.864\ 6) = 0.821\ 4$. This fraction is exceeded with $n = 6$ equal extractions.

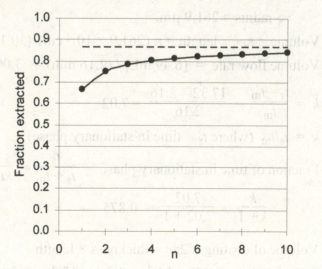

22-16. 1-C, 2-D, 3-A, 4-E, 5-B

22-17. The larger the partition coefficient, the greater the fraction of solute in the stationary phase, and the smaller the fraction that is moving through the column.

22-18. (a) $k = \dfrac{\text{time solute spends in stationary phase}}{\text{time solute spends in mobile phase}} = \dfrac{t_r - t_m}{t_m} = \dfrac{t_s}{t_m}$

(b) Fraction of time in mobile phase $= \dfrac{t_m}{t_m + t_s} = \dfrac{t_m}{t_m + kt_m} = \dfrac{1}{1 + k}$

(c) $R = \dfrac{t_m}{t_r} = \dfrac{t_m}{t_m + t_s} = \dfrac{1}{1 + k}.$ Parts (b) and (c) together tell us that

$$\dfrac{\text{time for solvent to pass through column}}{\text{time for solute to pass through column}} = \dfrac{\text{time spent by solute in mobile phase}}{\text{total time on column}}$$

22-19. (a) Volume per cm of length $= \pi r^2 \times \text{length} = \pi \left(\dfrac{0.461\ \text{cm}}{2}\right)^2 (1\ \text{cm}) = 0.167\ \text{mL}$

mobile phase volume $= (0.390)(0.167\ \text{mL}) = 0.065\ 1\ \text{mL per cm of column}$

linear flow rate $= u_x = \dfrac{1.13\ \text{ml/min}}{0.065\ 1\ \text{mL/cm}} = 17.4\ \text{cm/min}$

(b) $t_m = (10.3\ \text{cm}) / (17.4\ \text{cm/min}) = 0.592\ \text{min}$

(c) $k = \dfrac{t_r - t_m}{t_m} \Rightarrow t_r = kt_m + t_m = 10(0.592) + 0.592 = 6.51\ \text{min}$

22-20. (a) Linear flow rate $= (30.1\ \text{m}) / (2.16\ \text{min}) = 13.9\ \text{m/min}.$

Inner diameter of open tube $= 530\ \mu\text{m} - 2(3.1\ \mu\text{m}) = 523.8\ \mu\text{m}$

$\Rightarrow$ radius $= 261.9\ \mu\text{m}.$

Volume $= \pi r^2 \times \text{length} = \pi (261.9 \times 10^{-4}\ \text{cm})^2 (30.1 \times 10^2\ \text{cm}) = 6.49\ \text{mL}$

Volume flow rate $= (6.49\ \text{mL}) / (2.16\ \text{min}) = 3.00\ \text{mL/min}$

(b) $k = \dfrac{t_r - t_m}{t_m} = \dfrac{17.32 - 2.16}{2.16} = 7.02$

$k = t_s/t_m$ (where t_s = time in stationary phase)

Fraction of time in stationary phase $= \dfrac{t_s}{t_s + t_m} = \dfrac{kt_m}{kt_m + t_m} =$

$\dfrac{k}{k + 1} = \dfrac{7.02}{7.02 + 1} = 0.875$

(c) Volume of coating $\approx 2\pi r \times \text{thickness} \times \text{length}$

$= 2\pi [(261.9 + 1.55) \times 10^{-4}\ \text{cm}](3.1 \times 10^{-4}\ \text{cm})(30.1 \times 10^2\ \text{cm}) = 0.154\ \text{mL}$

$$k = K\frac{V_s}{V_m} \Rightarrow 7.02 = K\frac{0.154\ \text{mL}}{6.49\ \text{mL}} \Rightarrow K = \frac{c_s}{c_m} = 295$$

22-21. (a) $\dfrac{\text{Large load}}{\text{Small load}} = \left(\dfrac{\text{Large column radius}}{\text{Small column radius}}\right)^2$

$\dfrac{100\ \text{mg}}{4.0\ \text{mg}} = \left(\dfrac{\text{Large column diameter}}{0.85\ \text{cm diameter}}\right)^2 \Rightarrow$ large column diameter = 4.25 cm

Use a 40-cm-long column with a diameter near 4.25 cm.

(b) The linear flow rate should be the same. Since the cross-sectional area of the column is increased by a factor of 25, the volume flow rate should be increased by a factor of 25 $\Rightarrow u_v = 5.5$ mL/min.

(c) Volume of small column $= \pi r^2 \times \text{length} = \pi(0.85/2\ \text{cm})^2(40\ \text{cm}) = 22.7$ mL

Mobile phase volume = 35% of column volume = 7.94 mL

Linear flow $= \dfrac{40\ \text{cm}}{(7.94\ \text{mL})/(0.22\ \text{mL/min})} = 1.11$ cm/min for both columns

22-22. (a) $k = \dfrac{9.0 - 3.0}{3.0} = 2.0$

(b) Fraction of time solute is in mobile phase $= \dfrac{t_m}{t_r} = \dfrac{3.0}{9.0} = 0.33$

(c) $K = k\dfrac{V_m}{V_s} = (2.0)\dfrac{V_m}{0.10\ V_m} = 20$

22-23. Solvent volume per cm of column length $= (0.15)\pi\left(\dfrac{0.30\ \text{cm}}{2}\right)^2 = 0.010\,6$ mL/cm.

A volume flow rate of 0.20 mL/min corresponds to a linear flow rate of $\left(\dfrac{0.20\ \text{mL/min}}{0.010\,6\ \text{mL/cm}}\right) = 19$ cm/min.

22-24. $k = K\dfrac{V_s}{V_m} = 3\left(\dfrac{1}{5}\right) = \dfrac{3}{5}$ For $K = 30$, $k = 30\left(\dfrac{1}{5}\right) = 6$.

22-25. $k = \dfrac{V_r'}{V_m} = \dfrac{V_r - V_m}{V_m} = \dfrac{76.2 - 16.6}{16.6} = 3.59$

$K = k\dfrac{V_m}{V_s} = (3.59)\dfrac{16.6}{12.7} = 4.69$

22-26. $K = k\dfrac{V_m}{V_s}$

$k = \dfrac{t_r - t_m}{t_m} = \dfrac{433 - 63}{63} = 5.87$

$$\frac{V_m}{V_s} = \frac{\pi r^2 \times \text{length}}{2\pi r \times \text{thickness} \times \text{length}} = \frac{(103)^2}{2(103.25) \times 0.5} = 102.8$$

(In the numerator, r refers to the radius of the open tube $= \frac{1}{2}(207 - 1.0)$ μm $= 103$ μm. In the denominator, r is the radius at the center of the stationary phase, which is $103 + \frac{1}{2}(0.5) = 103.25$ μm.)

Therefore, the partition coefficient is $K = k\dfrac{V_m}{V_s} = 5.87\,(102.8) = 603$.

$$\text{Fraction of time in stationary phase} = \frac{t_s}{t_s + t_m} = \frac{kt_m}{kt_m + t_m} = \frac{k}{k+1}$$

$$= \frac{5.87}{5.87 + 1} = 0.854$$

22-27. (a) After 10 cycles, the compounds have passed through a length $10L$ containing $10N$ theoretical plates. We are told that $\gamma = 1.018$.

$$\text{resolution} = \frac{\sqrt{N}}{4}(\gamma - 1)$$

$$1.60 = \frac{\sqrt{10N}}{4}(1.018 - 1) \Rightarrow N = 1.2_6 \times 10^4$$

(b) Plate height $= H = L/N = 50 \text{ cm}/1.2_6 \times 10^4 = 4.0 \times 10^{-3} \text{ cm} = 40$ μm

(c) Resolution is proportional to $\sqrt{N}$ or $\sqrt{\text{number of cycles}}$

$$\frac{\text{resolution in 2 cycles}}{\text{resolution in 10 cycles}} = \sqrt{\frac{2}{10}} = 0.447$$

resolution in 2 cycles $= 0.447(\text{resolution in 10 cycles}) = 0.447(1.6) = 0.72$

(observed resolution $= 0.71$)

22-28. (a) Column 1 (sharper peaks)

(b) Column 2 (large plate height means fewer plates means broader peaks)

(c) Column 1 (less overlap between peaks because they are sharper)

(d) Neither (relative retention $(= t_r(B)/T_r(A))$ is equal for the two columns

(e) Compound B (longer retention time)

(f) Compound B (longer retention time means greater affinity for stationary phase)

(g) $\gamma = t_B/t_A = 10/8 = 1.25$

22-29. The linear rate at which solution goes past the stationary phase determines how completely the equilibrium between the two phases is established. This determines the size of the mass transfer term (Cu_x) in the van Deemter equation. The extent of longitudinal diffusion depends on the time spent on the column, which is inversely proportional to linear flow rate.

22-30. Smaller plate height gives less band spreading: 0.1 mm

22-31. Diffusion coefficients of gases are 10^4 times greater than those of liquids. Therefore, longitudinal diffusion occurs much faster in gas chromatography than in liquid chromatography.

22-32. The smaller the particle size, the more rapid is equilibration between mobile and stationary phases.

22-33. Minimum plate height is at 33 mL/min.

22-34. Silanization caps hydroxyl groups to which strong hydrogen bonding can occur.

22-35. Isotherms and band shapes are given in Figure 22-21. In overloading, the solute becomes more soluble in the stationary phase as solute concentration increases. This leaves little solute trailing behind the main band, and gives a non-Gaussian shape. Tailing occurs when small quantities of solute are retained more strongly than large quantities. The beginning of the band is abrupt, but the back part trails off slowly as the tightly bound solute is gradually eluted.

22-36. With 5.0 mg, the column may be overloaded. That is, the quantity of solute per unit length may be too great for the volume of stationary phase. This leads to the upper nonlinear isotherm in Figure 22-21, which broadens bands and decreases resolution.

22-37. Equation 22-26 says that the standard deviation of the band is proportional to $\sqrt{t}$. Here is what we know of the rate of diffusion:

time	standard deviation
t_1	$\sigma_1 = 1$
$t_2 = t_1 + 20$	$\sigma_2 = 2$
$t_3 = t_1 + 40$	$\sigma_3 = ?$

From the bandwidths at times t_1 and t_2, we can write

$$\frac{\sigma_2}{\sigma_1} = \sqrt{\frac{t_2}{t_1}} \Rightarrow \frac{2}{1} = \sqrt{\frac{t_1 + 20}{t_1}} \Rightarrow t_1 = \frac{20}{3} \text{ min}$$

For time t_3: $\dfrac{\sigma_3}{\sigma_1} = \sqrt{\dfrac{t_3}{t_1}} \Rightarrow \dfrac{\sigma_3}{1} = \sqrt{\dfrac{\dfrac{20}{3} + 40}{\dfrac{20}{3}}} \Rightarrow \sigma_3 = 2.65$ mm

22-38.   (a) $N = \dfrac{5.55\, t_r^2}{w_{1/2}^2} = \dfrac{5.55\,(9.0\ \text{min})^2}{(2.0\ \text{min})^2} = 1.1_2 \times 10^2$ plates

        (b) $(10\ \text{cm})/(1.1_2 \times 10^2\ \text{plates}) = 0.89$ mm

22-39.   (a) $N = \dfrac{41.7\,(t_r/w_{0.1})^2}{(A/B) + 1.25} = \dfrac{41.7\,(900\ \text{s}/44\ \text{s})^2}{(33\ \text{s}/11\ \text{s}) + (1.25)} = 4.1 \times 10^3$ plates

        (b)   To use the equation $N = (t_r/\sigma)^2$, we need to find the standard deviation of the

             peak. The width at 1/10 height is $22 + 22 = 44$ s, which we are told is equal

             to 4.297σ. Therefore, $\sigma = (44\ \text{s})/4.297 = 10.24$ s. $N = (t_r/\sigma)^2 =$

             $(900\ \text{s}/10.24\ \text{s})^2 = 7.72 \times 10^3$ plates.

             The equation for an asymmetric peak from (a) gives

             $N = \dfrac{41.7\,(t_r/w_{0.1})^2}{(A/B) + 1.25} = \dfrac{41.7\,(900\ \text{s}/44\ \text{s})^2}{(22\ \text{s}/22\ \text{s}) + (1.25)} = 7.75 \times 10^3$ plates

22-40.   Resolution $= \dfrac{\Delta t_r}{w} = \dfrac{5\ \text{min}}{6\ \text{min}} = 0.83$. This is most like the diagram for a

        resolution of 0.75.

22-41.   Since $w = 4V_r/\sqrt{N}$, w is proportional to V_r (if N is constant).

        $w_2/w_1 = V_2/V_1 = 127/49 \Rightarrow w_2 = (127/49)(4.0) = 10.4$ mL.

22-42. $\sigma_{obs}^2 = \left(\dfrac{w_{1/2}}{2.35}\right)^2 = \left(\dfrac{39.6}{2.35}\right)^2 = 283.96\ \text{s}^2$

        $\Delta t_{injection} = (0.40\ \text{mL})/(0.66\ \text{mL/min}) = 0.606\ \text{min} = 36.36$ s

        $\sigma_{injection}^2 = \dfrac{\Delta t_{injection}^2}{12} = \dfrac{36.36^2}{12} = 110.19\ \text{s}^2$

        $\Delta t_{detector} = (0.25\ \text{mL})/(0.66\ \text{mL/min}) = 22.73$ s

        $\sigma_{detector}^2 = (\Delta t)_{detector}^2/12 = 43.04\ \text{s}^2$

        $\sigma_{obs}^2 = \sigma_{column}^2 + \sigma_{injection}^2 + \sigma_{detector}^2$

        $283.96 = \sigma_{column}^2 + 110.19 + 43.04 \Rightarrow \sigma_{column} = 11.4_3$ s

        $w_{1/2} = 2.35\,\sigma_{column} = 26.9$ s

22-43. $\alpha = \dfrac{t'_{r2}}{t'_{r1}} = \dfrac{k_2}{k_1} = \dfrac{K_2}{K_1} = \dfrac{18}{15} = 1.2_0$

$k_2 = K_2 \dfrac{V_s}{V_m} = 18\left(\dfrac{1}{3.0}\right) = 6.0 \qquad k_1 = 15\left(\dfrac{1}{3.0}\right) = 5.0$

$k_1 = (t_1 - t_m)/t_m = t_1/t_m - 1 \implies t_1 = t_m(k_1 + 1) = (1.0 \text{ min})(5.0 + 1) = 6.0 \text{ min}$

$k_2 = (t_2 - t_m)/t_m \implies t_2 = t_m(k_2 + 1) = (1.0 \text{ min})(6.0 + 1) = 7.0 \text{ min}$

$\gamma = t_2/t_1 = 7.0/6.0 = 1.16_7$

$\text{Resolution} = \dfrac{\sqrt{N}}{4}(\gamma - 1)$

$1.5 = \dfrac{\sqrt{N}}{4}(1.16_7 - 1) \implies 1.3 \times 10^3 \text{ plates}$

22-44. (a) $\gamma = t_2/t_1 = 1.01, 1.05, \text{ or } 1.10$

$\text{Resolution} = \dfrac{\sqrt{N}}{4}(\gamma - 1) \implies N = \left(\dfrac{4 \times \text{resolution}}{\gamma - 1}\right)^2$

resolution	γ	N
2.0	1.01	640 000
2.0	1.05	25 600
2.00	1.10	6 400

(b) For the same kind of column, N can be increased by increasing the column length ($N \propto \sqrt{L}$). γ can be increased by changing solvent and/or stationary phase to change the partition coefficients of the two components.

22-45. (a) C_6HF_5: $t'_r = 12.98 - 1.06 = 11.92$ min. $k = 11.92/1.06 = 11.25$

C_6H_6: $t'_r = 13.20 - 1.06 = 12.14$ min. $k = 12.14/1.06 = 11.45$

(b) $\alpha = 12.14/11.92 = 1.018$

(c) $\gamma = t_2/t_1 = 13.20/12.98 = 1.017$

(d) $w_{1/2} (C_6HF_5) = 0.124$ min; $w_{1/2} (C_6H_6) = 0.121$ min

C_6HF_5: $N = \dfrac{5.55\, t_r^2}{w_{1/2}^2} = \dfrac{5.55\,(12.98)^2}{0.124^2} = 6.08 \times 10^4 \text{ plates}$

$\text{Plate height} = \dfrac{30.0 \text{ m}}{6.08 \times 10^4 \text{ plates}} = 0.493 \text{ mm}$

C_6H_6: $N = \dfrac{5.55\,(13.20)^2}{0.121^2} = 6.60 \times 10^4 \text{ plates}$

$$\text{Plate height} = \frac{30.0 \text{ m}}{6.60 \times 10^4 \text{ plates}} = 0.455 \text{ mm}$$

(e) w (C_6HF_5) = 0.220 min; w (C_6H_6) = 0.239 min

$$C_6HF_5: N = \frac{16 \, t_r^2}{w^2} = \frac{16 \, (12.98)^2}{0.220^2} = 5.57 \times 10^4 \text{ plates}$$

$$C_6H_6: N = \frac{16 \, (13.20)^2}{0.239^2} = 4.88 \times 10^4 \text{ plates}$$

(f) $\text{Resolution} = \dfrac{\Delta t_r}{w_{av}} = \dfrac{13.20 - 12.98}{0.229} = 0.96$

(g) $N = \sqrt{(5.57 \times 10^4)(4.88 \times 10^4)} = 5.21 \times 10^4 \text{ plates}$

$$\text{Resolution} = \frac{\sqrt{N}}{4}(\gamma - 1) = \frac{\sqrt{5.21 \times 10^4}}{4}(1.017 - 1) = 0.97$$

22-46. Initial concentration (m) = 10 nmol/(1.96 × 10⁻³ m²) = 5.09 × 10⁻⁶ mol/m². Diffusion will be symmetric about the origin. Only diffusion in the positive direction is computed below for $t = 60$ s. Other conditions in the graphs are obtained by changing t and the diffusion coefficient D.

	A	B	C
1	Diffusion problem		
2		x (m)	c(mol/m³)
3	moles =	0	4.637E-03
4	1.00E-08	0.0001	4.518E-03
5	diameter (m) =	0.0002	4.178E-03
6	0.05	0.0003	3.668E-03
7	x-sectional area (m²)	0.0004	3.057E-03
8	0.001963495	0.0005	2.418E-03
9	m (mol/m²)=	0.0006	1.816E-03
10	5.093E-06	0.0007	1.294E-03
11	D (m²/s) =	0.0008	8.758E-04
12	1.600E-09	0.0009	5.625E-04
13	t (s) =	0.001	3.430E-04
14	60	0.0012	1.090E-04
15		0.0014	2.815E-05
16		0.0016	5.901E-06
17		0.0018	1.004E-06
18	A10 = A4/A8	0.002	1.388E-07
19			
20	C3 = ($A$10/(SQRT(4*PI()*$A$12*$A$14)))		
21		*EXP(-(B3^2)/(4*$A$12*$A$14))	

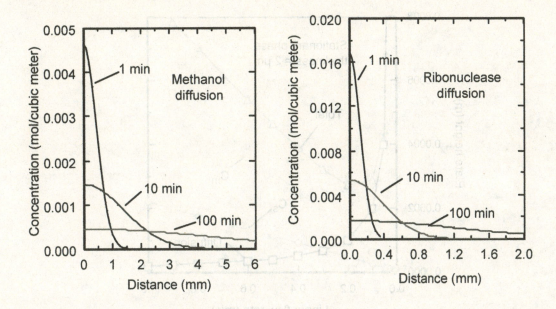

22-47.   Plate height $= H_D + H_{\text{mass transfer}} = \dfrac{B}{u_x} + (C_s + C_m)u_x$

$$= \frac{2D_m}{u_x} + \left(\frac{2kd^2}{3(k+1)^2 D_s} + \frac{1 + 6k + 11k^2 r^2}{24(k+1)^2 D_m} \right) u_x$$

Parameters for 0.25 µm thick stationary phase:

$D_m = 1.0 \times 10^{-5}$ m²/s $D_s = 1.0 \times 10^{-9}$ m²/s

$d = 2.5 \times 10^{-7}$ m $r = 12.5 \times 10^{-4}$ m

$k = 10$

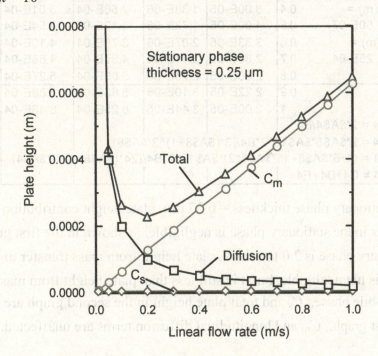

For stationary phase thickness = 0.25 µm, we see that contribution from mass transfer in the stationary phase is negligible, shown in the first graph. If the stationary phase is 5.0 µm thick, contributions from mass transfer in the stationary phase is much greater. Shown with mass transfer in the stationary phase, and overall plate height, in the second graph are greater than in the first graph. C_m and diffusion terms are unaffected.

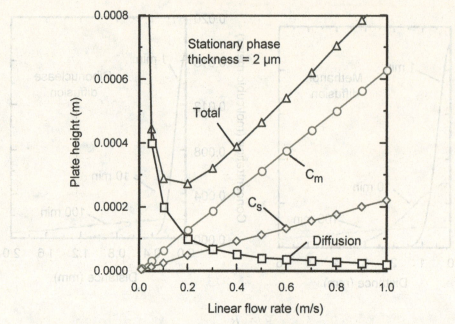

	A	B	C	D	E	F
1	Plate height calculation for 0.25-µm-thick stationary phase					
2				H(mass transfer)		
3	$D_m =$	u_x (m/s)	H(diffusion)	C_s term	C_m term	H (total)
4	0.00001	0.01	2.00E-03	3.44E-08	6.25E-06	2.01E-03
5	$D_s =$	0.05	4.00E-04	1.72E-07	3.12E-05	4.31E-04
6	1E-09	0.1	2.00E-04	3.44E-07	6.25E-05	2.63E-04
7	k =	0.2	1.00E-04	6.89E-07	1.25E-04	2.26E-04
8	10	0.3	6.67E-05	1.03E-06	1.87E-04	2.55E-04
9	d (m) =	0.4	5.00E-05	1.38E-06	2.50E-04	3.01E-04
10	2.50E-07	0.5	4.00E-05	1.72E-06	3.12E-04	3.54E-04
11	r (m) =	0.6	3.33E-05	2.07E-06	3.75E-04	4.10E-04
12	1.25E-04	0.7	2.86E-05	2.41E-06	4.37E-04	4.68E-04
13		0.8	2.50E-05	2.75E-06	5.00E-04	5.27E-04
14		0.9	2.22E-05	3.10E-06	5.62E-04	5.88E-04
15		1	2.00E-05	3.44E-06	6.25E-04	6.48E-04
16	C4 = 2*$A$4/B4					
17	D4 = 2*$A$8*$A$10^2*B4/(3*($A$8+1)^2*$A$6)					
18	E4 = (1+6*$A$8+11*$A$8^2)*$A$12^2*B4/(24*($A$8+1)^2*$A$4)					
19	F4 = C4+D4+E4					

For stationary phase thickness = 0.25 µm, plate height contribution from mass transfer in the stationary phase is negligible, as shown in the first graph. If the stationary phase is 2.0 µm thick, plate height from mass transfer in the stationary phase is not negligible, but it is still less than plate height from mass transfer in the mobile phase. C_s and total plate height in the second graph are greater than in the first graph. C_m and longitudinal diffusion terms are unaffected.

22-48. Inspection of Equation 4-3 shows that the general form of a Gaussian curve is $y = Ae^{-(x-x_0)^2/2\sigma^2}$, where A is a constant proportional to the area under the curve, x_0 is the abscissa of the center of the peak, and σ is the standard deviation. We can arbitrarily let $\sigma = 1$, which means that the width at the base ($w = 4\sigma$) is 4. A peak with an area of 1 centered at the origin is $y = 1*e^{-(x)^2/2}$. A curve of area 4 is $y = 4*e^{-(x-x_0)^2/2}$. The resolution is $\Delta x/w$. For a resolution of 0.5, $\Delta x = 0.5*w = 2$. That is, the second peak is centered at $x = 2$ if the resolution is 0.5. Its equation is $y = 4*e^{-(x-2)^2/2}$. Similarly, for a resolution of 1, $\Delta x = 1*w = 4$ and the second peak is centered at $x = 4$. For a resolution of 2, the second peak is centered at $x = 8$. The equations of the curves plotted below are:

Resolution = 0.5: $y = 1*e^{-(x)^2/2} + 4*e^{-(x-2)^2/2}$
Resolution = 1: $y = 1*e^{-(x)^2/2} + 4*e^{-(x-4)^2/2}$
Resolution = 2: $y = 1*e^{-(x)^2/2} + 4*e^{-(x-8)^2/2}$

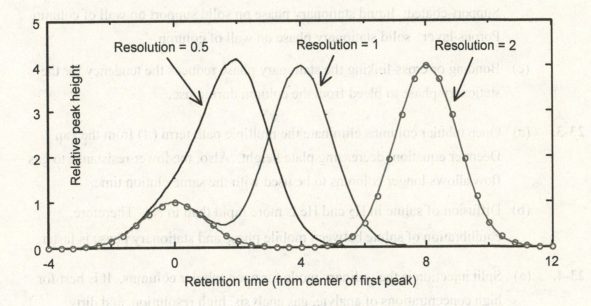

CHAPTER 23
GAS CHROMATOGRAPHY

23-1. (a) Low boiling solutes are separated well at low temperature, and the retention of high boiling solutes is reduced to a reasonable time at high temperature.

(b) Higher pressure gives higher flow rate. If pressure is increased during a separation, retention times of late-eluting peaks are reduced. The effect is the same as increasing temperature, but high temperatures are not required. Pressure programming reduces the likelihood of decomposing thermally sensitive compounds.

23-2. (a) Packed columns offer high sample capacity, while open tubular columns give better separation efficiency (smaller plate height), shorter analysis time, and increased sensitivity to small quantities of analyte.

(b) Wall-coated: liquid stationary phase bonded to the wall of column
Support-coated: liquid stationary phase on solid support on wall of column
Porous-layer: solid stationary phase on wall of column

(c) Bonding or cross-linking the stationary phase reduces the tendency for the stationary phase to bleed from the column during use.

23-3. (a) Open tubular columns eliminate the multiple path term (A) from the van Deemter equation, decreasing plate height. Also, the lower resistance to gas flow allows longer columns to be used with the same elution time.

(b) Diffusion of solute in H_2 and He is more rapid than in N_2. Therefore, equilibration of solute between mobile phase and stationary phase is faster.

23-4. (a) Split injection is the ordinary mode for open tubular columns. It is best for high concentrations of analyte, gas analysis, high resolution, and dirty samples (with an adsorbent packing in the injection liner). Splitless injection is useful for trace analysis (dilute solutions) and for compounds with moderate thermal stability. On-column injection is best for quantitative analysis and for thermally sensitive solutes that might decompose during a high-temperature injection.

(b) In solvent trapping, the initial column temperature is low enough to condense solvent at the beginning of the column. Solute is very soluble in the solvent and is trapped in a narrow band at the start of the column. In cold trapping, the initial column temperature is 150° lower than the boiling points of solutes, which condense in a narrow band at the start of the column. In both cases, elution occurs as the column temperature is raised.

23-5.   (a)   All analytes

(b)   Carbon atoms bearing hydrogen atoms

(c)   Molecules with halogens, CN, NO_2, conjugated C=O

(d)   P and S and other elements selected by wavelength

(e)   P and N (and also hydrocarbons)

(f)   Aromatic and unsaturated compounds

(g)   S

(h)   Most elements (selected individually by wavelength)

(i)   All analytes

23-6.   The thermal conductivity detector measures changes in the thermal conductivity of the gas stream exiting the column. Any substance other than the carrier gas will change the conductivity of the gas stream. Therefore, the detector responds to all analytes. The flame ionization detector burns eluate in an H_2/O_2 flame to create CH radicals from carbon atoms (except carbonyl and carboxyl carbons), which then go on to be ionized to a small extent in the flame: CH + O $\rightarrow$ CHO^+ + e^-. Most other kinds of molecules do not create ions in the flame and are not detected.

23-7.   A *reconstructed total ion chromatogram* is created by summing all ion intensities (above a selected value of m/z) in each mass spectrum at each time interval during a chromatography experiment. The technique responds to essentially everything eluted from the column and has no selectivity at all.
In *selected ion monitoring*, intensities at just one or a few values of m/z are plotted versus elution time. Only species with ions at those m/z values are detected, so the selectivity is much greater than that of the reconstructed total ion chromatogram. The signal-to-noise ratio is increased because ions are collected at each m/z for a longer time than would be allowed if the entire spectrum were being scanned.

Selected reaction monitoring is most selective. One ion from the first mass separator is passed through a collision cell, where it breaks into several product ions that are separated by a second mass separator. The intensities of one or a few of these product ions are plotted as a function of elution time. The selectivity is high because few species from the column produce the first selected ion and even fewer break into the same fragments in the collision cell. This technique is so selective that it can transform a poor chromatographic separation into a highly specific determination of one component with virtually no interference.

23-8.   Column (a): hexane < butanol < benzene < 2-pentanone < heptane < octane

Column (b): hexane < heptane < butanol < benzene < 2-pentanone < octane

Column (c): hexane < heptane < octane < benzene < 2-pentanone < butanol

23-9.   Column (a): 3, 1, 2, 4, 5, 6; Column (b): 3, 4, 1, 2, 5, 6; Column (c): 3, 4, 5, 6, 2, 1

23-10.   (a) $t'_r = 8.4 - 3.7 = 4.7$ min; $k = 4.7/3.7 = 1.3$

(b) $k = KV_s/V_m \Rightarrow K = (1.3)(1.4) = 1.8$

23-11. $I = 100 \left[8 + (9 - 8) \dfrac{\log(12.0) - \log(11.0)}{\log(14.0) - \log(11.0)} \right] = 836$

23-12. $\left. \begin{array}{l} \log(15.0) = \dfrac{a}{373} + b \\[2mm] \log(20.0) = \dfrac{a}{363} + b \end{array} \right\} \Rightarrow a = 1.69 \times 10^3 \text{ K} \qquad b = -3.36$

To solve for a, subtract one equation from the other to eliminate b. Once you have a, substitute it back into either equation and solve for b.

At 353 K: $\log t'_r = \dfrac{1.69_2 \times 10^3}{353} - 3.36 \Rightarrow t'_r = 27.1$ min

23-13.   Derivatization uses a chemical reaction to convert analyte into a form that is more convenient to separate or easier to detect. In Box 23-1, amino and carboxylate groups of amino acids were converted to covalent derivatives to make the molecules volatile enough to be separated by gas chromatography:

Amino acid

H_2N CO_2H R   H

F_3C O   N   H $CO_2CH_2CH_3$ R   H   Volatile derivative

23-14. (a) In solid-phase microextraction, analyte is extracted from a liquid or gas into a thin coating on a silica fiber extended from a syringe. After extraction, the fiber is withdrawn into the syringe. To inject sample into a chromatograph, the metal needle is inserted through the septum and the fiber is extended into the injection port. Analyte slowly evaporates from the fiber in the high-temperature port. Cold trapping is required to condense analyte at the start of the column during slow evaporation from the fiber. If cold trapping were not used, the peaks would be extremely broad because of the slow evaporation from the fiber. During solid-phase microextraction, analyte equilibrates between the unknown and the coating on the fiber. Only a fraction of analyte is extracted into the fiber.

(b) In stir-bar sorptive extraction, a thick coating on the outside of a glass-coated stirring bar is used in place of a thin coating on a fiber. After extraction, the bar is placed in a thermal desorption tube where analyte is vaporized and cold trapped for chromatography. The volume of the coating is ~100 times greater in stir-bar sorptive extraction, so the sensitivity is ~100 times higher.

23-15. The idea of purge and trap is to collect *all* of the analyte from the unknown and to inject *all* of the analyte into the chromatography column. Splitless injection is required so analyte is not lost during injection. Any unknown loss of analyte would lead to an error in quantitative analysis.

23-16. The order of decisions is: (1) goal of the analysis, (2) sample preparation method, (3) detector, (4) column, and (5) injection method.

23-17. (a) A thin stationary phase permits rapid equilibration of analyte between the mobile and stationary phases, which reduces the C term in the van Deemter equation. A thin stationary phase in a narrow-bore column gives small plate height and high resolution. In a wide-bore column, the large diameter of the column slows down the rate of mass transfer between the mobile and stationary phases (because it takes time for analyte to diffuse across the diameter of the column), which defeats the purpose of the thin stationary phase.

(b) Narrow-bore column: plate height = $1/(5\,000\text{ m}^{-1}) = 2.0 \times 10^{-4}$ m = 200 μm. The area of a length (ℓ) of the inside wall of the column is $\pi d\ell$, where d is the column diameter. The volume of stationary phase in this length is $\pi d\ell t$, where t is the thickness of the stationary phase. For d = 250 μm, ℓ = 200 μm, and t = 0.10 μm, the volume is $1.5_7 \times 10^4$ μm³. A density of 1.0 g/mL is 1.0 g/cm³ = 1.0 g/$(10^4$ μm$)^3$ = 1.0 g/10^{12} μm³ = 1 pg/μm³. The mass of stationary phase in one theoretical plate is $(1.5_7 \times 10^4$ μm³)(1 pg/μm³) = $1.5_7 \times 10^4$ pg. 1.0 % of this mass is = 0.16 ng.

Wide-bore column: For d = 530 μm, ℓ = 667 μm, and t = 5.0 μm, the volume is $5.5_5 \times 10^6$ μm³. Mass of stationary phase is $(5.5_5 \times 10^6$ μm³)(1 pg/μm³) = $5.5_5 \times 10^6$ pg. 1.0% of this mass is = 56 ng.

23-18. Use a narrower column or a longer column (doubling the length increases resolution by $\sqrt{2}$) or try a different stationary phase.

23-19. (a) The column on a chip is part of a system intended to be an autonomous environmental monitor. Therefore, it needs to be compact and to require little power and consumables. Air is selected as carrier gas because it can be taken from the atmosphere. Any other carrier gas would require a supply tank which would be heavy, bulky, and would run out of gas. Oxygen from air could degrade the column at elevated temperature. Therefore, the temperature must be kept below the point at which oxidation would occur. Air has impurities which must be removed by a filtration system. The filter is most likely a consumable which eventually needs replacement.

(b) The optimum velocity gives the lowest plate height. It is the minimum in each curve. Optimum velocity = 9.3 cm/s for air and 17.6 cm/s for H_2. Plate height at optimum velocity = 0.036 cm for air and 0.051 for H_2. (Values come from the original publication. You will probably measure somewhat different values from the figure.)

(c) Plates = column length/plate height = 3.0 m/0.036 cm = 8 300 for air and 5 900 for H_2

(d) Time = column length/optimum velocity = 3.0 m/9.3 cm/s = 32 s for air and 17 s for H_2

(e) The two terms describe broadening due to the finite time for solute to diffuse through the stationary phase and the mobile phase. If the stationary phase is sufficiently thin, the time for diffusion through the stationary phase (the C_s term) becomes negligible.

(f) Acceptable flow rates for H_2 are higher than for air because solutes diffuse through H_2 faster than they diffuse through air. With H_2 carrier, solutes can diffuse from the center of the column to the wall more rapidly than they can with air carrier.

23-20. (a) $S = [\text{pentanol}] = \dfrac{234 \text{ mg} / 88.15 \text{ g/mol}}{10.0 \text{ mL}} = 0.265_5 \text{ M}$

$X = [2,3\text{-dimethyl-2-butanol}] = \dfrac{237 \text{ mg} / 102.17 \text{ g/mol}}{10.0 \text{ mL}} = 0.232_0 \text{ M}$

$\dfrac{A_X}{[X]} = F\left(\dfrac{A_S}{[S]}\right) \Rightarrow \dfrac{1.00}{[0.232_0 \text{ M}]} = F\left(\dfrac{0.913}{[0.265_5 \text{ M}]}\right) \Rightarrow F = 1.25_3$

(b) I estimate the areas by measuring the height and $w_{1/2}$ in millimeters. Your answer will be different from mine if the figure size in your book is different from that in my manuscript. However, relative peak areas should be the same.

pentanol: height $= 40.1$ mm; $w_{1/2} = 3.7$ mm;

$\qquad$ area $= 1.064 \times$ peak height $\times w_{1/2} = 15_8$ mm^2

2,3-dimethyl-2-butanol:

$\qquad$ height $= 77.0$ mm; $w_{1/2} = 2.0$ mm; area $= 16_4$ mm^2

(c) $\dfrac{164}{2,3\text{-dimethyl-2-butanol}} = 1.25_3 \left(\dfrac{158}{[93.7 \text{ mM}]}\right)$

$\Rightarrow [2,3\text{-dimethyl-2-butanol}] = 77._6 \text{ mM}$

23-21. $\dfrac{A_X}{[X]} = F\left(\dfrac{A_S}{[S]}\right) \Rightarrow \dfrac{395}{[63 \text{ nM}]} = F\left(\dfrac{787}{[200 \text{ nM}]}\right) \Rightarrow F = 1.59$

The concentration of internal standard mixed with unknown is

$\dfrac{0.100 \text{ mL}}{10.00 \text{ mL}} (1.6 \times 10^{-5} \text{ M}) = 0.16 \text{ μM}$

$\dfrac{633}{[\text{iodoacetone}]} = 1.59 \left(\dfrac{520}{[0.16 \text{ μM}]}\right) \Rightarrow [\text{iodoacetone}] = 0.12_2 \text{ μM}$

$[\text{iodoacetone}]$ in original unknown $= \dfrac{10.00}{3.00} (0.12_2 \text{ μM}) = 0.41 \text{ μM}$

23-22. $I = 100 \left[(7 + (10 - 7) \dfrac{\log (20.0) - \log (12.6)}{\log (22.9) - \log (12.6)}\right] = 932$

23-23. (a) NaCl lowers the solubility of moderately nonpolar compounds, such as ethers, in water. Adding NaCl increases the fraction of the organic compounds that will be transferred to the extraction fiber.

(b) Selected ion monitoring is measuring ion abundance for m/z 73. Only three compounds in the extract have appreciable intensity at m/z 73.

(c) The base peak for both MTBE and TAME is at m/z 73. This mass corresponds to M-15 (loss of CH_3) for MTBE and M-29 (loss of C_2H_5) for TAME. Loss of the ethyl group bound to carbon in TAME suggests that the methyl group lost from MTBE is also bound to carbon, not to oxygen. If methyl bound to oxygen were easily lost from MTBE and TAME, we would expect to see the ethyl group bound to oxygen lost from ETBE. There is no significant peak at M-29 (m/z 73) in ETBE. The following structures are suggested:

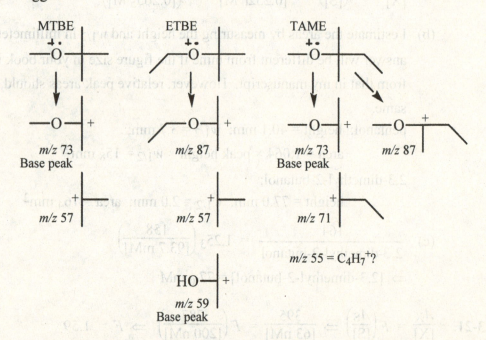

23-24. (a) The vial was heated to increase the vapor pressure of the analyte and the internal standard, so there would be enough in the gas phase (the headspace) to extract a significant quantity with the microextraction fiber.

(b) At 60°C the analyte and internal standard are cold trapped at the beginning of the column. Since desorption from the fiber takes many minutes, we do not want chromatography to begin until desorption is complete.

(c) H^+ pyrrolidine structure with N-CH_3 $C_5H_{10}N^+$, *m/z* 84

For 5-aminoquinoline, *m/z* 144 is the molecular ion, $C_9H_8NC_5H_{10}N_2^+$

(d)

	A	B	C	D	E
1	Least-Squares Spreadsheet				
2					
3		x	y		
4		12	0.056		
5		12	0.059		
6		51	0.402		
7		51	0.391		
8		102	0.684		
9	Highlight cells B16:C18	102	0.669		
10	Type "= LINEST(C4:C13,	157	1.011		
11	B4:B13,TRUE,TRUE)	157	1.063		
12	For PC, press	205	1.278		
13	CTRL+SHIFT+ENTER	205	1.355		
14	For Mac, press				
15	COMMAND+RETURN	LINEST output:			
16	m	0.006401	0.0222	b	
17	s_m	0.000185	0.0234	s_b	
18	R^2	0.9933	0.0409	s_y	
19					
20	n =		10	B20 = COUNT(B4:B13)	
21	Mean y =		0.6968	B21 = AVERAGE(C4:C13)	
22	$\Sigma(x_i - mean\ x)^2$ =		48554.4	B22 = DEVSQ(B4:B13)	
23					
24	Measured y =	1.25	Input		
25	k = Number of replicate measurements of y =	2	Input		
26	Derived x =	191.83	B26 = (B24-C16)/B16		
27	s_x =	5.54			
28	B27 = (C18/B16)*SQRT((1/B25)+(1/B20)+((B24-B21)^2)/(B16^2*B22))				

Least-squares parameters are computed in the block B16:C18. In cell B24, we insert the mean *y* value (1.25) for 2 replicate unknowns. The number of replicates is entered in cell B25. The derived value of *x* is computed in cell B26 and the uncertainty is computed with Equation 4-27 in cell B27.

Answers for the unknowns:

nonsmoker: 78 ± 5 μg/L

nonsmoker with smoking parents: 192 ± 6 μg/L

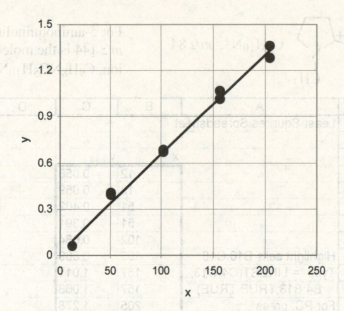

23-25. Nitrite: $[^{14}NO_2^-] = [^{15}NO_2^-](R - R_{blank}) = [80.0 \ \mu M](0.062 - 0.040) = 1.8 \ \mu M$

Nitrate: $[^{14}NO_3^-] = [^{15}NO_3^-](R - R_{blank}) = [800.0 \ \mu M](0.538 - 0.058) = 384 \ \mu M$

23-26. (a) The A term describing multiple flow paths is 0 for an open tubular column. Multiple paths arise in a packed column when liquid takes different paths through the column.

(b) $B = 2D_m$, where D_m is the diffusion coefficient of solute in the mobile phase.

(c) $C = C_s + C_m$

$$C_s = \frac{2k}{3(k+1)^2} \frac{d^2}{D_s} \qquad C_m = \frac{1 + 6k + 11k^2}{24(k+1)^2} \frac{r^2}{D_m}$$

where k = retention factor
   d = thickness of stationary phase
   r = column radius
   D_s = diffusion coefficient of solute in the stationary phase
   D_m = diffusion coefficient of solute in the mobile phase

(d) $H = B/u_x + Cu_x$ (u_x = linear velocity)

Plate height is a minimum at the optimum velocity:

$$\frac{dH}{du_x} = -\frac{B}{u_x^{\,2}} + C = 0 \Rightarrow u_x \text{ (optimum)} = \sqrt{\frac{B}{C}}$$

The minimum plate height is found by plugging this value of u_x (optimum) back into the van Deemter equation:

$$H_{\min} = B/u_x + Cu_x = B\sqrt{\frac{C}{B}} + C\sqrt{\frac{B}{C}} = 2\sqrt{BC} = 2\sqrt{B(C_s + C_m)}$$

$$H_{min} = 2\sqrt{(2D_m)\left(\frac{2k}{3(k+1)^2}\frac{d^2}{D_s} + \frac{1+6k+11k^2}{24(k+1)^2}\frac{r^2}{D_m}\right)}$$

$$H_{min} = 2\sqrt{\frac{4k}{3(k+1)^2}\frac{d^2 D_m}{D_s} + \frac{(1+6k+11k^2)\,2r^2}{24(k+1)^2}}$$

23-27. (a) As $k \to 0$, $H_{min}/r = \sqrt{1/3} = 0.58$

As $k \to \infty$, $H_{min}/r = \sqrt{\frac{1+6k+11k^2}{3(1+k)^2}} \to \sqrt{\frac{11k^2}{3k^2}} = \sqrt{\frac{11}{3}} = 1.9$

(b) As $k \to 0$, $H_{min} = 0.58\,r = 0.058$ mm

As $k \to \infty$, $H_{min} = 1.9\,r = 0.19$ mm

(c) For $k = 5.0$, $H_{min} = r\sqrt{\frac{1+6\cdot5.0+11\cdot25}{3(36)}} = 1.68\,r = 0.168$ mm

Number of plates $= \dfrac{50 \times 10^3 \text{ mm}}{0.168 \text{ mm/plate}} = 3.0 \times 10^5$

(d) $k = KV_s/V_m$, where V_s is the volume of stationary phase and V_m is the volume of mobile phase. For a length of column, ℓ, the volume of mobile phase is $\pi r^2 \ell$ and the volume of stationary phase is $2\pi r t \ell$. Substituting these volumes into the equation for k gives $k = K(2\pi r t \ell)/(\pi r^2 \ell) = 2tK/r$.

$$k = \frac{2\,(0.20\ \mu m)\,(1\,000)}{(100\ \mu m)} = 4.0$$

23-28. The van Deemter equation has the form

$$H = B/u_x + Cu_x = B/u_x + (C_s + C_m)u_x$$

$$B = 2D_m \qquad C_s = \frac{2k}{3(k+1)^2}\frac{d^2}{D_s} \qquad C_m = \frac{1+6k+11k^2}{24(k+1)^2}\frac{r^2}{D_m}$$

where $\qquad k = $ retention factor $= 8.0$

$d = $ thickness of stationary phase $= 3.0 \times 10^{-6}$ m

$r = $ column radius $= 2.65 \times 10^{-4}$ m

$D_s = $ diffusion coefficient of solute in the stationary phase

$D_m = $ diffusion coefficient of solute in the mobile phase

Experimentally, we find $H = (6.0 \times 10^{-5}\text{ m}^2/\text{s})/u_x + (2.09 \times 10^{-3}\text{ s})u_x$.

Therefore, $B = 2D_m = (6.0 \times 10^{-5}\text{ m}^2/\text{s})$, or $D_m = 3.0 \times 10^{-5}\text{ m}^2/\text{s}$.

From the second term of the experimental van Deemter equation, we know that

$$C_s + C_m = 2.09 \times 10^{-3}\text{ s} = \frac{2k}{3(k+1)^2}\frac{d^2}{D_s} + \frac{1+6k+11k^2}{24(k+1)^2}\frac{r^2}{D_m}$$

Inserting the known values of all parameters allows us to solve for D_s:

$$2.09 \times 10^{-3}\text{ s} =$$

$$\frac{2(8.0)}{3((8.0) + 1)^2} \quad \frac{(3.0 \times 10^{-6} \text{ m})^2}{D_s} + \frac{1 + 6(8.0) + 11(8.0)^2}{24((8.0) + 1)^2} \quad \frac{(2.65 \times 10^{-4} \text{ m})^2}{(3.0 \times 10^{-5} \text{ m}^2/\text{s})}$$

$$\Rightarrow D_s = 5.0 \times 10^{-10} \text{ m}^2/\text{s}$$

The diffusion coefficient in the mobile phase is $(3.0 \times 10^{-5} \text{ m}^2/\text{s})/(5.0 \times 10^{-10}$ $\text{m}^2/\text{s}) = 6.0 \times 10^4$ times greater than the diffusion coefficient in the stationary phase. This makes sense, because it is easier for solute to diffuse through He gas than through a viscous liquid phase.

23-29.   (a)

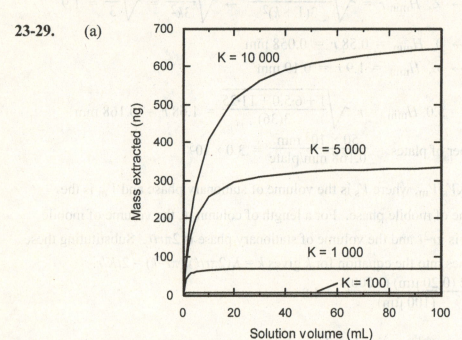

Mass of analyte extracted by solid phase microextraction			
K = partition	V_s (mL)	m (ng)	
coefficient =	0	0.000	
1.00E+02	1	6.455	
V_f = volume of	2	6.670	
film (mL) =	3	6.745	
6.90E-04	4	6.783	
C_o = initial	5	6.806	
concentration	10	6.853	
in solution	50	6.890	
(µg/mL) =	100	6.895	
0.1	1000	6.900	
C4 = 1000*($A$5*$A$8*$A$13*B4)/($A$5*$A$8+B4)			

(b) $m = \dfrac{KV_f c_o V_s}{KV_f + V_s}$ If $V_s \gg KV_f$, $m = KV_f c_o$

For $V_f = 6.9 \times 10^{-4}$ mL and $c_0 = 0.1$ µg/mL, $m \rightarrow (6.9 \times 10^{-5})(K)$ µg

For $K = 100$, $m \rightarrow 6.9$ ng, which agrees with the graph.

For $K = 10\,000$, $m \rightarrow 690$ ng, which is where the graph is heading, but it will require about 1 liter of solution to attain the limiting concentration in the fiber.

(c) The spreadsheet tells us that when $K = 100$, 6.85 ng have been extracted into the fiber and when $K = 10\,000$, 408 ng have been extracted into the fiber. The total analyte in 10.0 mL is (0.10 µg/mL)(10.0 mL) = 1.0 µg. The fraction extracted for $K = 100$ is 6.86 ng/1.0 µg = 0.006 9 (or 0.69%). The fraction extracted for $K = 10\,000$ is 0.41 (or 41%).

23-30. (a) For the formula $C_9H_4N_2Cl_6$,

rings + double bonds = $c - h/2 + n/2 + 1 = 9 - (4+6)/2 + 2/2 + 1 = 6$,

which agrees with the structure that has 2 rings + 4 double bonds.

(b) Nominal mass = integer mass of the species with the most abundant isotope of each of the constituent atoms. For $C_9H_4N_2Cl_6$, nominal mass = $(9 \times 12) + (4 \times 1) + (2 \times 14) + (6 \times 35) = 350$.

(c) The sequence m/z 350, 315, 280, 245, and 210 corresponds to successive losses of mass 35 Da. A logical assignment is $C_9H_4N_2Cl_6^+$, $C_9H_4N_2Cl_5^+$, $C_9H_4N_2Cl_4^+$, $C_9H_4N_2Cl_3^+$, $C_9H_4N_2Cl_2^+$.

(d) Here is the spreadsheet for 5 Cl atoms:

	A	B	C	D	E
1	Isotopic abundance from binomial distribution				
2					
3	$^{35}Cl =$	0.7577	natural abundance		
4	$^{37}Cl =$	0.2423	natural abundance		
5	n =	5			
6					Relative
7	^{35}Cl	Formula	Mass	Abundance	abundance
8	5	$^{35}Cl_5{}^{37}Cl_0$	M	0.24974	62.54
9	4	$^{35}Cl_4{}^{37}Cl_1$	M+2	0.39931	100.00
10	3	$^{35}Cl_3{}^{37}Cl_2$	M+4	0.25539	63.96
11	2	$^{35}Cl_2{}^{37}Cl_3$	M+6	0.08167	20.45
12	1	$^{35}Cl_1{}^{37}Cl_4$	M+8	0.01306	3.27
13	0	$^{35}Cl_0{}^{37}Cl_5$	M+10	0.00084	0.21
14	D8 = BINOMDIST(A8,$B$5,$B$3,FALSE)				
15	E8 = 100*D8/MAX($D$8:$D$13)				

And here are the results for species with 6, 5, 4, 3, 2, and 1 Cl atom. The predicted patterns are in reasonable agreement with the observed amplitudes of the clusters of peaks at m/z 350, 315, 280, 245, and 210.

	Predicted relative abundance					
	Cl_6	Cl_5	Cl_4	Cl_3	Cl_2	Cl_1
M	52.12	62.54	78.18	100.00	100.00	100.00
M+2	100.00	100.00	100.00	95.94	63.96	31.98
M+4	79.95	63.96	47.97	30.68	10.23	
M+6	34.09	20.45	10.23	3.27		
M+8	8.18	3.27	0.82			
M+10	1.05	0.21				
M+12	0.06					

CHAPTER 24
HIGH-PERFORMANCE LIQUID CHROMATOGRAPHY

24-1. (a) In reversed-phase chromatography, the solutes are nonpolar and more soluble in a nonpolar mobile phase. In normal-phase chromatography, the solutes are polar and more soluble in a polar mobile phase.

(b) A gradient of increasing pressure gives increasing solvent density, which gives increasing eluent strength in supercritical fluid chromatography.

24-2. Solvent is competing with solute for adsorption sites. The strength of the solvent-adsorbent interaction is independent of solute.

24-3. In hydrophilic interaction chromatography, solute equilibrates between the mobile phase and an aqueous layer on the surface of the polar stationary phase. The more water in the eluent, the better can eluent compete with the stationary aqueous layer to dissolve polar solute and elute it from the column.

24-4. (a) Small particles give increased resistance to flow. High pressure is required to obtain a usable flow rate.

(b) A bonded stationary phase is covalently attached to the support.

24-5. (a) $L(\text{cm}) \approx \dfrac{N d_{\mathrm{p}}(\mu\text{m})}{3\,000}$

If $N = 1.0 \times 10^4$ and $d_{\mathrm{p}} = 10.0\ \mu\text{m}$, $L = 33$ cm

$d_{\mathrm{p}} = 5.0\ \mu\text{m} \Rightarrow L = 17$ cm; $\qquad d_{\mathrm{p}} = 3.0\ \mu\text{m} \Rightarrow L = 10$ cm

$d_{\mathrm{p}} = 1.5\ \mu\text{m} \Rightarrow L = 5$ cm

(b) Efficiency increases because solute equilibrates between phases more rapidly if the thicknesses of both phases are smaller. This effect decreases the C term in the van Deemter equation. Also, migration paths between small particles are more uniform, decreasing the multiple path (A) term.

24-6. Plates $(N) = (15\ \text{cm})/(5.0 \times 10^{-4}\ \text{cm/plate}) = 3.0 \times 10^4$

$N = \dfrac{5.55\, t_{\mathrm{r}}^2}{w_{1/2}^2} \Rightarrow w_{1/2} = t_{\mathrm{r}} \sqrt{\dfrac{5.55}{N}} = (10.0\ \text{min})\sqrt{\dfrac{5.55}{3.0 \times 10^4}} = 0.13_6\ \text{min}$

If plate height = 25 μm, plates = 6 000 and $w_{1/2} = 0.30_4$ min

311

24-7. Silica dissolves above pH 8 and the siloxane bond to the stationary phase hydrolyzes below pH 2. Bulky isobutyl groups hinder the approach of H_3O^+ to the Si–O–Si bond, so the rate of acid-catalyzed hydrolysis is decreased.

24-8. The high concentration of additive binds to the sites on the stationary phase that would otherwise hold on tightly to solutes and cause tailing.

24-9. (a) Your sketch should look like Figure 22-14, in which the asymmetry factor is $A/B = 1.8$, measured at one tenth of the peak height.

(b) Tailing of amines might be eliminated by adding 30 mM triethylamine to the mobile phase. Tailing of acidic compounds might be eliminated by adding 30 mM ammonium acetate. For unknown mixtures, 30 mM triethylammonium acetate is useful. If tailing persists, 10 mM dimethyloctylamine or dimethyloctylammonium acetate might be effective. Tailing could also be caused by a clogged frit which you might be able to clean by washing with reversed flow.

24-10. (a)

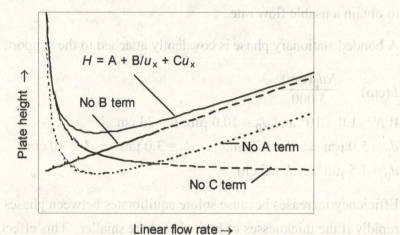

(b) For 1.8-μm particle size, the experimental van Deemter curve looks almost like the curve with no C term in (a) (that is, finite equilibration time ≈ 0). When particle size is small enough, equilibration between the mobile and stationary phases is very rapid and this process contributes little to peak broadening. The experimental curve for 1.8-μm particles levels off at a smaller plate height than the curves for 5- and 3.5-μm particles. This behavior suggests that the A term (multiple flow paths) is smaller for the smaller particles.

(c) A superficially porous particle has a thin porous shell on a solid inner core. Solute only needs to diffuse short distances into the thin shell, so equilibration occurs on a time scale similar to that of smaller particles. However, the overall diameter of the superficially porous particle is not small, so its resistance to fluid flow is not as high as that of a small particle.

24-11. (a) $N = \dfrac{5.55\, t_r^2}{w_{1/2}^2} = \dfrac{5.55\,(4.70\ \text{min})^2}{(0.28\ \text{min})^2} = 1\,560$ for l enantiomer

$N = \dfrac{5.55\,(5.37\ \text{min})^2}{(0.35\ \text{min})^2} = 1\,310$ for d enantiomer

(b) $w_{1/2\text{av}} = \tfrac{1}{2}(0.28\ \text{min} + 0.35\ \text{min}) = 0.31_5\ \text{min}$

$\text{Resolution} = \dfrac{0.589\Delta t_r}{w_{1/2\text{av}}} = \dfrac{0.589\,(5.37\ \text{min} - 4.70\ \text{min})}{0.315\ \text{min}} = 1.25$

(c) Unadjusted relative retention: $\gamma = (5.37\ \text{min})/(4.70\ \text{min}) = 1.14_3$

Average $N = \tfrac{1}{2}(1\,560 + 1\,310) = 1\,435$

$\text{Resolution} = \dfrac{\sqrt{N}}{4}(\gamma - 1) = \dfrac{\sqrt{1\,435}}{4}(1.14_3 - 1) = 1.35$

24-12. (a) $P \propto \dfrac{1}{d_p^2}$ For two difference conditions (1 and 2),

$\dfrac{P_2}{P_1} \propto \dfrac{d_1^2}{d_2^2} = \left(\dfrac{3\ \mu\text{m}}{0.7\ \mu\text{m}}\right)^2 = 18.$ Pressure must be 18 times greater.

(b) $u_x \propto P$, so if pressure is increased by a factor of 10, then linear velocity should increase by a factor of 10.

(c) Mass transfer between the mobile and stationary phase is faster for small particles than for large particles. The optimum velocity for maximum efficiency (highest plate number) increases as the rate of mass transfer increases. In the example cited, the high flow rate is closer to the optimum flow rate than is the low flow rate.

24-13. (a) Bonded reversed-phase chromatography

(b) Bonded normal-phase chromatography (Dioxane is closer to ethyl acetate than to chloroform in eluent strength.)

(c) Ion-exchange or ion chromatography

(d) Molecular-exclusion chromatography

(e) Ion-exchange chromatography

(f) Molecular-exclusion chromatography

24-14. 10-µm-diameter spheres: volume $= \frac{4}{3}\pi r^3 = \frac{4}{3}\pi(5 \times 10^{-4}\text{ cm})^3 = 5.24 \times 10^{-10}\text{ cm}^3$

Mass of one sphere $= (5.24 \times 10^{-10}\text{ mL})(2.2\text{ g/mL}) = 1.15 \times 10^{-9}\text{ g}$

Number of particles in 1 g $= 1\text{ g} / (1.15 \times 10^{-9}\text{ g/particle}) = 8.68 \times 10^8$

Surface area of one particle $= 4\pi r^2 = 4\pi(5 \times 10^{-6}\text{ m})^2 = 3.14 \times 10^{-10}\text{ m}^2$

Surface area of 8.68×10^8 particles $= 0.27\text{ m}^2$

Since the observed surface area is 300 m², the particles must have highly irregular shapes or be porous.

24-15. (a) Since the nonpolar compounds should become more soluble in the mobile phase, the retention time will be shorter in 90% methanol.

(b) At pH 3, the predominant forms are neutral RCO_2H and cationic RNH_3^+. The amine will be eluted first, since RNH_3^+ is insoluble in the nonpolar stationary phase.

24-16. (a) Unretained component travels at the solvent velocity, u_x.

$$u_x = \frac{\text{column length}}{\text{transit time}} = \frac{4\,400\text{ mm}}{(41.7\text{ min})(60\text{ s/min})} = 1.76\text{ mm/s}$$

(b) $k = \dfrac{t_r - t_m}{t_m} = \dfrac{188.1\text{ min} - 41.7\text{ min}}{41.7\text{ min}} = 3.51$

(c) $N = \dfrac{5.55\,t_r^2}{w_{1/2}^2} = \dfrac{5.55\,(188.1\text{ min})^2}{(1.01\text{ min})^2} = 192\,000$

$H = \dfrac{4\,400\text{ mm}}{192\,000} = 22.9\text{ µm}$

(d) Resolution $= \dfrac{0.589\Delta t_r}{w_{1/2av}} = \dfrac{0.589(1.01\text{ min})}{1.01\text{ min}} = 0.589$

(e) $\alpha = \dfrac{t_{r2}'}{t_{r1}'} = \dfrac{194.3\text{ min} - 41.7\text{ min}}{193.3\text{ min} - 41.7\text{ min}} = 1.006_6$

$\gamma = \dfrac{t_{r2}}{t_{r1}} = \dfrac{194.3\text{ min}}{193.3\text{ min}} = 1.005_2$

(f) Resolution $= \dfrac{\sqrt{N}}{4}(\gamma - 1)$

$1.000 = \dfrac{\sqrt{N}}{4}(1.0052 - 1) \Rightarrow N = 5.9_2 \times 10^5$

A column length of 440 cm gave N $= 1.92 \times 10^5$ plates. To obtain $5.9_2 \times 10^5$ plates, the column must be longer by a factor of $\dfrac{5.9_2 \times 10^5 \text{ plates}}{1.92 \times 10^5 \text{ plates}} = 3.0_8$.

Required length $= (3.0_8)(4.40 \text{ m}) = 13.6 \text{ m}$

(g) Slow the flow rate to possibly decrease H and thereby increase N. Change the solvent to change the relative retention.

(h) Resolution $= \dfrac{\sqrt{N}}{4}(\gamma - 1) = \dfrac{\sqrt{192\,000}}{4}(1.008_3 - 1) = 0.91$

24-17. (a) On (R,R)-stationary phase, (S)-gimatecan is eluted at 6.10 min. On (S,S)-stationary phase, (S)-gimatecan is retained more strongly and is eluted at 6.96 min. (R)-gimatecan *must have the exact opposite behavior*. It will be eluted at 6.96 min from (R,R)-stationary phase and at 6.10 min from (S,S)-stationary phase.

(b) With (S,S)-stationary phase, we observe a small peak at 6.10 min for (R)-gimatecan. This peak is well separated from the front of the big (S)-gimatecan peak centered at 6.96 min, so the two areas can be integrated and compared with each other. With (R,R)-stationary phase, we see the (S)-gimatecan peak at 6.10 min with no evidence of the minor (R)-gimatecan peak at 6.96 min. The minor peak is lost beneath the tail of (S)-gimatecan. Chromatography on each enatiomer of the stationary phase enables us to unambiguously locate where each enantiomer of gimatecan is eluted, even though we do not have a standard sample of (R)-gimatecan.

(c) For the (S,S)-stationary phase, we have the following information:

(S)-gimatecan: $t_r = 6.96$ min $k = 1.50$

(R)-gimatecan: $t_r = 6.10$ min $k = 1.22$

The definition of retention factor is $k = (t_r - t_m)/t_m$. Inserting $k = 1.50$ and $t_r = 6.96$ min for (S)-gimatecan gives $t_m = 2.78_4$ min. We should get the same value of t_m for (R)-gimatecan by inserting $k = 1.22$ and $t_r = 6.10$ min into the equation $k = (t_r - t_m)/t_m$. In fact, this pair of numbers gives $t_m = 2.74_8$ min. The difference is from experimental error plus roundoff. Let's take the average value $t_m = 2.76_6$ min.

The adjusted retention time is $t_r' = t_r - t_m$.

(S)-gimatecan: $t_r' = t_r - t_m = 6.96 - 2.76_8 = 4.19_2$ min

(R)-gimatecan: $t_r' = t_r - t_m = 6.10 - 2.76_8 = 3.33_2$ min

Relative retention: $\alpha = \dfrac{t_{r2}'}{t_{r1}'} = \dfrac{4.19_2 \text{ min}}{3.33_2 \text{ min}} = 1.25_8$

Unadjusted relative retention: $\gamma = \dfrac{t_{r2}}{t_{r1}} = \dfrac{6.96 \text{ min}}{6.10 \text{ min}} = 1.14_1$

(d) Resolution $= \dfrac{\sqrt{N}}{4}(\gamma - 1) = \dfrac{\sqrt{6\,800}}{4}(1.14_1 - 1) = 2.91$ which is more than

adequate for "baseline" separation. Tailing of the peaks creates a little

overlap, but it should not be very serious for an equal mixture of the

enantiomers.

24-18. Peak areas will be proportional to molar absorptivity, since the number of moles

of A and B are equal.

$$\frac{\text{Area of A}}{\text{Area of B}} = \frac{2.26 \times 10^4}{1.68 \times 10^4} = \frac{1.064 \times h_A w_{1/2}}{1.064 \times h_B w_{1/2}} = \frac{(128)(10.1)}{h_B (7.6)}$$

$$\Rightarrow h_B = 126 \text{ mm}$$

24-19. Acetophenone is neutral at all pH values. Its retention is nearly unaffected by

pH. For salicylic acid, we expect the neutral molecule, HA, to have some affinity

for the C_8 nonpolar stationary phase and the ion, A^-, to have little affinity for C_8.

Salicylic acid is predominantly HA below pH 2.97 and A^- above pH 2.97. At pH

3, there is nearly a 1:1 mixture of HA and A^-, which is moderately retained on the

nonpolar column. At pH 5 and 7, more than 99% of the molecules are A^-, so

retention is weak (small retention factor).

Ionic forms of nicotine ought to have low affinity for the nonpolar stationary

phase and the neutral molecule would have some affinity. Abbreviating nicotine

as B, the form B is dominant above pH = pK_2 = 7.85. BH^+ is dominant between

pH 3.15 and 7.85. BH_2^{2+} is dominant below pH 3.15. B does not become appreciable until pH $\approx$ 7, so the retention factor is low below pH 7 and increases at pH 7.

24-20. (a) $V_m \approx 0.5\, L\, d_c^2 = 0.5(5.0 \text{ cm})(0.46 \text{ cm})^2 = 0.53 \text{ cm}^3 = 0.53 \text{ mL}$

$t_m = V_m/F = (0.53 \text{ mL})/(1.4 \text{ mL/min}) = 0.38 \text{ min for column A}$

$= (0.53 \text{ mL})/(2.0 \text{ mL/min}) = 0.26 \text{ min for column B}$

(b) Morphine 3-β-D-glucuronide is more polar than morphine because of the added hydroxyl groups and the carboxylic acid. The more polar compound is less retained by the nonpolar reversed-phase column.

(c) Bare silica is a polar, hydrophilic surface. Morphine should not be retained as strongly as the more polar morphine 3-β-D-glucuronide. The gradient goes to increasing H_2O for increasing polarity (that is, increasing solvent strength) to remove the more strongly adsorbed, more polar compound.

(d) $k = \dfrac{t_r - t_m}{t_m} = \dfrac{1.5 - 0.65}{0.65} = 1.3$ for morphine 3-β-D-glucuronide

$k = \dfrac{t_r - t_m}{t_m} = \dfrac{2.8 - 0.65}{0.65} = 3.3$ for morphine

(e) $V_m = F\, t_m = (2.0 \text{ mL/min})(0.50 \text{ min}) = 1.0 \text{ mL}$

$k^* = \dfrac{t_G\, F}{\Delta\Phi\, V_m\, S} = \dfrac{(5.0 \text{ min})(2.0 \text{ mL/min})}{(0.4)(1.0 \text{ mL})(4)} = 6.2$

24-21. (a) Electrical power = current × voltage. Current is the rate of flow of charge through a circuit. It is analogous to the rate of flow of liquid through a column. Voltage is the potential difference driving charge through the wire. It is analogous to the pressure difference driving liquid through a column.

(b) $1 \text{ mL} = 1 \text{ cm}^3 = (10^{-2} \text{ m})^3 = 10^{-6} \text{ m}^3$

$1 \text{ mL/min} = 10^{-6} \text{ m}^3/60 \text{ s} = 1.67 \times 10^{-8} \text{ m}^3/\text{s}.$

$3\,500 \text{ bar} = 3\,500 \times 10^5 \text{ Pa} = 3.5 \times 10^8 \text{ Pa}$

power = volume flow rate × pressure drop

$= (1.67 \times 10^{-8} \text{ m}^3/\text{s})(3.5 \times 10^8 \text{ Pa}) = 5.8 \text{ W}$

24-22.   (a)

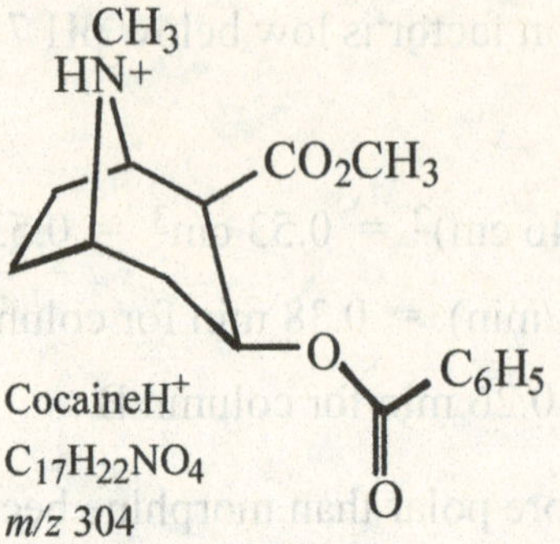

CH3
HN+

CO2CH3

O   C6H5

CocaineH+
C17H22NO4
m/z 304

C

O

(b)   The C6H5CO2 group has a mass of 121 Da. Subtracting 121 from 304 gives
183 Da. The peak at *m/z* 182 probably represents cocaine minus
C6H5CO2H. The structure might be the one below or some rearranged form
of it.

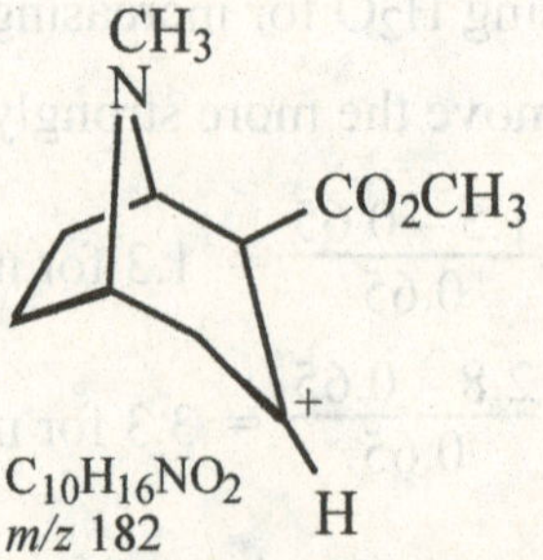

CH3
N

CO2CH3

+

C10H16NO2
m/z 182   H

(c)   The ion at *m/z* 304 was selected by mass filter Q1. Its isotopic partner
containing ^{13}C at *m/z* 305 was blocked by Q1. Because the species at *m/z*
304 is isotopically pure, there is no ^{13}C-containing partner for the
collisionally activated dissociation product at *m/z* 182.

(d)   For selected reaction monitoring, the mass filter Q1 selects just *m/z* 304,
which eliminates components of plasma that do not give a signal at *m/z* 304.
Then this ion is passed to the collision cell, in which it breaks into a major
fragment at *m/z* 182 which passes through Q3. Few other components in the
plasma that give a signal at *m/z* 304 also break into a fragment at *m/z* 282.
The 2-step selection process essentially eliminates everything else in the
sample and produces just one clean peak in the chromatogram.

(e) The phenyl group must be labeled with deuterium because the labeled product gives the same fragment at m/z 182 as unlabeled cocaine.

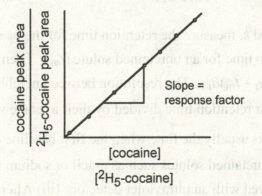

$C_{17}D_5H_{17}NO_4$
m/z 309

$C_{10}H_{16}NO_2$
m/z 182

(f) First, we need to construct a calibration curve to get the response factor for cocaine compared to $^2H_5$-cocaine. We expect this response factor to be close to 1.00. We would prepare a series of solutions with known concentration ratios [cocaine]/[$^2H_5$-cocaine] and measure the area of each chromatographic peak in the chromatography/atmospheric chemical ionization/selected reaction monitoring experiment. A graph would be constructed, in which [peak area of cocaine]/[peak area of $^2H_5$-cocaine] is plotted versus [cocaine]/[$^2H_5$-cocaine]. The slope of this line is the response factor.

For quantitative analysis, a known amount of the internal standard $^2H_5$-cocaine is injected into the plasma. From the calibration curve, the relative peak areas tell us the relative concentrations of cocaine and the internal standard. From the known quantity of internal standard injected into the plasma, we can calculate the quantity of cocaine.

24-23. (a) Atmospheric pressure chemical ionization gives a prominent peak at *m/z* 234, which must be MH⁺. The peak at *m/z* 84 is probably the fragment $C_5H_{10}N^+$, which might have the structure shown below.

In selected reaction monitoring, *m/z* 234 is selected by mass filter Q1 and *m/z* 84 is selected by mass filter Q3 in a triple quadrupole spectrometer.

(b) Deuterated internal standard has the formula $C_{14}H_{16}{}^2H_3O_2N$, with a nominal mass of 236. The protonated molecule is *m/z* 237. Cleavage of the C-C bond gives the same $C_5H_{10}N^+$ fragment as unlabeled Ritalin. The transition to monitor is *m/z* 237 → 84.

24-24. (a) To find *k*, measure the retention time for the peak of interest (t_r) and the elution time for an unretained solute (t_m). Then use the formula $k = (t_r - t_m)/t_m$. The resolution between neighboring peaks is the difference in their retention time divided by their average width at the baseline.

(b) (i) t_m is usually the time when the first baseline disturbance is observed. (ii) Unretained solutes such as uracil or sodium nitrate could be run and observed with an ultraviolet detector. (iii) Alternatively, the formula $t_m \approx Ld_c^2/(2F)$ can be used, where *L* is the length of the column (cm), d_c is the column diameter (cm), and *F* is the flow rate (mL/min).

(c) $t_m \approx Ld_c^2/(2F) = (15)(0.46)^2/(2 \cdot 1.5) = 1.0_6$ min

t_m does not depend on particle size. The estimate is 1.0_6 min for both 5.0- and 3.5-μm particles.

24-25. *Dead volume* is the volume of the system (not including the chromatography column) from the point of injection to the point of detection. *Dwell volume* is the volume of the system from the point of mixing solvents to the beginning of the column. Excessive dead volume causes peak broadening by longitudinal diffusion. In gradient elution, dwell volume determines the time from the initiation of a gradient until the gradient reaches the column. The greater the dwell volume, the more the delay between initiating a gradient and the actual increase of solvent strength on the column.

24-26. A rugged procedure should not be seriously affected by gradual deterioration of the column, *small* variations in solvent composition, pH, and temperature, or use of a different batch of the same stationary phase. A procedure should be rugged so that inevitable, small variations in conditions do not substantially affect the outcome of the separation.

24-27. $0.5 \leq k \leq 20$; resolution ≥ 2; operating pressure ≤ 15 MPa; $0.9 \leq$ asymmetry factor ≤ 1.5

24-28. Run a wide gradient (such as 5%B to 100%B) in a gradient time, t_G selected to produce $k^* \approx 5$ in Equation 24-10. Measure the difference in retention time (Δt) between the first and last peaks eluted. Use a gradient if $\Delta t/t_G > 0.25$ and use isocratic elution if $\Delta t/t_G < 0.25$.

24-29. The first steps are to (1) determine the goal of the analysis, (2) select a method of sample preparation, and (3) choose a detector that allows you to observe the desired analytes in the mixture. The next step could be a wide gradient elution to determine whether or not an isocratic or gradient separation is more appropriate. If the isocratic separation is chosen, %B is varied until criteria for a good separation are met. If adequate resolution is not attained, you can try different organic solvents. If adequate resolution is still not attained, you can use a slower flow rate, a longer column, smaller particles, or a different stationary phase.

24-30. To use two organic solvents (A and B), the optimum concentration of A is first found to get the best separation while keeping all retention factors in the range 0.5–20. If adequate separation does not result, then the same procedure is carried out with solvent B. If adequate separation is still not attained, a 1:1 mixture of the best compositions of A and B should be tried. If it looks promising, other mixtures of the optimum concentrations of A and B can be tried.

24-31. Chromatography is conducted with four conditions: (A) high %B, low T, (B) high %B, high T, (C) low %B, high T, and (D) low %B, low T. Based on the appearance of the chromatograms, combinations between the points A, B, C, and D can be explored for further improvement in the separation.

24-32. Peak 5 has a retention time (t_r) of 11.0 min for 50% B. The retention factor is $k = (t_r - t_m)/t_m = (11.0 - 2.7)/2.7 = 3.1$. When B is reduced to 40%, the rule of three predicts $k = 3(3.1) = 9.3$. Rearranging the definition of retention factor, we find $t_r = t_m k + t_m = t_m(k + 1)$. We predict for 40% B $t_r = t_m(k + 1) = (2.7)(9.3 + 1) = 27.8$ min. The observed retention time at 40% B is 20.2 min.

24-33. (a)

%B	Retention time (min)		
	peak 6	peak 7	peak 8
90	4.4	4.4	4.9
80	4.5	4.5	5.1
70	5.6	5.6	7.3
60	8.2	8.2	12.2
50	13.1	13.6	24.5
40	24.8	27.5	65.1
35	37.6	44.2	125.2

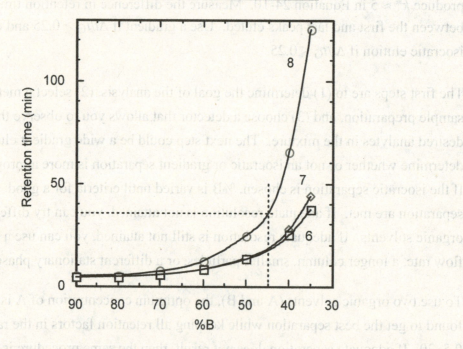

At 45% B, we could estimate that Peak 8 will be eluted halfway between the times for 40% B and 50% B, which is about 45 min. The fit to the curve above suggests that 36 min is a more realistic estimate.

(b) The table shows the calculation of retention factor k for Peaks 6-8.

				$t_m =$	2.7 min				
	retention time t_r (min)			retention factor $k = (t_r - t_m)/t_m$			log k		
Φ	Peak 6	Peak 7	Peak 8	Peak 6	Peak 7	Peak 8	Peak 6	Peak 7	Peak 8
0.9	4.4	4.4	4.9	0.630	0.630	0.815	-0.201	-0.201	-0.089
0.8	4.5	4.5	5.1	0.667	0.667	0.889	-0.176	-0.176	-0.051
0.7	5.6	5.6	7.3	1.074	1.074	1.704	0.031	0.031	0.231
0.6	8.2	8.2	12.2	2.037	2.037	3.519	0.309	0.309	0.546
0.5	13.1	13.6	24.5	3.852	4.037	8.074	0.586	0.606	0.907
0.4	24.8	27.5	65.1	8.185	9.185	23.111	0.913	0.963	1.364
0.35	37.6	44.2	125.2	12.926	15.370	45.370	1.111	1.187	1.657

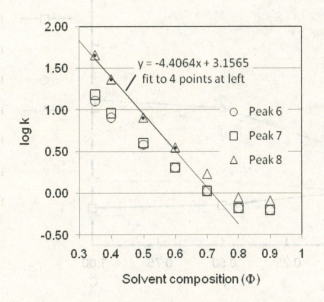

The first obvious point is that log k versus Φ does not follow a straight line over a wide range of solvent composition. The straight line going through the four points for Peak 8 from $\Phi = 0.35$ to 0.6 is log $k = -4.4064\Phi + 3.1565$. At $\Phi = 0.45$, we compute log $k = 1.1736$ and $k = 14.92$. We compute $t_r = t_m(k+1) = 43.0$ min. If we had only taken the first three points ($\Phi = 0.35$ to 0.5), we would find log $k = -4.9364\Phi + 3.3660$. At $\Phi = 0.45$, we compute log $k = 1.11447$, $k = 13.95$, and $t_r = 40.4$ min.

24-34.   (a)

Solvent composition		Retention times (min) for Peaks 1-7						
		1	2	3	4	5	6	7
B	0.0	8.0	8.0	11.5	13.8	12.8	12.8	37.0
F	0.5	6.0	9.6	5.0	20.5	17.3	22.0	16.0
C	1.0	5.5	9.6	4.3	21.7	23.6	16.5	13.6

Predicted positions (by linear interpolation):

0.25	7.0	8.80	8.25	17.15	15.05	17.40	26.50
0.75	5.75	9.6	4.65	21.10	20.45	19.75	14.80

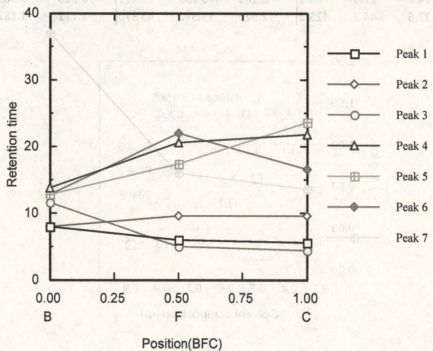

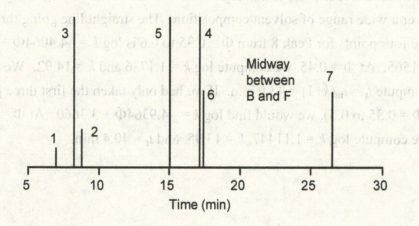

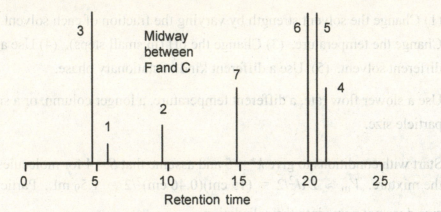

(b)   B: 40% methanol/60% buffer

C: 32% tetrahydrofuran/68% buffer

F: 20% methanol/16% tetrahydrofuran/63% buffer

Between B and F: 30% methanol/8% tetrahydrofuran/62% buffer

Between F and C: 10% methanol/24% tetrahydrofuran/66% buffer

24-35.   D: 25% acetonitrile/30% methanol/45% buffer

E: 25% acetonitrile/20% tetrahydrofuran/55% buffer

F: 30% methanol/20% tetrahydrofuran/50% buffer

G: 16.7% acetonitrile/20% methanol/13.3% tetrahydrofuran/50% buffer

24-36.   In the nomograph in Figure 24-26, a vertical line at 48% methanol intersects the acetonitrile line at 38%.

24-37.   (a)   Lower solvent strength usually increases the difference in retention between different compounds. Use a lower percentage of acetonitrile.

(b)   In normal-phase chromatography, solvent strength increases as the solvent becomes more polar, which means increasing the methyl t-butyl ether concentration. We need a higher concentration of hexane to lower the solvent strength, increase the retention times, and probably improve resolution.

24-38.   (a) $\Delta t/t_G = 19/60 = 0.32$. Because $\Delta t/t_G > 0.25$, gradient elution is suggested.

(b)   At $t = 22$ min, the solvent composition entering the column can be calculated by linear interpolation: $5 + \frac{22}{60}(100 - 5) = 39.8\%$. At 41 minutes, the composition is $5 + \frac{41}{60}(100 - 5) = 69.9\%$. A reasonable gradient for the second experiment is from 40 to 70% acetonitrile in 60 min.

24-39.   (a) (1) Change the solvent strength by varying the fraction of each solvent. (2) Change the temperature. (3) Change the pH (in small steps). (4) Use a different solvent. (5) Use a different kind of stationary phase.

(b) Use a slower flow rate, a different temperature, a longer column, or a smaller particle size.

24-40.   (a) Start with conditions to give $k^* = 5$ and assume that $S = 4$ for molecules in the mixture. $V_m \approx L\, d_c^2/2 = (15\text{ cm})(0.46\text{ cm})^2/2 = 1.5_9\text{ mL}$. Particle size does not come into the calculation.

$$t_G = \frac{k^*\,\Delta\Phi\,V_m\,S}{F} = \frac{(5)(0.9)(1.59\text{ mL})(4)}{(1.0\text{ mL/min})} = 29\text{ min}$$

(b) $k^* = \dfrac{t_G\,F}{\Delta\Phi\,V_m\,S} = \dfrac{(11.5\text{ min})(1.0\text{ mL/min})}{(0.14)(1.59\text{ mL})(4)} = 12.9$

The large column has the same length as the small column, but the diameter is increased from 0.46 to 1.0 cm. The volume increases by a factor of $(1.0/0.46)^2 = 4.7$. Therefore, we increase the flow rate and the sample loading by a factor of 4.7. Flow rate = 4.7 mL/min and sample load = 4.7 mg. The gradient time is unchanged at 11.5 min. For the large column, $V_m \approx L\, d_c^2/2 = (15\text{ cm})(1.0\text{ cm})^2/2 = 7.5\text{ mL}$ and

$$k^* = \frac{t_G\,F}{\Delta\Phi\,V_m\,S} = \frac{(11.5\text{ min})(4.7\text{ mL/min})}{(0.14)(7.5\text{ mL})(4)} = 12.9$$

CHAPTER 25
CHROMATOGRAPHIC METHODS AND CAPILLARY ELECTROPHORESIS

25-1. The separator column separates ions by ion exchange, while the suppressor exchanges the counterion to reduce the conductivity of eluent. After separating cations in the cation-exchange column, the suppressor must exchange the anion for OH⁻, which makes H_2O from the HCl eluent.

25-2. Increased cross-linking gives decreased swelling, increased exchange capacity and selectivity, but longer equilibration time.

25-3. Deionized water has been passed through ion-exchangers to convert cations to H^+ and anions to OH⁻, making H_2O. Nonionic impurities (e. g., organic compounds) are not removed by this process, but can be removed by activated carbon.

25-4. One way is to wash extensively with NaOH a column containing a weighed amount of resin to load all ion-exchange sites with OH⁻. After a thorough washing with water to remove excess NaOH, the column can be eluted with a large quantity of aqueous NaCl to displace OH⁻. Eluate is then titrated with standard HCl to determine the moles of displaced OH⁻.

25-5. (a) As pH is lowered the protein becomes protonated, so the magnitude of the negative charge decreases. The protein becomes less strongly retained.

(b) As the ionic strength of eluent is increased, the protein will be displaced from the gel by solute ions.

25-6. Particles pass through 200 mesh (75 μm) sieve and are retained by 400 mesh (38 μm) sieve. 200/400 mesh particles are smaller than 100/200 mesh particles.

25-7. The pK_a values are: NH_4^+ (9.24), $CH_3NH_3^+$ (10.64), $(CH_3)_2NH_2^+$ (10.77), and $(CH_3)_3NH^+$ (9.80). If the four ammonium ions are adsorbed on a cation exchange resin at, say, pH 7, they might be separated by elution with a gradient of increasing pH. The anticipated order of elution is $NH_3 < (CH_3)_3N < CH_3NH_2 < (CH_3)_2NH$. We should not be surprised if the elution order were different, since steric and hydrogen bonding effects could be significant determinants of the selectivity coefficients. It is also possible that elution with a constant pH (of, say, 8) might separate all four species from each other.

25-8. (a) $[Cl^-]_i ([Cl^-]_i + [R^-]_i) = [Cl^-]_o^2$

$[Cl^-]_i ([Cl^-]_i + 3.0) = (0.10)^2 \Rightarrow [Cl^-]_i = 0.003\,33$ M

$\Rightarrow [Cl^-]_o/[Cl^-]_i = 0.10/0.003\,3 = 30$

(b) Using $[Cl^-]_o = 1.0$ in (a) gives $[Cl^-]_o/[Cl^-]_i = 1.0/0.30 = 3.3$

(c) As $[Cl^-]_o$ increases, the fraction of $[Cl^-]_i$ increases.

25-9. The sum of anion charge in the spreadsheet is $-0.001\,59$ M, and the sum of cation charge is $0.002\,02$ M. Either some of the ion concentrations are inaccurate, or there are other ions in the pondwater that were not detected. For example, there could be large organic anions derived from living matter (such as humic acid from plants) that are not detected in this experiment.

	A	B	C	D	E	F
1	Ion	Formula mass	Concentration		Ion	Charge
2		(g/mol)	(µg/mL)	(mol/L)	charge	(mol/L)
3	Fluoride	18.998	0.26	1.37E-05	-1	-1.37E-05
4	Chloride	35.453	43.6	1.23E-03	-1	-1.23E-03
5	Nitrate	62.005	5.5	8.87E-05	-1	-8.87E-05
6	Sulfate	96.064	12.6	1.31E-04	-2	-2.62E-04
7						
8			Sum of anion charge =			-0.00159
9						
10	Sodium	22.990	2.8	1.22E-04	1	1.22E-04
11	Ammonium	18.038	0.2	1.11E-05	1	1.11E-05
12	Potassium	39.098	3.5	8.95E-05	1	8.95E-05
13	Magnesium	24.305	7.3	3.00E-04	2	6.01E-04
14	Calcium	40.078	24.0	5.99E-04	2	1.20E-03
15						
16			Sum of cation charge =			0.00202

25-10. (a) The hydrophilic stationary phase is a zwitterion with fixed positive and negative charges. Anions are retained by positive charges and cations are retained by negative charges. In hydrophilic interaction chromatography, the stationary phase is polar and there is thought to be a thin layer of aqueous phase on the surface of the stationary phase. Solvent must be made more polar to compete with the stationary phase to elute polar solutes. Eluent strength is increased when the acetonitrile content is decreased.

(b) Eluent strength increases in hydrophilic interaction chromatography as the fraction of aqueous phase increases. With 20 vol% acetonitrile / 80 vol% aqueous buffer, the eluent strength is high and both ions are eluted rapidly, without an opportunity to be separated. The eluent strength of 40 vol% acetonitrile / 60 vol% aqueous buffer is lower, so the ions are eluted slower and more selectively.

25-11. Hydrophobic regions of the protein are less soluble in water as the salt concentration in the water increases. This decrease in solubility of nonpolar substances in water with increasing salt concentration is known as "salting out." By decreasing the salt concentration, the protein becomes more soluble in the aqueous phase and can be eluted from the column. Eluent strength increases as the salt concentration decreases.

25-12. At pH 2 (0.01 M HCl), TCA is more dissociated than DCA, which is more dissociated than MCA. The greater the average charge of the compound, the more it is excluded from the ion-exchange resin and the more rapidly it is eluted.

25-13. (a) Sodium octyl sulfate dissolved in the stationary phase forms an ion-pair with NE or DHBA. Other ions in the eluent compete with NE or DHBA, and slowly elute them from the column by ion exchange.

(b) Construct a graph of (peak height ratio) vs. (added concentration of NE). The x-intercept gives [NE] = 29 ng/mL.

Added NE	signal
0	0.298
12	0.414
24	0.554
36	0.664
48	0.792

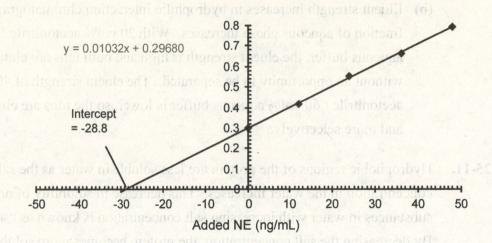

y = 0.01032x + 0.29680

Intercept
= -28.8

Added NE (ng/mL)

25-14. This is an example of *indirect detection*. Eluent contains naphthalenetrisulfonate, which absorbs at 280 nm. Charge balance dictates that when one of the analyte anions is emerging from the column, there must be less naphthalenetrisulfonate anion emerging. Since analytes do not absorb as strongly at 280 nm, the absorbance is negative with respect to the steady baseline.

25-15. (a) K^+ in reservoir = $(0.75)(1.5 \text{ L})(2.0 \frac{\text{mol } K_2PO_4}{\text{L}})(2 \frac{\text{mol } K^+}{\text{mol } K_2PO_4}) = 4.5 \text{ mol}$

Flow rate = $(20 \times 10^{-3} \frac{\text{mol KOH}}{\text{L}})(0.001 \ 0 \frac{\text{L}}{\text{min}}) = 2.0 \times 10^{-5} \frac{\text{mol KOH}}{\text{min}}$

Time available = $\dfrac{4.5 \text{ mol } K}{2.0 \times 10^{-5} \frac{\text{mol KOH}}{\text{min}}} = 2.25 \times 10^5 \text{ min}$

$\dfrac{2.25 \times 10^5 \text{ min}}{60 \text{ min/h}} = 3.8 \times 10^2 \text{ h}$

(b) A flow of 5.0 mM KOH at 1.0 mL/min provides

$(5.0 \times 10^{-3} \text{mol KOH/L})(0.001 \ 0 \text{ L/min}) = 5.0 \times 10^{-6} \text{ mol KOH/min}$

$\dfrac{5.0 \times 10^{-6} \text{mol KOH/min}}{60 \text{ s/min}} = 8.33 \times 10^{-8} \text{ mol KOH/s}$

One electron provides one OH^- at the cathode, so the current must provide 8.33×10^{-8} mol e^-/s. We multiply by the Faraday constant to convert moles of electrons into coulombs:

$(8.33 \times 10^{-8} \text{ mol } e^-/s)(9.648 \ 5 \times 10^4 \text{ C/mol } e^-)$

$= 8.0 \times 10^{-3} \text{ C/s} = 8.0 \times 10^{-3} \text{ A} = 8.0 \text{ mA}.$

To produce 0.10 M KOH at 1.0 mL/min requires 20 times as much current, because the concentration of KOH is 20 times higher than 5.0 mM. The current at the end of the gradient will be (20)(8.0 mA) = 160 mA = 0.16 A.

25-16.

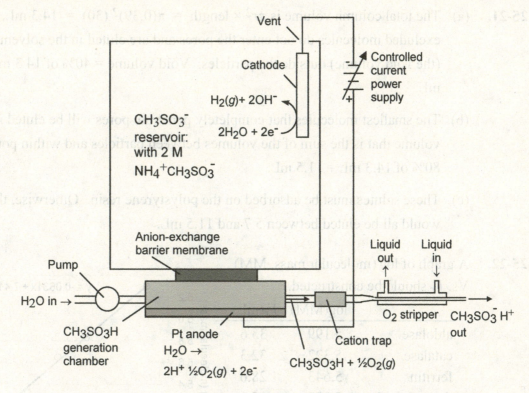

25-17. (a) There is a range in which retention volume is logarithmically related to molecular mass. The unknown is compared to a series of standards of known molecular mass.

(b) FM 10^5 is near the middle range of the 10 μm pore size column.

25-18. (a) $V_t = \pi(0.80 \text{ cm})^2 (20.0 \text{ cm}) = 40.2 \text{ mL}$

(b) $K_{av} = \dfrac{V_r - V_0}{V_m - V_0} = \dfrac{27.4 - 18.1}{35.8 - 18.1} = 0.53$

25-19. Ferritin maximum is in tube 22 (= 22 × 0.65 mL) = 14.3 mL = V_0

Ferric citrate maximum is in tube 84 (= 84 × 0.65 mL) = 54.6 mL = V_m

Does this value of V_m make sense? The total column volume is $V_t = \pi r^2 \times$ length = $\pi(0.75 \text{ cm})^2(37 \text{ cm}) = 65.4 \text{ mL}$, so $V_m = 54.6 \text{ mL}$ is plausible.

Transferrin maximum = tube 32 = 20.8 mL $\Rightarrow K_{av} = \dfrac{20.8 - 14.3}{54.6 - 14.3} = 0.16$

25-20. (a) The vertical line begins at log (molecular mass) $\approx 3.3 \Rightarrow$ mass = $10^{3.3}$ = 2 000 Da.

(b) A vertical line at 6.5 mL intersects the 10-nm calibration line at log (molecular mass) $\approx 2.5 \Rightarrow$ mass = $10^{2.5}$ = 300 Da.

25-21. (a) The total column volume is $\pi r^2 \times$ length $= \pi(0.39)^2 (30) = 14.3$ mL. Totally excluded molecules do not enter the pores and are eluted in the solvent volume (the void volume) outside the particles. Void volume $= 40\%$ of 14.3 mL $= 5.7$ mL.

(b) The smallest molecules that completely penetrate pores will be eluted in a volume that is the sum of the volumes between particles and within pores $= 80\%$ of 14.3 mL $= 11.5$ mL.

(c) These solutes must be adsorbed on the polystyrene resin. Otherwise, they would all be eluted between 5.7 and 11.5 mL.

25-22. A graph of log (molecular mass, MM) Vs. V_r should be constructed.

	log(MM)	V_r(mL)
aldolase	5.199	35.6
catalase	5.322	32.3
ferritin	5.643	28.6
thyroglobulin	5.825	25.1
Blue Dextran	6.301	17.7
unknown	?	30.3

The equation of the graph of K_{av} vs. log (MM) is $y = -0.063\,1\,x + 7.416$. Inserting $x = 30.3$ gives $y = $ log (MM) $= 5.50 \Rightarrow$ molecular mass $= 320\,000$

25-23. Electroosmosis is the bulk flow of fluid in a capillary caused by migration of the dominant ion in the diffuse part of the double layer toward the anode or cathode.

25-24. At pH 10, the wall of the bare capillary is negatively charged with $-\text{Si}-\text{O}^-$ groups and there is strong electroosmotic flow toward the cathode. At pH 2.5, the wall is nearly neutral with $-\text{Si}-\text{OH}$ groups and there is almost no electroosmotic flow. The few $-\text{Si}-\text{O}^-$ groups left give slight flow toward the cathode. The aminopropyl capillary also has positive flow at pH 10, but the rate is only about half as great as that of the bare capillary. The negative charge might be reduced because there are fewer $-\text{Si}-\text{O}^-$ groups (because some of them have been converted to $-\text{Si}-\text{CH}_2\text{CH}_2\text{CH}_2\text{NH}_2$) or because some of the aminopropyl groups are protonated ($-\text{Si}-\text{CH}_2\text{CH}_2\text{CH}_2\text{NH}_2^+$) at pH 10. At pH 2.5, all the aminopropyl groups are protonated. The net charge on the wall is *positive* and the flow is *reversed*.

25-25. Arginine is the only amino acid listed with a positively charged side chain. All of the derivatized amino acids have a negative charge because the fluorescent group and the terminal carboxyl group are both negative. Arginine is least negative, so its electrophoretic mobility toward the anode is slowest and its net migration toward the cathode (from electroosmosis) is fastest.

25-26. Under ideal conditions, longitudinal diffusion is the principle source of zone broadening. Even under ideal conditions, the finite length of the injected sample and, possibly, the finite length of the detector contribute to zone broadening. In real electrophoresis, adsorption on the capillary wall and irregular flow paths due to imperfections in the capillary could contribute to zone broadening. For an experimental study of zone broadening, see D. Xiao, T. V. Le, and M. J. Wirth, "Surface Modification of the Channels of Poly(dimethylsiloxane) Microfluidic Chips with Polyacrylamide for Fast Electrophoretic Separations of Proteins," *Anal. Chem.* **2004**, *76*, 2055.

25-27. (a) At pH 2.8, electroosmotic flow will be very small. Anionic analyte will migrate from negative to positive polarity with little effect from the reverse electroosmotic flow.

(b) The conductivity of the buffer needs to be higher than the conductivity of the sample so that the sample will stack. At lower buffer concentration, analyte bands will be broader and resolution of heparin from its impurities would be diminished.

(c) High buffer concentration gives high conductivity, high current, and high heat generation. The narrow column reduces the current and the heat generation and makes it easier to cool the entire volume inside the capillary.

(d) Li^+ has lower mobility than Na^+, so the conductivity of lithium phosphate solution will be lower than the conductivity of sodium phosphate solution at the same pH. The lower the conductivity, the higher the electric field required to generate the same current.

High field strength reduces the migration time to shorten the analysis. Also, according to Equation 25-14, the number of plates increases in proportion to applied voltage.

25-28. (a) Volume = cross-sectional area × length

100×10^{-9} cm^3 (= 100 pL) = $(12 \times 10^{-4}$ cm$)(50 \times 10^{-4}$ cm$)$(length)

$\Rightarrow$ length = 0.167 mm

(b) Time = distance/speed

Width of injection band (in seconds) = $\Delta t = \dfrac{\text{band length}}{\text{speed}} = \dfrac{0.167 \text{ mm}}{24 \text{ mm} / 8 \text{ s}} =$ 0.055 7 s

$\sigma_{\text{injection}} = \Delta t / \sqrt{12} = 0.016$ s

(c) $\sigma_{\text{diffusion}} = \sqrt{2Dt} = \sqrt{2(1.0 \times 10^{-8} \text{ m}^2/\text{s})(8 \text{ s})} = 0.000\,40$ s

(d) $\sigma_{\text{total}}^2 = \sigma_{\text{diffusion}}^2 + \sigma_{\text{injection}}^2 = (0.004 \text{ s})^2 + (0.016 \text{ s})^2$

$\Rightarrow \sigma_{\text{total}} = 0.016$ s

$w = 4\sigma_{\text{total}} = 0.064$ s

25-29. Electroosmotic flow can be reduced by (a) lowering the pH, so the charge on the capillary wall is reduced; (b) adding ions such as $^+H_3NCH_2CH_2CH_2NH_3^+$ that adhere to the capillary wall and effectively neutralize its charge; and (c) covalently attaching silanes with neutral, hydrophilic substituents to the Si—O$^-$ groups on the walls. A cationic surfactant can form a bilayer, such as that shown in Figure 25-24, which effectively reverses the charge on the wall.

25-30. In the absence of micelles, neutral molecules are all swept through the capillary at the electroosmotic velocity. Negatively charged micelles swim upstream with some electrophoretic velocity, so they take longer than neutral molecules to reach the detector. A neutral molecule spends some time free in solution and some time dissolved in the micelles. Therefore, the net velocity of the neutral molecule is reduced from the electroosmotic velocity. Because different neutral molecules have different partition coefficients between the solution and the micelles, each type of neutral molecule has its own net migration speed. We say that micellar electrokinetic chromatography is a form of chromatography because the micelles behave as a "stationary" phase in the capillary because their concentration is uniform throughout the capillary. Analyte partitions between the mobile phase and the micelles as the analyte travels through the capillary.

25-31. (a) Volume of sample = cross-sectional area × length

= πr^2(length) = $\pi(25 \times 10^{-6} \text{ m})^2(0.006\,0 \text{ m}) = 1.18 \times 10^{-11} \text{ m}^3$

$$\Delta P = \frac{128\eta L_t(\text{Volume})}{t\pi d^4} = \frac{128(0.001\ 0\ \text{kg/(m·s)})(0.600\ \text{m})(1.18\times10^{-11}\ \text{m}^3)}{(4.0\ \text{s})\pi(50\times10^{-6}\ \text{m})^4}$$

$$= 1.15 \times 10^4\ \text{Pa}\ (= 1.15 \times 10^4\ \text{kg/(m·s}^2))$$

(b) $\Delta P = h\rho g \Rightarrow h = \dfrac{\Delta P}{\rho g} = \dfrac{1.15 \times 10^4\ \text{kg/(m·s}^2)}{(1\ 000\ \text{kg/m}^3)(9.8\ \text{m/s}^2)} = 1.17\ \text{m}$

Since the column is only 0.6 m long, we cannot raise the inlet to 1.17 m. Instead, we could use pressure at the inlet (1.15×10^4 Pa = 0.114 atm) or an equivalent vacuum at the outlet.

25-32. (a) Volume = $\pi r^2(\text{length}) = \pi(12.5 \times 10^{-6}\ \text{m})^2(0.006\ 0\ \text{m}) = 2.95 \times 10^{-12}\ \text{m}^3 = 2.95\text{nL}$. Moles = $(10.0 \times 10^{-6}\ \text{M})(2.95 \times 10^{-9}\ \text{L}) = 29.5\ \text{fmol}$.

(b) Moles injected $= \mu_{app}\left(E\dfrac{\kappa_b}{\kappa_s}\right)t\pi r^2 C = \mu_{app}\left(\dfrac{V}{L_t}\dfrac{\kappa_b}{\kappa_s}\right)t\pi r^2 C$

In order for the units to work out, we need to express the concentration, C, in mol/m^3: $(10.0 \times 10^{-6}\ \text{mol/L})(1\ 000\ \text{L/m}^3) = 1.00 \times 10^{-2}\ \text{mol/m}^3$

$$V = \frac{(\text{moles})L_t(\kappa_s/\kappa_b)}{\mu_{app}\ t\pi r^2 C}$$

$$= \frac{(29.5 \times 10^{-15}\ \text{mol})(0.600\ \text{m})(1/10)}{(3.0 \times 10^{-8}\ \text{m}^2/(\text{V·s}))(4.0\ \text{s})\pi(12.5 \times 10^{-6}\ \text{m})^2(1.00 \times 10^{-2}\ \text{mol/m}^3)}$$

$$= 3.00 \times 10^3\ \text{V}$$

25-33. Electrophoretic peak: $N = \dfrac{16\ t_r^2}{w^2} = \dfrac{16\ (6.08\ \text{min})^2}{(0.080\ \text{min})^2} = 9.2 \times 10^4\ \text{plates}$

Chromatographic peak: $N \approx \dfrac{41.7(t_r/w_{0.1})^2}{(A/B + 1.25)}$

$$= \frac{41.7(6.03\ \text{min}/0.37\ \text{min})^2}{(1.45 + 1.25)} = 4.1 \times 10^3\ \text{plates}$$

(According to my measurements, both plate counts are about 1/3 lower than the values labeled in the figure from the original source.)

25-34. (a) Fumarate is a longer molecule than maleate, so we guess that fumarate has a greater friction coefficient than maleate. Electrophoretic mobility is (charge)/(friction coefficient). Both ions have the same charge, so we predict that maleate will have the greater electrophoretic mobility.

(b) Since maleate moves upstream faster than fumarate, fumarate is eluted first.

(c) Since the anions move faster than the endosmotic flow, the faster anion (maleate) is eluted first.

25-35. (a) pH 2: $u_{neutral} = \mu_{eo}E = \left(1.3 \times 10^{-8} \dfrac{m^2}{V \cdot s}\right)\left(\dfrac{27 \times 10^3\ V}{0.62\ m}\right) = 5.6_6 \times 10^{-4}$ m/s

Migration time $= (0.52\ m)/(5.6_6 \times 10^{-4}\ m/s) = 9.2 \times 10^2$ s

pH 12: $u_{neutral} = \mu_{eo}E = \left(8.1 \times 10^{-8} \dfrac{m^2}{V \cdot s}\right)\left(\dfrac{27 \times 10^3\ V}{0.62\ m}\right) = 3.5_3 \times 10^{-3}$ m/s

Migration time $= (0.52\ m)/(3.5_3 \times 10^{-3}\ m/s) = 1.4_7 \times 10^2$ s

(b) pH 2: $\mu_{app} = \mu_{ep} + \mu_{eo} = (-1.6 + 1.3) \times 10^{-8} \dfrac{m^2}{V \cdot s} = -0.3 \times 10^{-8} \dfrac{m^2}{V \cdot s}$

The anion will not migrate toward the detector at pH 2.

pH 12: $\mu_{app} = \mu_{ep} + \mu_{eo} = (-1.6 + 8.1) \times 10^{-8} \dfrac{m^2}{V \cdot s} = 6.5 \times 10^{-8} \dfrac{m^2}{V \cdot s}$

$u_{anion} = \mu_{app}E = \left(6.5 \times 10^{-8} \dfrac{m^2}{V \cdot s}\right)\left(\dfrac{27 \times 10^3\ V}{0.62\ m}\right) = 2.8_3 \times 10^{-3}$ m/s

Migration time $= (0.52\ m)/(2.8_3 \times 10^{-3}\ m/s) = 1.8_4 \times 10^2$ s

25-36. (a) The net speed of an ion moving through the capillary by electroosmosis plus electrophoresis is proportional to electric field ($u_{net} = \mu_{app}E$), which, in turn, is proportional to voltage. Increasing voltage by 120 kV/28 kV = 4.3 should increase the speed by 4.3 and decrease the migration time by 4.3. Peak 1 has a migration time of 211.3 min at 28 kV and 54.36 min at 120 kV. The ratio is 211.3 min/54.36 min = 3.9.

(b) Plate count is proportional to voltage ($N = \dfrac{\mu_{app}V}{2D}\dfrac{L_d}{L_t}$). Increasing voltage by a factor of 4.3 should increase the plate count by 4.3.

(c) Bandwidth is proportional to the $1/\sqrt{N}$ ($N = L_d^2/\sigma^2 \Rightarrow \sigma = L_d/\sqrt{N}$). Increasing voltage by 4.3 should increase N by 4.3 and decrease bandwidth by $1/\sqrt{4.3} = 0.48$. Bandwidth at 120 kV should be 48% as great as bandwidth at 28 kV.

(d) Increasing voltage makes the ions move faster, which gives them less time to diffuse apart. Therefore, the bandwidth is reduced and resolution is increased.

25-37. At low voltage (low electric field), the number of plates increases in proportion to voltage, as predicted by Equation 25-14. Above ~25 000 V/m, the capillary is probably overheating, which produces band broadening and decreases the number of plates.

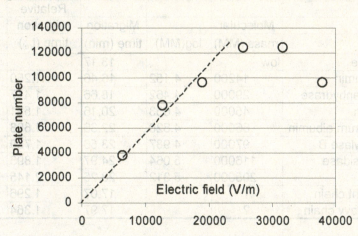

25-38. $N = \dfrac{5.55\, t_r^2}{w_{1/2}^2} = \dfrac{5.55\, (39.9\text{ min})^2}{(0.81\text{ min})^2} = 1.3_5 \times 10^4$ plates

Plate height = 0.400 m$/(1.3_5 \times 10^4$ plates$) = 30$ μm

25-39. $t = \dfrac{L}{u_{\text{net}}} = \dfrac{L}{\mu_{\text{app}} E}$ (t = migration time, L = length, u = speed, E = field)

$\Rightarrow \mu_{\text{app}} = \dfrac{L}{tE} = \dfrac{L/E}{17.12}$ for Cl$^-$ and $\mu_{\text{app}} = \dfrac{L/E}{17.78}$ for I$^-$

Therefore, we can write that the difference in mobilities is

$\Delta\mu_{\text{app}}(\text{I-Cl}) = \dfrac{L/E}{17.12} - \dfrac{L/E}{17.78}$ (L/E is an unknown constant)

But we know that $\Delta\mu_{\text{app}}(\text{I-Cl}) = [\mu_{\text{eo}} + \mu_{\text{ep}}(\text{I}^-)] - [\mu_{\text{eo}} + \mu_{\text{ep}}(\text{Cl}^-)] = \mu_{\text{ep}}(\text{I}^-) - \mu_{\text{ep}}(\text{Cl}^-) = 0.05 \times 10^{-8}$ m^2/(s·V) in Table 14-1.

For the difference between Cl$^-$ and Br$^-$ we can say

$\Delta\mu_{\text{app}}(\text{Br-Cl}) = \dfrac{L/E}{17.12} - \dfrac{L/E}{x}$

and we know that $\Delta\mu_{\text{app}}(\text{Br-Cl}) = 0.22 \times 10^{-8}$ m^2/(s·V) in Table 14-1.

Therefore, we can set up a proportion:

$\dfrac{\Delta\mu_{\text{app}}(\text{Br-Cl})}{\Delta\mu_{\text{app}}(\text{I-Cl})} = \dfrac{0.22}{0.05} = \dfrac{\dfrac{L/E}{17.12} - \dfrac{L/E}{x}}{\dfrac{L/E}{17.12} - \dfrac{L/E}{17.78}} \Rightarrow x = 20.5$ min

The observed migration time is 19.6 min. Considering the small number of significant digits in the $\Delta\mu$ values, this is a reasonable discrepancy.

25-40.

	A	B	C	D	E	F
1	Molecular mass by SDS/capillary gel electrophoresis					
2					Relative	
3		Molecular		Migration	migration	
4	Protein	mass (MM)	log(MM)	time (min)	time (t_{rel})	$1/t_{rel}$
5	Marker dye	low		13.17		
6	a-Lactalbumin	14200	4.152	16.46	1.250	0.8001
7	Carbonic anhydrase	29000	4.462	18.66	1.417	0.7058
8	Ovalbumin	45000	4.653	20.16	1.531	0.6533
9	Bovine serum albumin	66000	4.820	22.36	1.698	0.5890
10	Phosphorylase B	97000	4.987	23.56	1.789	0.5590
11	b-Galactosidase	116000	5.064	24.97	1.896	0.5274
12	Myosin	205000	5.312	28.25	2.145	0.4662
13	Ferritin light chain	?		17.07	1.296	0.7715
14	Ferritin heavy chain	?		17.97	1.364	0.7329

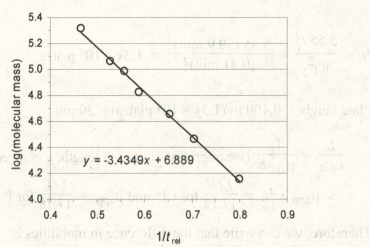

$$\log(\text{MM}) = (-3.434\,9)/t_{rel} + 6.889 = 4.239 \text{ for } t_{rel} = 1.296 \text{ (ferritin light chain)}$$
$$= 4.372 \text{ for } t_{rel} = 1.364 \text{ (ferritin heavy chain)}$$
$$\text{Molecular mass} = 10^{\log(\text{MM})} = 17\,300 \text{ (ferritin light chain)}$$
$$= 23\,500 \text{ (ferritin heavy chain)}$$

Molecular masses observed from amino acid sequences are 19 766 and 21 099 Da

25-41.

	A	B	C	D	E	F
1	Protein charge ladder	Charge (Δz)	Migration	Apparent	Electro-	($\mu n/\mu o$)-1
2		relative to	time (s)	mobility	phoretic	
3	Total length	native		m^2/(Vxs)	mobility	
4	of column	protein			μn	
5	Lt (m) =	0	343.0	6.27E-08	-7.02E-09	0.00
6	0.840	-1	355.4	6.05E-08	-9.20E-09	0.31
7	Distance to	-2	368.2	5.84E-08	-1.13E-08	0.61
8	detector	-3	382.2	5.63E-08	-1.34E-08	0.92
9	Ld (m) =	-4	395.5	5.44E-08	-1.53E-08	1.18
10	0.640	-5	409.1	5.26E-08	-1.71E-08	1.44
11	Voltage (V) =	-6	424.9	5.06E-08	-1.91E-08	1.72
12	25000	-7	438.5	4.90E-08	-2.07E-08	1.94
13	Field (V/Lt) =	-8	453.0	4.75E-08	-2.22E-08	2.17
14	2.98E+04	-9	467.0	4.60E-08	-2.37E-08	2.37
15	Migration time of	-10	482.0	4.46E-08	-2.51E-08	2.57
16	neutral marker (s) =	-11	496.4	4.33E-08	-2.64E-08	2.76
17	308.5	-12	510.1	4.22E-08	-2.75E-08	2.93
18	Electroosmotic	-13	524.1	4.10E-08	-2.87E-08	3.09
19	mobility (m^2/(Vxs)) =	-14	536.9	4.00E-08	-2.97E-08	3.23
20	6.97E-08	-15	551.4	3.90E-08	-3.07E-08	3.38
21		-16	565.1	3.81E-08	-3.16E-08	3.51
22	A20 =(A10/A17)/A14	-17	577.4	3.72E-08	-3.25E-08	3.63
23	D5 = ($A$10/C5)/$A$14	-18	588.5	3.65E-08	-3.32E-08	3.73
24	E5 = D5-$A$20					
25	F5 = (E5/$E$5)-1			slope from points 0 to -2 =		-3.27874
26	F25 = SLOPE(B5:B7,F5:F7)			slope from points 0 to -3 =		-3.28115
27				slope from points 0 to -4 =		-3.36122

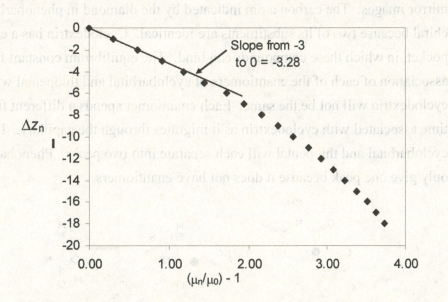

The graph is curved, most likely because the shape and friction coefficient of the protein change somewhat as the degree of acetylation increases. The first 4 points (from $x = 0$ to $x = -3$) lie on a straight line with a slope of -3.28. This slope is the charge of the unmodified protein, $z_0 = -3.28$. There is no reason why z_0 should be an integer. At any given pH, such as pH 8.3 in this experiment, the native protein is likely to have a fractional average charge because of different amounts of ionization in different species that are all in equilibrium.

25-42. SO_4^{2-}:　　$\mu_{ep} = -8.27 \times 10^{-8}$ m^2/(s·V) in Table 14-1

$$\mu_{app} = \mu_{eo} + \mu_{ep} = 16.1 \times 10^{-8} - 8.27 \times 10^{-8} = 7.8_3 \times 10^{-8} \text{ m}^2/(\text{s·V})$$

Br$^-$:　　$\mu_{ep} = -8.13 \times 10^{-8}$ m^2/(s·V) in Table 14-1

$$\mu_{app} = \mu_{eo} + \mu_{ep} = 16.1 \times 10^{-8} - 8.13 \times 10^{-8} = 7.9_7 \times 10^{-8} \text{ m}^2/(\text{s·V})$$

$$\mu_{av} = \tfrac{1}{2}(7.8_3 + 7.9_7 \times 10^{-8}) = 7.9_0 \times 10^{-8} \text{ m}^2/(\text{s·V})$$

$$\Delta\mu = (8.27 - 8.13) \times 10^{-8} = 0.14 \times 10^{-8} \text{ m}^2/(\text{s·V})$$

$$N = \left(4 \,(\text{Resolution})\frac{\mu_{av}}{\Delta\mu}\right)^2 = \left(4\,(2.0)\frac{7.9_0}{0.14}\right)^2 = 2.0 \times 10^5 \text{ plates}$$

25-43. In the absence of micelles, the expected order of elution is cations before neutrals before anions: thiamine < (niacinamide + riboflavin) < niacin. Since thiamine is eluted last, it must be most soluble in the micelles.

25-44. Carbon atoms labeled with black circles in cyclobarbital and thiopental are chiral, with four different substitutents. These compounds are not superimposable on their mirror images. The carbon atom indicated by the diamond in phenobarbital is not chiral because two of its substituents are identical. Cyclodextrin has a chiral pocket, in which these compounds can bind. The equilibrium constant for association of each of the enantiomers of cyclobarbital and thiopental with cyclodextrin will not be the same. Each enantiomer spends a different fraction of time associated with cyclodextrin as it migrates through the capillary. Therefore, cyclobarbital and thiopental will each separate into two peaks. Phenobarbital will only give one peak because it does not have enantiomers.

Cyclobarbital　　Thiopental　　Phenobarbital

25-45. (a) Plate height rises sharply at low velocity because bands broaden by diffusion when they spend more time in the capillary. This is the effect of the B term in the van Deemter equation, and it always operates in capillary electrophoresis. Plate height rises gradually at high velocity because solutes require a finite time to equilibrate with the micelles on the column. This is the effect of the C term in the van Deemter equation, and it is absent in capillary electrophoresis but present to a small extent in micellar electrokinetic capillary chromatography.

(b) There should be no irregular flow paths because the micelles are nanosized structures in solution. The large A term most likely arises from extra-column effects, such as the finite size of the injection plug and the finite width of the detector zone.

25-46. For the acid H_2A, the average charge is $\alpha_{HA^-} + 2\alpha_{A^{2-}}$, where α is the fraction in each form. From our study of acids and bases, we know that

$$\alpha_{HA^-} = \frac{K_1[H^+]}{[H^+]^2 + K_1[H^+] + K_1K_2} \text{ and } \alpha_{A^{2-}} = \frac{K_1K_2}{[H^+]^2 + K_1[H^+] + K_1K_2}$$

where K_1 and K_2 are acid dissociation constants of H_2A. The following spreadsheet finds the average charge of malonic acid (H_2M) and phthalic acid (H_2P) and finds that the maximum difference between them occurs at pH 5.55.

Charge Difference Between Malonic and Phthalic Acids

	A	B	C	D	E	F	G	H	I	J
1	Malonic:			Alpha	Alpha	Alpha	Alpha	Average charges		Charge
2	K1 =	pH	[H+]	HM-	M2-	HP-	P2-	Malonate	Phthalate	Difference
3	1.42E-03	5.52	3.0E-06	0.600	0.399	0.436	0.563	-1.398	-1.562	-0.16392
4	K2 =	5.53	3.0E-06	0.594	0.405	0.430	0.569	-1.403	-1.567	-0.16405
5	2.01E-06	5.54	2.9E-06	0.589	0.410	0.425	0.574	-1.409	-1.573	-0.16413
6	Phthalic:	5.55	2.8E-06	0.583	0.416	0.419	0.580	-1.415	-1.579	-0.16418
7	K1 =	5.56	2.8E-06	0.577	0.421	0.413	0.585	-1.420	-1.584	-0.16417
8	1.12E-03	5.57	2.7E-06	0.572	0.427	0.408	0.591	-1.426	-1.590	-0.16413
9	K2 =	5.58	2.6E-06	0.566	0.433	0.402	0.597	-1.432	-1.596	-0.16404
10	3.90E-06	5.59	2.6E-06	0.561	0.438	0.397	0.602	-1.437	-1.601	-0.16391
11										
12	D3 = $A$3*C3/(C3^2+$A$3*C3+$A$3*$A$5)						C3 = 10^-B3			
13	E3 = $A$3*$A$5/(C3^2+$A$3*C3+$A$3*$A$5)						H3 = -D3-2*E3			
14	F3 = $A$8*C3/(C3^2+$A$8*C3+$A$8*$A$10)						I3 = -F3-2*G3			
15	G3 = $A$8*$A$10/(C3^2+$A$8*C3+$A$8*$A$10)						J3 =I3-H3			

25-47. (a) $^{18}\alpha = \dfrac{K/R}{(K/R) + [H^+]} = \dfrac{K}{K + R[H^+]}$

Approximating $\bar{\alpha}$ as $^{16}\alpha$, we can write

$$\frac{\Delta\alpha}{\sqrt{\bar{\alpha}}} \approx \frac{\dfrac{K}{K+[H^+]} - \dfrac{K}{K+R[H^+]}}{\sqrt{\dfrac{K}{K+[H^+]}}}$$

$$\frac{\Delta\alpha}{\sqrt{\bar{\alpha}}} \approx \frac{(R-1)\,K\,[H^+]}{(K+[H^+])(K+R[H^+])}\,\frac{\sqrt{K+[H^+]}}{\sqrt{K}} = \frac{(R-1)\sqrt{K}\,[H^+]}{\sqrt{K+[H^+]}\,(K+R\,[H^+])}$$

(b) The function $\dfrac{\Delta\alpha}{\sqrt{\bar{\alpha}}}$ has the form $\dfrac{u}{v}$, where u and v are functions of $[H^+]$:

$$u = (R-1)\sqrt{K}\,[H^+] \qquad v = \sqrt{K+[H^+]}\,(K+R\,[H^+])$$

The derivative of u/v is $\dfrac{d\left(\dfrac{u}{v}\right)}{d\,[H^+]} = \dfrac{-v\,\dfrac{du}{d\,[H^+]} + u\,\dfrac{dv}{d\,[H^+]}}{v^2}$.

Setting the derivative equal to zero gives

$$-v\,\frac{du}{d\,[H^+]} + u\,\frac{dv}{d\,[H^+]} = 0 \;\Rightarrow\; v\,\frac{du}{d\,[H^+]} = u\,\frac{dv}{d\,[H^+]} \qquad (A)$$

The derivatives are

$$\frac{du}{d\,[H^+]} = (R-1)\sqrt{K}$$

$$\frac{dv}{d\,[H^+]} = (K+R[H^+])\tfrac{1}{2}(K+[H^+])^{-1/2} + \sqrt{K+[H^+]}\,(R)$$

Inserting u and v and the two derivatives into Equation A gives an equation that can be solved for $[H^+]$.

$$v\left(\frac{du}{d\,[H^+]}\right) = u\left(\frac{dv}{d\,[H^+]}\right)$$

$$\sqrt{K+[H^+]}\,(K+R[H^+])\left\{(R-1)\sqrt{K}\right\}$$

$$= (R-1)\sqrt{K}[H^+]\left\{K+R[H^+])\tfrac{1}{2}(K+[H^+])^{-1/2} + \sqrt{K+[H^+]}\,(R)\right\}$$

Solving for $[H^+]$ gives — after many lines of algebra —

$$[H+] = \frac{K + K\sqrt{1 + 8R}}{2R}$$

(c) Setting $R = 1$ gives $[H^+] = \dfrac{K + K\sqrt{9}}{2} = 2K$

$$-\log [H^+] = -\log K - \log 2$$

$$\text{pH} = \text{p}K - 0.30$$

25-48. (a) In ion mobility spectrometry, gaseous ions are generated by irradiating analyte plus reagent gas (such as acetone in air) with high energy electrons (β emission) from radioactive ^{63}Ni. Periodically, ions are admitted into a drift tube by a short voltage pulse applied to an electronic gate (a grid). In the drift tube, ions experience a constant electric field that causes either cations or anions to migrate from the gate to a detector at the other end of the tube. The time to reach the detector is the drift time. Ions drift at a constant speed governed by the driving force of the electric field and the retarding force of friction (drag) by the atmosphere of gas (usually dry air) in the drift tube. Also, gas in the drift tube flows from the detector to the source, further decreasing the migration speed of an ion.

The electric field in ion mobility spectrometry causes ions to migrate from the source to the detector, just as the electric field in electrophoresis causes ions to migrate. Drift time in ion mobility spectrometry is the same quantity as migration time in electrophoresis. The mobility of an ion in liquid or in gas is governed by the charge-to-size ratio. The greater the charge and the smaller the size, the greater the mobility. In liquid or gas, the retarding force is caused by collisions with solvent or gas molecules.

(b)

	A	B	C	D	E	F
1	Ion Mobility Spectrometry					
2						
3	k =		Volts	t_d (s)	$w_{1/2}$ (s)	N
4	1.38065E-23	J/K	100	5.0000	2.68E-01	1.94E+03
5	e =		1000	0.5000	8.47E-03	1.94E+04
6	1.60218E-19	C	2000	0.2500	2.99E-03	3.87E+04
7	T =		3000	0.1667	1.63E-03	5.80E+04
8	300	K	4000	0.1250	1.06E-03	7.73E+04
9	μ =		5000	0.1000	7.59E-04	9.64E+04
10	0.00008	m^2/(sV)	6000	0.0833	5.78E-04	1.15E+05
11	z =		7000	0.0714	4.60E-04	1.34E+05
12	1		8000	0.0625	3.77E-04	1.52E+05
13	L =		9000	0.0556	3.18E-04	1.70E+05
14	0.2	m	10000	0.0500	2.72E-04	1.87E+05
15	t_g =		12000	0.0417	2.10E-04	2.19E+05
16	5.00E-05	s	14000	0.0357	1.69E-04	2.47E+05
17	16kT(ln2)/ez =		16000	0.0313	1.41E-04	2.71E+05
18	2.86707E-01		18000	0.0278	1.22E-04	2.90E+05
19			20000	0.0250	1.07E-04	3.03E+05
20	D4 = $A$14^2/($A$10*C4)					
21	E4 = SQRT($A$16^2+($A$18/C4)*D4^2)					
22	F4 = 5.55*(D4/E4)^2					

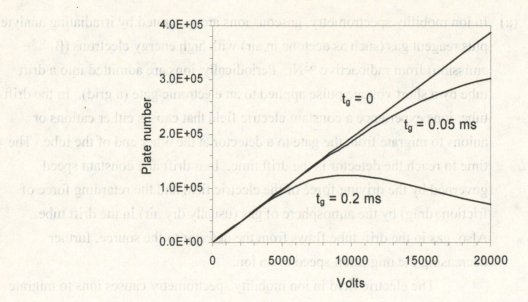

Increasing V increases N by decreasing the drift time, and therefore decreasing the time for diffusion to broaden the peak. Increasing the time that the ion gate is open increases the initial width of the peak, and therefore decreases the plate number. The peak can never be narrower than the pulse that is admitted by the gate. At high voltage, the effect of t_g on plate number overwhelms the effect of t_d. The disadvantage of using a short gate opening time is that fewer ions are admitted to the drift cell and the signal will be weaker.

(c) Decreasing T increases N because diffusional broadening decreases with decreasing temperature.

(d) $N = 5.55 \, (t_d/w_{1/2})^2 = 5.55 \, (0.024\,925 \text{ s}/0.000\,154 \text{ s})^2 = 1.45 \times 10^5$ plates

Theoretical $w_{1/2}^2 = t_g^2 + \left(\dfrac{16kT \ln 2}{Vez}\right) t_d^2$

$= (5.0 \times 10^{-5} \text{ s})^2 + \left(\dfrac{16(1.38 \times 10^{-23} \text{ J/K})(300 \text{ K}) \ln 2}{(12\,500 \text{ V})(1.602 \times 10^{-19} \text{ C})(1)}\right)(0.0249\,25 \text{ s})^2$

$= 1.674 \times 10^{-8} \text{ s}^2$

Theoretical $N = 5.55 \, (t_d^2/w_{1/2}^2) = 5.55 \, (0.024\,925 \text{ s})^2/(1.674 \times 10^{-8} \text{ s}^2)$

$= 2.06 \times 10^5$ plates

(e) $R = (\sqrt{N}/4)(\gamma - 1) = \dfrac{\sqrt{80\,000}}{4}\left(\dfrac{22.5}{22.0} - 1\right) = 1.6$

26-1. (a) In adsorption, a substance becomes bound to the surface of another substance. In absorption, a substance is taken up inside another substance.

(b) An inclusion is an impurity that occupies lattice sites in a crystal. An occlusion is an impurity trapped inside a pocket in a growing crystal.

26-2. An ideal gravimetric precipitate should be insoluble, easily filterable, pure, and possess a known, constant composition.

26-3. High relative supersaturation often leads to formation of colloidal product with a large amount of impurities.

26-4. Relative supersaturation can be decreased by increasing temperature (for most solutions), mixing well during addition of precipitant, and using dilute reagents. Homogeneous precipitation is also an excellent way to control relative supersaturation.

26-5. Washing with electrolyte preserves the electric double layer and prevents peptization.

26-6. HNO_3 evaporates during drying. $NaNO_3$ is nonvolatile and will lead to a high mass for the precipitate.

26-7. During the first precipitation, the concentration of unwanted species in the solution is high, giving a relatively high concentration of impurities in the precipitate. In the reprecipitation, the level of solution impurities is reduced, giving a purer precipitate.

26-8. In thermogravimetric analysis, the mass of a sample is measured as the sample is heated. The mass lost during decomposition provides some information about the composition of the sample.

26-9. A quartz crystal microbalance consists of a specially cut, thin, disk-shape slice of quartz with gold electrodes on each of the two faces. Application of an oscillating electric field causes the crystal to oscillate at a characteristic frequency. Binding of small masses to the gold electrodes increases the mass of the system and lowers the oscillation frequency. From the change in frequency, we can deduce how much mass was bound.

26-10. $\dfrac{0.2146 \text{ g AgBr}}{187.772 \text{ g AgBr/mol}} = 1.1429 \times 10^{-3}$ mol AgBr

$[\text{NaBr}] = \dfrac{1.1429 \times 10^{-3} \text{ mol}}{50.00 \times 10^{-3} \text{ L}} = 0.02286$ M

26-11. $\dfrac{0.104 \text{ g CeO}_2}{172.114 \text{ g CeO}_2/\text{mol}} = 6.043 \times 10^{-4}$ mol $CeO_2 = 6.043 \times 10^{-4}$ mol Ce

$= 0.08466$ g Ce

weight % Ce $= \dfrac{0.08466 \text{ g}}{4.37 \text{ g}} \times 100 = 1.94$ wt%

26-12. Formula mass of AgCl = 107.8 + 35.4 = 143.2

mol AgCl $= \dfrac{0.08890 \text{ g AgCl}}{143.2 \text{ g AgCl/mol AgCl}} = 6.208_1 \times 10^{-4}$ mol

mol radium $= \dfrac{6.208_1 \times 10^{-4} \text{ mol Cl}}{2 \text{ mol Cl/mol Ra}} = 3.104_0 \times 10^{-4}$ mol

$3.104_0 \times 10^{-4}$ mol $\text{RaCl}_2 = \dfrac{0.09192 \text{ g RaCl}}{x \text{ g RaCl}_2/\text{mol RaCl}_2}$

$x = \dfrac{0.09192 \text{ g RaCl}}{3.104_0 \times 10^{-4} \text{ mol RaCl}_2} = 296.1_3$ g $\text{RaCl}_2/\text{mol RaCl}_2$

formula mass of RaCl_2 = atomic mass of Ra + 2(35.4 g/mol) = 296.1_3 g/mol

$\Rightarrow$ atomic mass of Ra = 225.3 g/mol

26-13. One mole of product (206.240 g) comes from one mole of piperazine (86.136 g). Grams of piperazine in sample =
(0.7129 g of piperazine / g of sample) × (0.05002 g of sample) = 0.03566.
Mass of product $= \left(\dfrac{206.240}{86.136}\right)(0.03566) = 0.08538$ g.

26-14. 2.500 g bis(dimethylglyoximate) nickel (II) = 8.6532×10^{-3} mol Ni =
0.50785 g Ni = 50.79% Ni.

26-15. Formula masses: $\text{CaC}_{14}\text{H}_{10}\text{O}_6 \cdot \text{H}_2\text{O}$ (332.32), $CaCO_3$ (100.09), CaO (56.08). At 550°, $\text{CaC}_{14}\text{H}_{10}\text{O}_6 \cdot \text{H}_2\text{O}$ is converted to $CaCO_3$ (calcium carbonate).
332.32 g of starting material will produce 100.09 g of CaO.

Mass at $550° = (100.09/332.32)(0.635\ 6\ g) = 0.191\ 4\ g$. At $1\ 000°C$, the product is CaO (calcium oxide) and the mass is $(56.08/332.32)(0.635\ 6\ g) = 0.107\ 3\ g$.

26-16. $2.378\ mg\ CO_2\ /\ (44.010\ g/mol) = 5.403\ 3 \times 10^{-5}\ mol\ CO_2 = 5.403\ 3 \times 10^{-5}\ mol\ C$

$= 6.490\ 0 \times 10^{-4}\ g\ C$

$ppm\ C = 10^6\ (6.490\ 0 \times 10^{-4}\ /\ 6.234) = 104.1\ ppm$

26-17. 2.07% of $0.998\ 4\ g = 0.020\ 67\ g$ of $Ni = 3.521 \times 10^{-4}$ mol of Ni.

This requires $(2)(3.521 \times 10^{-4})$ mol of $DMG = 0.081\ 77\ g$.

A 50.0% excess is $(1.5)(0.081\ 77\ g) = 0.122\ 7\ g$. The mass of solution containing $0.122\ 7\ g$ is $0.122\ 7\ g\ DMG\ /\ (0.021\ 5\ g\ DMG/g\ solution) = 5.705\ g$ of solution.

The volume of solution is $5.705\ g/(0.790\ g/mL) = 7.22\ mL$.

26-18. Moles of Fe in product (Fe_2O_3) = moles of Fe in sample.

Because 1 mole of (Fe_2O_3) contains 2 moles of Fe, we can write the equation

$\dfrac{2\ (0.264\ g)}{159.69\ g/mol} = 3.306 \times 10^{-3}$ mol of Fe.

This many moles of Fe equals $0.919\ 2\ g$ of $FeSO_4 \cdot 7\ H_2O$. Because we analyzed just $2.998\ g$ out of $22.131\ g$ of tablets, the $FeSO_4 \cdot 7\ H_2O$ in the $22.131\ g$ sample is $\dfrac{22.131\ g}{2.998\ g}\ (0.919\ 2\ g) = 6.786\ g$. This is the $FeSO_4 \cdot 7\ H_2O$ content of 20 tablets.

The content in one tablet is $(6.786\ g)/20 = 0.339\ g$.

26-19. (a) Mass of product $(CaCO_3) = 18.546\ 7\ g - 18.231\ 1\ g = 0.315\ 6\ g$

$mol\ CaCO_3 = \left(\dfrac{0.315\ 6\ g}{100.087\ g/mol}\right) = 3.153 \times 10^{-3}\ mol$

The product contains $3.153\ mmol\ Ca = (3.153 \times 10^{-3}\ mol)(40.078\ g/mol) = 0.1264\ g\ Ca$.

$wt\%\ Ca = \dfrac{0.1264\ g\ Ca}{0.632\ 4\ g\ mineral} \times 100 = 19.98\ \%$

(b) The solutions are heated before mixing to increase the solubility of the product that will precipitate. If the solution is less supersaturated during the precipitation, crystals form more slowly and grow to be larger and purer than if they precipitate rapidly. The larger crystals are easier to filter.

(c) $(NH_4)_2C_2O_4$ provides oxalate ion to prevent CaC_2O_4 from redissolving. Also, the ammonium and oxalate ions provide an ionic atmosphere that prevents the precipitate from peptizing (breaking into colloidal particles).

(d) $AgNO_3$ solution is added to the filtrate to test for Cl^- in the filtrate. If Cl^- is present, $AgCl(s)$ will precipitate when Ag^+ is added. The source of Cl^- is the HCl used to dissolve the mineral. All the original solution needs to be washed away, so no extra material is present that would increase the mass of final product, which should be pure $CaCO_3(s)$.

26-20. (a) $70 \text{ kg} \left(\dfrac{6.3 \text{ g P}}{\text{kg}} \right) = 441 \text{ g P}$ in 8.00×10^3 L. This corresponds to

$$\frac{441 \text{ g P}}{8.00 \times 10^3 \text{L}} = 0.055_1 \text{ g/L or } 5.5_1 \text{ mg/100 mL.}$$

(b) Fraction of P in one formula mass is $\dfrac{2(30.974)}{3\,596.46} = 1.722\%$.

P in $0.338\,7$ g of $P_2O_5 \cdot 24\,MoO_3 = (0.017\,22)(0.338\,7) = 5.834$ mg

This is near the amount expected from a dissolved man.

26-21. Let x = mass of NH_4Cl and y = mass of K_2CO_3.

For the first part, 1/4 of the sample (25 mL) gave 0.617 g of precipitate containing both products:

$$\frac{1}{4}\left[\underbrace{\overbrace{\left(\frac{x}{53.492}\right)}^{\text{mol } NH_4Cl} (337.27)}_{\text{g } \phi_4BNH_4} + \underbrace{\overbrace{\left(\frac{2y}{138.21}\right)}^{\text{mol } K_2CO_3 \times 2} (358.33)}_{\text{g } \phi_4BK} \right] = 0.617 \text{ g} \qquad (\phi = \text{phenyl} = C_6H_5)$$

We multiplied moles of K_2CO_3 by 2 because one mole of K_2CO_3 gives 2 moles of ϕ_4BK. In the second part, half of the sample (50 mL) gave 0.554 g of ϕ_4BK:

$$\frac{1}{2}\underbrace{\overbrace{\left(\frac{2y}{138.21}\right)}^{\text{mol } K_2CO_3 \times 2}(358.33)}_{\text{g } \phi_4BK} = 0.554 \text{ g} \implies y = 0.213_7 \text{ g} = 14.5 \text{ wt\% } K_2CO_3$$

Putting this value of y into the first equation gives $x = 0.215_7$ g $= 14.6$ wt% NH_4Cl

26-22. $\underbrace{Fe_2O_3 + Al_2O_3}_{2.0\dot{1}9 \text{ g}} \xrightarrow[H_2]{\text{Heat}} \underbrace{Fe + Al_2O_3}_{1.7\dot{7}4 \text{ g}}$

The mass of oxygen lost is $2.019 - 1.774 = 0.245$ g, which equals 0.01531 moles of oxygen atoms. For every 3 moles of oxygen there is 1 mole of Fe_2O_3, so moles of $Fe_2O_3 = \frac{1}{3}(0.01531) = 0.005105$ mol of Fe_2O_3. This much Fe_2O_3 equals 0.815 g, which is 40.4 wt% of the original sample.

26-23. Let x = g of $FeSO_4 \cdot (NH_4)_2 SO_4 \cdot 6H_2O$ and y = g of $FeCl_2 \cdot 6H_2O$.

We can say that $x + y = 0.5485$ g. The moles of Fe in the final product (Fe_2O_3) must equal the moles of Fe in the sample.

The moles of Fe in $Fe_2O_3 = 2$ (moles of Fe_2O_3) $= 2\left(\dfrac{0.1678}{159.69}\right) = 0.0021016$ mol.

Mol Fe in $FeSO_4 \cdot (NH_4)_2 SO_4 \cdot 6H_2O = x / 392.13$ and
mol Fe in $FeCl_2 \cdot 6H_2O = y / 234.84$.

$$0.0021016 = \frac{x}{392.13} + \frac{y}{234.84} \tag{1}$$

Substituting $x = 0.5485 - y$ into Eq. (1) gives $y = 0.41146$ g of $FeCl_2 \cdot 6H_2O$.

Mass of Cl $= 2\left(\dfrac{35.453}{234.84}\right)(0.41146) = 0.12423$ g $= 22.65$ wt%

26-24. (a) Let x = mass of $AgNO_3$ and $(0.4321 - x)$ = mass of $Hg_2(NO_3)_2$ in unknown. Each mol of $AgNO_3$ gives $1/3$ mol $Ag_3[Co(CN)_6]$ and each mol of $Hg_2(NO_3)_2$ gives $1/3$ mol $(Hg_2)_3[Co(CN)_6]_2$. Mass of both products must equal 0.4515 g:

$$\underbrace{\frac{1}{3}\left(\frac{x}{169.873}\right)(538.643)}_{\text{mass of } Ag_3Co(CN)_6} + \underbrace{\frac{1}{3}\left(\frac{0.4321-x}{525.19}\right)(1633.62)}_{\text{mass of } (Hg_2)_3[Co(CN)_6]_2} = 0.4515$$

$\Rightarrow x = 0.1731$ g $= 40.05$ wt%

(b) 0.30% error in 0.4515 g $= \pm 0.00135$ g. This changes the equation of (a) to:

$\dfrac{1}{3}\left(\dfrac{x}{169.873}\right)(538.643) + \dfrac{1}{3}\left(\dfrac{0.4321-x}{525.19}\right)(1633.62) = 0.4515\ (\pm 0.00135)$

$1.056952\,x + 0.448020 - 1.036844\,x = 0.4515\ (\pm 0.00135)$

$0.020109\,x = 0.4515\ (\pm 0.00135) - 0.448020$

$0.020109\,x = 0.003480\ (\pm 0.00135)$

$x = \dfrac{0.003480(\pm 0.00135)}{0.020109} = \dfrac{0.003480(\pm 38.8\%)}{0.020109} = 0.17$ g $\pm 39\%$

26-25. (a) Balanced equation for overall (31.8%) mass loss:

$$Y_2(OH)_5Cl \cdot xH_2O \xrightarrow{\text{31.8\% mass loss}} Y_2O_3 + \underbrace{xH_2O + 2H_2O}_{} + HCl$$

FM 298.30 + x(18.015)   FM 225.81   FM (2+x)(18.015))   FM 36.461

$$\underbrace{(2 + x)(18.015) + 36.461}_{\text{mass lost}} = \underbrace{(0.318)[298.30 + x(18.015)]}_{\text{31.8\% of original mass}} \Rightarrow x = 1.82$$

(b) Logical molecular units that could be lost are H_2O and HCl. At ~8.1% mass loss, the product is $Y_2(OH)_5Cl$. Loss of 2 more H_2O would give a total mass loss of

$$\frac{1.82H_2O + 2H_2O}{Y_2(OH)_5Cl \cdot 1.82H_2O} = \frac{68.82}{331.09} = 20.8\%$$

Loss of HCl from $Y_2(OH)_5Cl$ would give a total mass loss of

$$\frac{1.82H_2O + HCl}{Y_2(OH)_5Cl \cdot 1.82H_2O} = \frac{69.25}{331.09} = 20.9\%$$

The composition at the ~19.2% plateau could be either $Y_2O_2(OH)Cl$ (from loss of $2H_2O$) or $Y_2O(OH)_4$ (from loss of HCl).

26-26. (a) $\alpha = \dfrac{\text{mass of } KPO_3}{\text{mass of } K(D_xH_{1-x})_2PO_4} = \dfrac{118.070\,3}{136.085\,3 + 2.012\,55x}$

Cross-multiply: $(136.085\,3)\alpha + (2.012\,55)\alpha x = 118.070\,3$

$(2.012\,55)\alpha x = 118.070\,3 - (136.085\,3)\alpha$

Divide by $(2.012\,55)\alpha$:

$$x = \frac{118.070\,3}{(2.012\,55)\alpha} - \frac{(136.085\,3)\alpha}{(2.012\,55)\alpha} \qquad\qquad x = \frac{58.667\,0}{\alpha} - 67.618\,3$$

For fully deuterated material, $x = 1$ and $\alpha = \dfrac{58.667\,0}{x + 67.618\,3} = 0.854\,976$

(b) $x = \dfrac{58.667\,0}{0.856\,7_7} - 67.618\,3 = 0.856\,3_2$

(c) For the function $x = f(\alpha)$, we can write

$$e_x = \sqrt{\left(\frac{\partial F}{\partial \alpha}\right)^2 e_\alpha{}^2}$$

For $x = \dfrac{58.667\,0}{\alpha} - 67.618\,3$, $\dfrac{\partial F}{\partial \alpha} = -\dfrac{58.667\,0}{\alpha^2}$ giving

$$e_x = \sqrt{\left(-\frac{58.667\,0}{\alpha^2}\right)^2 e_\alpha{}^2} = \frac{58.667\,0\, e_\alpha}{\alpha^2}$$

(d) For $e_\alpha = 0.000\ 1$, $e_x = \dfrac{(58.667\ 0)(0.000\ 1)}{(0.856\ 7_7)^2} = 0.008$

D:H stoichiometry $= x \pm e_x = 0.856 \pm 0.008$

If e_x were 0.001, then $e_x = \dfrac{(58.667\ 0)(0.001)}{(0.856\ 7_7)^2} = 0.08$ and

D:H stoichiometry $= 0.86 \pm 0.08$

26-27. (a) Formula mass of $YBa_2Cu_3O_{7-x} = 666.19 - (16.00)\,x$

mmol of $YBa_2Cu_3O_{7-x}$ in experiment $= \dfrac{34.397\ \text{mg}}{[666.19 - (16.00)x]\text{mg/mmol}}$

mmol of oxygen atoms lost in experiment $= \dfrac{(34.397 - 31.661)\ \text{mg}}{16.00\ \text{mg/mmol}}$

$= 0.171\ 00$ mmol

From the stoichiometry of the reaction, we can write

$$\dfrac{\text{mmol oxygen atoms lost}}{\text{mmol } YBa_2Cu_3O_{7-x}} = \dfrac{3.5 - x}{1}$$

$$\dfrac{0.171\ 00}{34.397\ /\ [666.19 - (16.00)x]} = 3.5 - x \Rightarrow x = 0.204\ 2$$

(without regard to significant figures)

(b) Now let the uncertainty in each mass be 0.002 mg and let all atomic and molecular masses have negligible uncertainty.

The mmol of oxygen atoms lost are:

$$\dfrac{[34.397(\pm0.002) - 31.661(\pm0.002)]\text{mg}}{16.00\ \text{mg}\,/\,\text{mmol}} = \dfrac{2.736(\pm0.002\ 8)}{16.00}$$

$$= 0.171\ 00\ (\pm0.102\%)$$

The relative error in the mass of starting material is $\dfrac{0.002}{34.397} = 0.005\ 8\%$

The master equation becomes

$$\dfrac{0.171\ 00\ (\pm0.102\%)}{34.397\ (\pm0.005\ 8\%)/[666.19 - (16.00)\,x]} = 3.5 - x$$

$$0.171\ 00\ (\pm0.102\%)[666.19 - (16.00)x] = (3.5 - x)[34.397\ (\pm0.005\ 8\%)]$$

$$113.918\ (\pm0.116) - [2.736\ (\pm0.002\ 79)]\,x$$
$$= 120.389\ 5\ (\pm0.006\ 98) - [34.397\ (\pm0.002)]\,x$$

$$[31.66\ (\pm0.003\ 46)]\,x = 6.471\ 5\ (\pm0.116)$$

$$= 0.204\ 4\ (\pm1.79\%) = 0.204 \pm 0.004$$

26-28.

Neutral DTPA has 2 carboxylic acid protons and 3 ammonium protons. We are not given the pK_a values, but, by analogy with EDTA, we expect carboxyl pK_a values to be below ~3 and ammonium pK_a values to be above ~6. At pH 14, we expect all the acidic protons of DTPA to be dissociated, so the predominant species will be DTPA^{5-}. At pH 3-4, the nitrogen atoms should all be protonated, but the carboxyl groups should be all (or mostly) deprotonated. The predominant species is probably DTPA^{2-}.

For HSO_4^-, $pK_a = 2.0$. At pH 14 and at pH 3, sulfate should mainly be in the form SO_4^{2-}.

At pH 14, DTPA^{5-} is apparently a strong enough ligand to chelate Ba^{2+} and dissolve $BaSO_4(s)$. At pH 3-4, DTPA^{2-} is not a strong enough ligand to dissolve $BaSO_4(s)$. Another way to say the same thing is that H^+ at a concentration of 10^{-3}-10^{-4} M competes with Ba^{2+} for binding sites on DTPA, but H^+ at a concentration of 10^{-14} M does not compete with Ba^{2+} for binding sites on DTPA.

26-29. In *combustion*, a substance is heated in the presence of excess O_2 to convert carbon to CO_2 and hydrogen to H_2O. In *pyrolysis*, the substance is decomposed by heating in the absence of added O_2. All oxygen in the sample is converted to CO by passage through a suitable catalyst.

26-30. WO_3 catalyzes the complete combustion of C to CO_2 in the presence of excess O_2. Cu converts SO_3 to SO_2 and removes excess O_2.

26-31. The tin capsule melts and is oxidized to SnO_2 to liberate heat and crack the sample. Tin uses the available oxygen immediately, ensures that sample oxidation occurs in the gas phase, and acts as an oxidation catalyst.

26-32. By dropping the sample in before very much O_2 is present, pyrolysis of the sample to give gaseous products occurs prior to oxidation. This minimizes the formation of nitrogen oxides.

26-33.
$$C_6H_5CO_2H + \frac{15}{2}O_2 \rightarrow 7\,CO_2 + 3\,H_2O$$
FM 122.123   44.010   18.015

One mole of $C_6H_5CO_2H$ gives 7 moles of CO_2 and 3 moles of H_2O.

4.635 mg of benzoic acid $= 0.037\,95$ mmol, which gives $0.265\,7$ mmol CO_2

($= 11.69$ mg CO_2) and $0.113\,9$ mmol H_2O ($= 2.051$ mg H_2O).

26-34. $C_8H_7NO_2SBrCl + 9\frac{1}{4}O_2 \rightarrow 8CO_2 + \frac{5}{2}H_2O + \frac{1}{2}N_2 + SO_2 + HBr + HCl$

26-35.   100 g of compound contains 46.21 g C, 9.02 g H, 13.74 g N, and 31.03 g O. The atomic ratios are $C:H:N:O =$

$$\frac{46.21\text{ g}}{12.010\,7\text{ g/mol}} : \frac{9.02\text{ g}}{1.007\,94\text{ g/mol}} : \frac{13.74\text{ g}}{14.006\,74\text{ g/mol}} : \frac{31.03\text{ g}}{15.999\,4\text{ g/mol}}$$
$$= 3.847 : 8.94_9 : 0.981\,0 : 1.940$$

Dividing by the smallest factor ($0.981\,0$) gives the ratios $C:H:N:O = 3.922 :$ $9.12 : 1 : 1.978$. The empirical formula is probably $C_4H_9NO_2$.

26-36.
$$C_6H_{12} + C_2H_4O \rightarrow CO_2 + H_2O$$
FM 84.159   44.053   44.010

Let $x =$ mg of C_6H_{12} and $y =$ mg of C_2H_4O

$$x + y = 7.290.$$

We also know that moles of $CO_2 = 6$ (moles of C_6H_{12}) + 2 (moles of C_2H_4O), by conservation of carbon atoms.

$$6\left(\frac{x}{84.159}\right) + 2\left(\frac{y}{44.053}\right) = \frac{21.999}{44.010}$$

Making the substitution $x = 7.290 - y$ allows us to solve for y.

$$y = 0.767\text{ mg} = 10.5\text{ wt\%}.$$

26-37.   The atomic ratio $H:C$ is

$$\frac{\left(\dfrac{6.76 \pm 0.12\text{ g}}{1.007\,94\text{ g/mol}}\right)}{\left(\dfrac{71.17 \pm 0.41\text{ g}}{12.010\,7\text{ g/mol}}\right)} = \frac{6.707 \pm 0.119}{5.926 \pm 0.034\,1} = \frac{6.707 \pm 1.78\%}{5.926 \pm 0.576\%} = 1.132 \pm 0.021$$

If we define the stoichiometry coefficient for C to be 8, then the stoichiometry coefficient for H is $8(1.132 \pm 0.021) = 9.06 \pm 0.17$.

The atomic ratio N:C is

$$\frac{\left(\dfrac{10.34 \pm 0.08 \text{ g}}{14.006\,74 \text{ g/mol}}\right)}{\left(\dfrac{71.17 \pm 0.41 \text{ g}}{12.010\,7 \text{ g/mol}}\right)} = \frac{0.738\,2 \pm 0.005\,7}{5.926 \pm 0.034\,1} = \frac{0.738\,2 \pm 0.774\%}{5.925 \pm 0.576\%}$$

$$= 0.124\,6 \pm 0.001\,2$$

If we define the stoichiometry coefficient for C to be 8, then the stoichiometry coefficient for N is $8(0.124\,6 \pm 0.001\,2) = 0.996\,8 \pm 0.009\,6$.

The empirical formula is reasonably expressed as $C_8H_{9.06\pm0.17}N_{0.997\pm0.010}$.

26-38. The reaction between H_2SO_4 and NaOH can be written

$$H_2SO_4 + 2NaOH \rightarrow 2H_2O + Na_2SO_4$$

One mole of H_2SO_4 requires two moles of NaOH. In 3.01 mL of 0.015 76 M NaOH there are $(0.003\,01 \text{ L})(0.015\,76 \text{ mol/L}) = 4.74_4 \times 10^{-5}$ mol of NaOH. The moles of H_2SO_4 must have been $(\frac{1}{2})(4.74_4 \times 10^{-5}) = 2.37_2 \times 10^{-5}$ mol. Because one mole of H_2SO_4 contains one mole of S, there must have been $2.37_2 \times 10^{-5}$ mol of S $(= 0.760_6$ mg). The percentage of S in the sample is

$$\frac{0.760_6 \text{ mg S}}{6.123 \text{ mg sample}} \times 100 = 12.4 \text{ wt\%}.$$

26-39. (a) Experiment 1: $\bar{x} = 10.16_0$ µmol Cl⁻ $s = 2.70_7$ µmol Cl⁻

95% confidence interval $= \bar{x} \pm \dfrac{ts}{\sqrt{n}}$

$$= 10.16_0 \pm \frac{(2.262)(2.70_7)}{\sqrt{10}} = 10.16_0 \pm 1.93_6 \text{ µmol Cl}^-$$

Experiment 2: $\bar{x} = 10.77_0$ µmol Cl⁻ $s = 3.20_5$ µmol Cl⁻

95% confidence interval $= \bar{x} \pm \dfrac{ts}{\sqrt{n}}$

$$= 10.77_0 \pm \frac{(2.262)(3.20_5)}{\sqrt{10}} = 10.77_0 \pm 2.29_3 \text{ µmol Cl}^-$$

(b) $s_{pooled} = \sqrt{\dfrac{s_1^2\,(n_1 - 1) + s_2^2\,(n_2 - 1)}{n_1 + n_2 - 2}} = \sqrt{\dfrac{2.70_7^2\,(10 - 1) + 3.20_5^2\,(10 - 1)}{10 + 10 - 2}}$

$$= 2.96_6$$

$$t_{calculated} = \frac{|\bar{x}_1 - \bar{x}_2|}{s_{pooled}} \sqrt{\frac{n_1 n_2}{n_1 + n_2}} = \frac{|10.16_0 - 10.77_0|}{2.96_6} \sqrt{\frac{(10)(10)}{10 + 10}}$$

$$= 0.46_0 < t_{tabulated} \text{ for 18 degrees of freedom for}$$

95% confidence level (or even for 50% confidence level)

Therefore, the **difference is not significant**. The result means that addition of excess Cl^- prior to precipitation does not lead to additional coprecipitation of Cl^- under the conditions of these experiments. (In general, under other conditions we would expect extra Cl^- to lead to extra coprecipitation.)

(c)   10.0 mg of SO_4^{2-} = 0.104_{10} mmol = 24.2_{95} mg $BaSO_4$

(d)   In Experiment 1, the precipitate includes an additional 10.16_0 μmol Cl^- = 5.08 μmol $BaCl_2$ = 1.05_8 mg $BaCl_2$. The increase in mass is $(1.05_8)/(24.2_{95})$ = 4.35%. This represents a large error in the analysis.

26-40.   (i) $I^-(excess) + Ag^+ \rightarrow AgI(s)$ $[Ag^+] = K_{sp}$ (for AgI) / $[I^-]$

(ii)   A stoichiometric quantity of Ag^+ has been added that would be just equivalent to I^-, if no Cl^- were present. Instead, a tiny amount of AgCl precipitates and a slight amount of I^- remains in solution.

(iii) $Cl^-(excess) + Ag^+ \rightarrow AgCl(s)$ $[Ag^+] = K_{sp}$ (for AgCl) / $[Cl^-]$

(iv)   Virtually all I^- and Cl^- have precipitated.
$[Ag^+] \approx [Cl^-] \Rightarrow [Ag^+] = \sqrt{K_{sp} \text{ (for AgCl)}}$

(v)   There is excess Ag^+ delivered from the buret.
$$[Ag^+] = [Ag^+]_{titrant} \cdot \left(\frac{\text{volume added past 2nd equivalence point}}{\text{total volume}} \right)$$

26-41.   At V_e, moles of Ag^+ = moles of I^-
$(V_e \text{ mL})(0.051\,1 \text{ M}) = (25.0 \text{ mL})(0.082\,3 \text{ M}) \Rightarrow V_e = 40.26 \text{ mL}$
When V_{Ag^+} = 39.00 mL, $[I^-] = \dfrac{40.26 - 39.00}{40.26} (0.082\,30) \left(\dfrac{25.00}{25.00 + 39.00} \right)$

$= 1.006 \times 10^{-3}$ M. $[Ag^+] = K_{sp}/[I^-] = 8.3 \times 10^{-14}$ M $\Rightarrow pAg^+ = 13.08$.
When V_{Ag^+} = V_e, $[Ag^+][I^-] = x^2 = K_{sp} \Rightarrow x = [Ag^+] = 9.1 \times 10^{-9}$ M
$\Rightarrow pAg^+ = 8.04$.
When V_{Ag^+} = 44.30 mL, there is an excess of $(44.30 - 40.26) = 4.04$ mL of Ag^+. $[Ag^+] = \left(\dfrac{4.04}{25.00 + 44.30} \right) (0.051\,10) = 2.98 \times 10^{-3}$ M $\Rightarrow pAg^+ = 2.53$.

26-42.   At the equivalence point, $[Ag^+][I^-] = K_{sp} \Rightarrow (x)(x) = 8.3 \times 10^{-17}$
$\Rightarrow [Ag^+] = 9.1 \times 10^{-9}$ M. The concentration of Cl^- in the titration solution is the initial concentration (0.050 0 M) corrected for dilution from an initial volume of 40.00 mL up to ~63.85 mL at the equivalence point:

$$[Cl^-] = (0.050\,0 \text{ M}) \left(\frac{40.00}{63.85} \right) = 0.031\,3 \text{ M}$$

Is AgCl solubility exceeded? The reaction quotient is $Q = [Ag^+][Cl^-] = (9.1 \times 10^{-9})(0.031\ 3) = 2.8 \times 10^{-10}$, which is greater than K_{sp} for AgCl ($= 1.8 \times 10^{-10}$). Therefore, AgCl begins to precipitate before AgI finishes precipitating. If $[Cl^-]$ were ~2 times lower, AgCl would not precipitate prematurely.

26-43. Moles of Ca^{2+} = moles of $C_2O_4^{2-}$

$(V_e)(0.025\ 7\ M) = (25.00\ mL)(0.031\ 1\ M) \Rightarrow V_e = 30.25\ mL$

(a) The fraction of $C_2O_4^{2-}$ remaining when 10.00 mL of Ca^{2+} have been added is

$(30.25 - 10.00)/(30.25) = 0.669\ 4$.

$[C_2O_4^{2-}] = (0.669\ 4)(0.031\ 10\ M)\left(\dfrac{25.00}{35.00}\right) = 0.014\ 87\ M$

$[Ca^{2+}] = K_{sp}/[C_2O_4^{2-}] = (1.3 \times 10^{-8})/(0.014\ 87) = 8.7 \times 10^{-7}$

$\Rightarrow pCa^{2+} = -\log(8.7 \times 10^{-7}) = 6.06$

(b) At the equivalence point, there are equal numbers of moles of Ca^{2+} and $C_2O_4^{2-}$ dissolved. Call each concentration x:

$[Ca^{2+}][C_2O_4^{2-}] = (x)(x) = K_{sp} \Rightarrow x = \sqrt{K_{sp}} = 1.1_4 \times 10^{-4}\ M$

$pCa^{2+} = -\log(1.1_4 \times 10^{-4}) = 3.94$

(c) $[Ca^{2+}] = (0.025\ 70\ M)\left(\dfrac{35.00 - 30.25}{60.00}\right) = 0.002\ 03_5\ M.$ $pCa^{2+} = 2.69$

26-44. Equilibrium constants for ion pair formation:

$$\dfrac{[AgX(aq)]}{[Ag^+][X^-]} = \begin{cases} 10^{3.31} & (X = Cl) \\ 10^{4.6} & (X = Br) \\ 10^{6.6} & (X = I) \end{cases}$$

Calling the ion pair formation constant K_f, we can write

$[AgX(aq)] = K_f[Ag^+][X^-]$. But the product $[Ag^+][X^-]$ is just K_{sp}. So,

$[AgX(aq)] = K_f K_{sp}$. Putting in the values $K_{sp} = 10^{-9.74}$ for AgCl, $10^{-12.30}$ for AgBr, and $10^{-16.08}$ for AgI gives

$[AgCl(aq)] = 10^{3.31}10^{-9.74} = 10^{-6.43}\ M = 370\ nM$

$[AgBr(aq)] = 10^{4.6}10^{-12.30} = 10^{-7.70}\ M = 20\ nM$

$[AgI(aq)] = 10^{6.6}10^{-16.08} = 10^{-9.48}\ M = 0.33\ nM$

26-45. FM of NaCl = 58.443. FM of KBr = 119.002.   48.40 mL of 0.048 37 M $Ag^+ = 2.341\ 1$ mmol. This must equal the mmol of $(Cl^- + Br^-)$. Let x = mass of NaCl and y = mass of KBr. $x + y = 0.238\ 6$ g.

$$\underbrace{\frac{x}{58.443}}_{\text{mol Cl}^-} + \underbrace{\frac{y}{119.002}}_{\text{mol Br}^-} = 2.341\,1 \times 10^{-3} \text{ mol}$$

Substituting $x = 0.238\,6 - y$ gives $y = 0.200\,0$ g of KBr $= 1.681$ mmol of KBr $= 1.681$ mmol of Br $= 0.134\,3$ g of Br $= 56.28\%$ of the sample.

26-46. $\text{mmol of BrCH}_2\text{CH}_2\text{CH}_2\text{CH}_2\text{Cl} = \dfrac{82.67 \text{ mg}}{171.46 \text{ mg/mmol}} = 0.482\,2$ mmol

There will be 0.482 2 mmol of Cl$^-$ and 0.482 2 mmol of Br$^-$ liberated by reaction with CH$_3$O$^-$Na$^+$.

$$\text{Ag}^+ \text{ required for Br}^- = \frac{0.482\,2 \text{ mmol}}{0.025\,70 \text{ mmol/mL}} = 18.76 \text{ mL}$$

The same amount of Ag$^+$ is required to react with Cl$^-$, so the second equivalence point is at $18.76 + 18.76 = 37.52$ mL.

26-47.
$$\underset{\text{Analyte}}{\text{M}^+} + \underset{\text{Titrant}}{\text{X}^-} \rightleftharpoons \text{MX}(s)$$
$$\qquad C_M^o, V_M^o \qquad C_X^o, V_X$$

Mass balance for M: $\qquad C_M^o V_M^o = [\text{M}^+](V_M^o + V_X) + \text{mol MX}(s)$

Mass balance for X: $\qquad C_X^o V_X = [\text{X}^-](V_M^o + V_X) + \text{mol MX}(s)$

Equating mol MX(s) from both mass balances gives

$$C_M^o V_M^o - [\text{M}^+](V_M^o + V_X) = C_X^o V_X - [\text{X}^-](V_M^o + V_X)$$

which can be rearranged to $\quad V_X = V_M^o \left(\dfrac{C_M^o - [\text{M}^+] + [\text{X}^-]}{C_X^o + [\text{M}^+] - [\text{X}^-]} \right)$

26-48. Your graph should look like the figure in the text.

26-49. Mass balance for M: $C_M^o V_M = [\text{M}^{m+}](V_M + V_X^o) + x\{\text{mol M}_x\text{X}_m(s)\}$

Mass balance for X: $C_X^o V_X^o = [\text{X}^{x-}](V_M + V_X^o) + m\{\text{mol M}_x\text{X}_m(s)\}$

Equating mol M$_x$X$_m$ from the two equations gives

$$\frac{1}{x}\{C_M^o V_M - [\text{M}^{m+}](V_M + V_X^o)\} = \frac{1}{m}\{C_X^o V_X^o - [\text{X}^{x-}](V_M + V_X^o)\}$$

which can be rearranged to the required form.

26-50.   Titration of chromate with Ag^+:

	A	B	C	D	E	F	G	H	I
1	K_{sp} =	pM	[M]	[X]	V_M				
2	1.2E-12	5.46	3.47E-06	9.98E-02	0.013		C2 = 10^-B2		
3	C_M =	5.4	3.98E-06	7.57E-02	1.932		D2 = $A$2/C2^2		
4	0.1	5.3	5.01E-06	4.78E-02	5.342		E2 = $A$8*(2*$A$6+C2-2*D2)		
5	C_X =	5.2	6.31E-06	3.01E-02	8.717			/($A$4-C2+2*D2)	
6	0.1	5.1	7.94E-06	1.90E-02	11.734				
7	V_X =	5	1.00E-05	1.20E-02	14.195				
8	10	4.9	1.26E-05	7.57E-03	16.057				
9		4.8	1.58E-05	4.78E-03	17.388				
10		4.6	2.51E-05	1.90E-03	18.908				
11		4.4	3.98E-05	7.57E-04	19.564				
12		4	1.00E-04	1.20E-04	19.958				
13		3.6	2.51E-04	1.90E-05	20.064				
14		3.2	6.31E-04	3.01E-06	20.189				
15		2.8	1.58E-03	4.78E-07	20.483				
16		2.4	3.98E-03	7.57E-08	21.244				
17		2.3	5.01E-03	4.78E-08	21.583				
18		2.2	6.31E-03	3.01E-08	22.020				
19		2.1	7.94E-03	1.90E-08	22.589				
20		2	1.00E-02	1.20E-08	23.333				
21		1.9	1.26E-02	7.57E-09	24.321				
22		1.8	1.58E-02	4.78E-09	25.650				

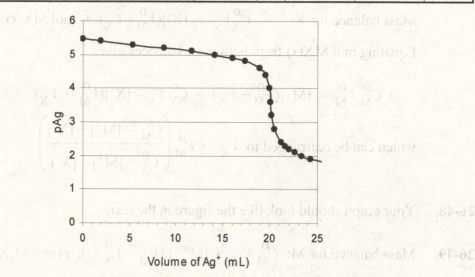

26-51.   Consider the titration of C^+ (in a flask) by A^- (from a buret). Before the equivalence point, there is excess C^+ in solution. Selective adsorption of C^+ on the CA crystal surface gives the crystal a positive charge. After the equivalence point, there is excess A^- in solution. Selective adsorption of A^- on the CA crystal surface gives it a negative charge.

26-52. Beyond the equivalence point, there is excess $Fe(CN)_6^{4-}$ in solution. Selective adsorption of this ion by the precipitate will give the particles a negative charge.

26-53. A known excess of Ag^+ is added to form AgI (s). In the presence of Fe^{3+}, the excess Ag^+ is titrated with standard SCN^- to precipitate $AgSCN(s)$. When Ag^+ is consumed, SCN^- reacts with Fe^{3+} to form the red complex, $FeSCN^{2+}$.

26-54. 50.00 mL of 0.365 0 M $AgNO_3$ = 18.25 mmol of Ag^+
37.60 mL of 0.287 0 M KSCN = 10.79 mmol of SCN^-
Difference = 18.25 − 10.79 = 7.46 mmol of I^- = 947 mg of I^-

27-1. There is no point analyzing a sample if you do not know that it was selected in a sensible way and stored so that its composition did not change after it was taken.

27-2. "Analytical quality" refers to the accuracy and precision of the method applied to the sample that was analyzed. High quality means that the analysis is accurate and precise. "Data quality" means that the sample that was analyzed is representative and appropriate for the question being asked and that the analytical quality is adequate for the intended purpose. If an accurate and precise analysis is performed on an unrepresentative or contaminated or decomposed sample, the results are meaningless.

27-3. (a) $s_o^2 = s_a^2 + s_s^2 = 3^2 + 4^2 \Rightarrow s_o = 5\%$.

(b) $s_s^2 = s_o^2 - s_a^2 = 4^2 - 3^2 \Rightarrow s_s = 2.6\%$.

27-4. $mR^2 = K_s$. $m(6^2) = 36 \text{ g} \Rightarrow m = 1.0 \text{ g}$

27-5. Pass the powder through a 120 mesh sieve and then through a 170 mesh sieve. Sample retained by 170 mesh sieve has a size between 90 and 125 μm. It would be called 120/170 mesh.

27-6. 11.0×10^2 g will contain 10^6 total particles, since 11.0 g contains 10^4 particles. $n_{KCl} = np = (10^6)(0.01) = 10^4$.
Relative standard deviation $= \sqrt{npq}/n_{KCl} = \sqrt{(10^6)(0.01)(0.99)}/10^4 = 0.99\%$.

27-7. (a) $\sqrt{(10^3)(0.5)(0.5)} = 15.8$.

(b) We are looking for the value of z, whose area is 0.45 (since the area from $-z$ to $+z$ is 0.90). The value lies between $z = 1.6$ and 1.7, whose areas are 0.445 2 and 0.455 4, respectively. Linear interpolation:
$$\frac{z - 1.6}{1.7 - 1.6} = \frac{0.45 - 0.445\,2}{0.455\,4 - 0.445\,2} \Rightarrow z = 1.647.$$

(c) Since $z = (x - \bar{x})/s$, $x = \bar{x} \pm zs = 500 \pm (1.647)(15.8) = 500 \pm 26$.
The range 474–526 will be observed 90% of the time.

27-8. Use Equation 27-7, with $s_s = 0.05$ and $e = 0.04$. The initial value of t for 95% confidence in Table 4-2 is 1.960. $n = t^2 s_s^2 / e^2 = 6.0$ For $n = 6$, there are 5 degrees of freedom, so $t = 2.571$, which gives $n = 10.3$. For 9 degrees of freedom, $t = 2.262$, which gives $n = 8.0$. Continuing, we find $t = 2.365 \Rightarrow n = 8.74$. This gives $t = 2.306 \Rightarrow n = 8.30$. Use <u>8 samples</u>. For 90% confidence, the initial t is 1.645 in Table 4-2 and the same series of calculations gives $n = $ <u>6 samples</u>.

27-9. (a) $mR^2 = K_S$. For $R = 2$ and $K_S = 20$ g, we find $m = 5.0$ g.

(b) Use Equation 27-7 with $s_s = 0.02$ and $e = 0.015$. The initial value of t for 90% confidence in Table 4-2 is 1.645. $n = t^2 s_s^2 / e^2 = 4.8$

For $n = 5$, there are 4 degrees of freedom, so $t = 2.132$, which gives $n = 8.1$. For 7 degrees of freedom, $t = 1.895$, which gives $n = 6.4$. Continuing, we find $t = 2.015 \Rightarrow n = 7.2$. This gives $t = 1.943 \Rightarrow n = 6.7$. Use <u>7 samples</u>.

27-10.

	A	B	C	D
1	Evaluation of the relation $mR^2 = K_s$			
2				
3	m (pg)	R %	R^2	$K_s = mR^2$
4	57	0.057	0.00325	0.185
5	68	0.069	0.00476	0.324
6	110	0.049	0.00240	0.264
7	110	0.045	0.00203	0.223
8	506	0.035	0.00123	0.620
9	515	0.027	0.00073	0.375
10	916	0.018	0.00032	0.297
11	955	0.022	0.00048	0.462
12			average	0.344
13			std dev	0.141

Average value of $K_s =$

0.34 ± 0.14 pg

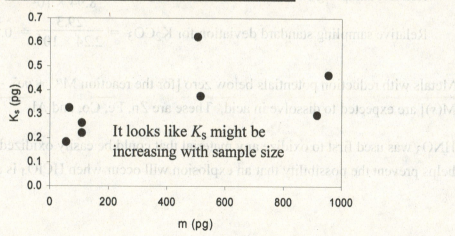

It looks like K_s might be increasing with sample size

The confidence interval is $\pm ts/\sqrt{n}$, where t is Student's t, s is the standard deviation, and n is the number of replicate measurements. If n is the same for all points, then t is the same for all points and the confidence interval is proportional to s. The equation in the text is expressed in terms of s. If the confidence interval is proportional to s, then the same equation should hold for the confidence interval.

27-11. (a) Volume $= (4/3)\pi r^3$, where $r = 0.075$ mm $= 7.5 \times 10^{-3}$ cm.

Volume $= 1.767 \times 10^{-6}$ mL.

Na_2CO_3 mass $= (1.767 \times 10^{-6}$ mL$)(2.532$ g/mL$) = 4.474 \times 10^{-6}$ g.

K_2CO_3 mass $= (1.767 \times 10^{-6}$ mL$)(2.428$ g/mL$) = 4.291 \times 10^{-6}$ g.

Number of particles of $Na_2CO_3 = (4.00$ g$)/(4.474 \times 10^{-6}$ g/particle$)$

$= 8.941 \times 10^5$.

Number of particles of $K_2CO_3 = (96.00$ g$)/(4.291 \times 10^{-6}$ g/particle$)$

$= 2.237 \times 10^7$.

The fraction of each type (which we will need for part c) is

$p_{Na_2CO_3} = (8.941 \times 10^5)/(8.941 \times 10^5 + 2.237 \times 10^7) = 0.0384$

$q_{K_2CO_3} = (2.237 \times 10^7)/(8.941 \times 10^5 + 2.237 \times 10^7) = 0.962$.

(b) Total number of particles in 0.100 g is $n = 2.326 \times 10^4$.

(c) Expected number of Na_2CO_3 particles in 0.100 g is 1/1000 of number in 100 grams $= 8.94 \times 10^2$.

Expected number of K_2CO_3 particles in 0.100 g is 1/1000 of number in 100 grams $= 2.24 \times 10^4$.

Sampling standard deviation $= \sqrt{npq} = \sqrt{(2.326 \times 10^4)(0.0384)(0.962)}$

$= 29.3$.

Relative sampling standard deviation for $Na_2CO_3 = \dfrac{29.3}{8.94 \times 10^2} = 3.28\%$.

Relative sampling standard deviation for $K_2CO_3 = \dfrac{29.3}{2.24 \times 10^4} = 0.131\%$.

27-12. Metals with reduction potentials below zero [for the reaction $M^{n+} + ne^- \rightarrow M(s)$] are expected to dissolve in acid. These are Zn, Fe, Co, and Al.

27-13. HNO_3 was used first to oxidize any material that could be easily oxidized. This helps prevent the possibility that an explosion will occur when $HClO_4$ is added.

27-14.   Barbital has a higher affinity for the octadecyl phase than for water, so it is retained by the column. The drug dissolves readily in acetone/chloroform, which elutes it from the column.

27-15.   Cocaine is an amine base. It will be a cation at low pH and neutral in ammonia. The cation at pH 2 is retained by the cation-exchange resin. The neutral molecule is easily eluted by methanol. Benzoylecgonine has an amine and carboxylate functionality. At pH 2, the amine will be protonated and the carboxylic acid should be neutral, so the molecule will be retained by the cation exchange column. At elevated pH, the amine will be neutral and the carboxylate will be negative. The anion is not retained by the cation-exchanger and is eluted with methanol.

27-16.   The product gas stream is passed through an anion-exchange column, on which SO_2 is absorbed by the following reactions:

$$SO_2 + H_2O \ \rightarrow \ H_2SO_3$$

$$2Resin^+OH^- + H_2SO_3 \ \rightarrow \ (Resin^+)_2SO_3^{2-} + H_2O$$

The sulfite is eluted with Na_2CO_3/H_2O_2, which oxidizes it to sulfate that can be measured by ion chromatography.

27-17.   Large particle size allows sample to drain through the solid-phase extraction column without applying high pressure. In chromatography, small particle size increases the efficiency of separation, but high pressure is necessary to force solvent through the column.

27-18.   (a) Solid-phase extraction retains acrylamide while passing many other components in the aqueous extract of the french fries. We want to remove as many other components as possible to simplify the chromatographic analysis. The strong acid of the ion-exchange resin protonates acrylamide and retains it by ionic attraction:

$$R{-}SO_3H \ + \ CH_2{=}CHCONH_2 \ \rightarrow \ R{-}SO_3^- \ + \ CH_2{=}CHCONH_3^+$$

(b) There are many ultraviolet-absorbing components in addition to acrylamide in the acrylamide fraction obtained from the ion-exchange column. Ultraviolet absorbance is not specific for acrylamide.

(c) For acrylamide, m/z 72 is selected by the mass filter Q1 of the mass spectrometer. This ion dissociates by collisions in Q2. The product m/z 55 is selected in Q3 for passage to the detector.

$$CH_2=CHCONH_3^+ \rightarrow CH_2=CHC\equiv O^+$$
$$\text{m/z 72} \qquad\qquad \text{m/z 55}$$

$$CD_2=CDCONH_3^+ \rightarrow CD_2=CDC\equiv O^+$$
$$\text{m/z 75} \qquad\qquad \text{m/z 58}$$

(d) Even though many compounds are applied to the chromatography column, acrylamide is the only one with m/z 72 that gives a significant reaction product at m/z 55.

(e) Acrylamide gives one peak by selected reaction monitoring of the transition m/z 72→55. The internal standard gives just one peak for $^2H_3$-acrylamide monitored by the transition m/z 75→58 with the same retention time as acrylamide. The transition m/z 72→55 does not respond to the internal standard, and the transition m/z 75→58 does not respond to unlabeled acrylamide. We know the concentration of internal standard added to the aqueous extract of the french fries, and we measure the integrated area of the m/z 75→58 peak for the internal standard. We also measure the integrated area of the m/z 72→55 peak for acrylamide. The concentration of acrylamide in the aqueous extract is found by the proportion

$$\frac{[\text{acrylamide}]}{[\text{internal standard}]} = \frac{[\text{area of m/z 72→55 peak}]}{[\text{area of m/z 75→58 peak}]}$$

(f) With ultraviolet absorption, the internal standard appears at the same elution time as acrylamide. The molar absorptivity of deuterated internal standard is probably very similar to that of acrylamide, so equal concentrations of internal standard and acrylamide contribute approximately the same integrated area in the chromatogram. With selected reaction monitoring by mass spectrometry, the detector sees either acrylamide or the internal standard, with no interference from the other, even though they are eluted at the same time.

(g) The internal standard is mixed with the aqueous extract from the french fries prior to solid-phase extraction. We expect little isotope effect on the binding of acrylamide to the solid-phase sorbent or the HPLC stationary phase. Therefore, the fraction of acrylamide and the fraction of internal standard that bind to and are recovered from the solid-phase extraction column are equal. Even though neither one is bound or eluted quantitatively, equal fractions of each are bound and eluted. The ratio of acrylamide and internal standard should remain constant throughout the entire procedure.

27-19. (a) Highest concentration of Ni $\approx$ 80 ng/mL. A 10 mL sample contains 800 ng Ni = 1.36×10^{-8} mol Ni. To this is added 50 μg Ga = 7.17×10^{-7} mol Ga. Atomic ratio Ga/Ni = $(7.17 \times 10^{-7})/(1.36 \times 10^{-8})$ = 53.

(b) Apparently all the Ni is in solution because filtration does not decrease its total concentration. Since filtration removes most of the Fe, it must be present as a suspension of solid particles.

27-20. One-fourth of the sample (25 mL out of 100 mL) required
$(0.011\,44$ M$)\,(0.032\,49$ L$)$ = 3.717×10^{-4} mol EDTA $\Rightarrow (3.717 \times 10^{-4})\,(4)$ = 1.487×10^{-3} mol Ba^{2+} in sample = $0.204\,2$ g Ba = 64.90 wt%.

27-21. (a) From the acid dissociation constants of Cr(III), we see that the dominant forms at pH 8 are Cr(OH)$_2^+$ and Cr(OH)$_3(aq)$. The dominant form of Cr(VI) is CrO$_4^{2-}$.

(b) The anion exchanger retains the anion, CrO$_4^{2-}$, but permits the Cr(OH)$_2^+$ cation and neutral Cr(OH)$_3(aq)$ to pass through, thereby separating Cr(VI) from Cr(III).

(c) A "weakly basic" anion exchanger contains a protonated amine ($-^+$NHR$_2$) that might lose its positive change in basic solution. A "strongly basic" anion exchanger ($-^+$NR$_3$) is a stable cation in basic solution.

(d) CrO$_4^{2-}$ is eluted from the anion exchanger when the concentration of sulfate in the buffer is increased from 0.05 M in step 3 to 0.5 M in step 4.

27-22. One possible cost-saving scheme is to monitor wells 8, 11, 12, and 13 individually, but to pool samples from the other sites. For example, a composite sample could be made with equal volumes from wells 1, 2, 3, and 4. Other composites could be constructed from (5, 6, 7), (9, 10), (14, 15, 16, 17), and (18, 19, 20, 21). If no warning level of analyte is found in a composite sample, we would assume that each well in that composite is free of the analyte. If analyte is found in a composite sample, then each contributor to the composite would be separately analyzed. The disadvantage of pooling samples from n wells is that the sensitivity of the analysis for analyte in any one well is reduced by $1/n$.